Uncertainty Quantification of Electromagnetic Devices, Circuits, and Systems

The ACES Series on Computational and Numerical Modelling in Electrical Engineering

Andrew F. Peterson, PhD – Series Editor
The volumes in this series will encompass the development and application of numerical techniques to electrical and electronic systems, including the modelling of electromagnetic phenomena over all frequency ranges and closely related techniques for acoustic and optical analysis. The scope includes the use of computation for engineering design and optimization, as well as the application of commercial modelling tools to practical problems. The series will include titles for senior undergraduate and postgraduate education, research monographs for reference, and practitioner guides and handbooks.

Titles in the Series

K. Warnick, **"Numerical Methods for Engineering,"** 2010.
W. Yu, X. Yang and W. Li, **"VALU, AVX and GPU Acceleration Techniques for Parallel FDTD Methods,"** 2014.
A.Z. Elsherbeni, P. Nayeri and C.J. Reddy, **"Antenna Analysis and Design Using FEKO Electromagnetic Simulation Software,"** 2014.
A.Z. Elsherbeni and V. Demir, **"The Finite-Difference Time-Domain Method in Electromagnetics with MATLAB® Simulations, 2nd Edition,"** 2015.
M. Bakr, A.Z. Elsherbeni and V. Demir, **"Adjoint Sensitivity Analysis of High Frequency Structures with MATLAB®,"** 2017.
O. Ergul, **"New Trends in Computational Electromagnetics,"** 2019.
D. Werner, **"Nanoantennas and Plasmonics: Modelling, design and fabrication,"** 2020.
K. Kobayashi and P.D. Smith, **"Advances in Mathematical Methods for Electromagnetics,"** 2020.

Uncertainty Quantification of Electromagnetic Devices, Circuits, and Systems

Edited by
Sourajeet Roy

The Institution of Engineering and Technology

Published by SciTech Publishing, an imprint of The Institution of Engineering and Technology, London, United Kingdom

The Institution of Engineering and Technology is registered as a Charity in England & Wales (no. 211014) and Scotland (no. SC038698).

First published 2021

The Institution of Engineering and Technology
Michael Faraday House
Six Hills Way, Stevenage
Herts SG1 2AY, United Kingdom

www.theiet.org

British Library Cataloguing in Publication Data
A catalogue record for this product is available from the British Library

ISBN 978-1-83953-171-2 (hardback)
ISBN 978-1-83953-172-9 (PDF)

Typeset in India by MPS Limited

Contents

About the editor

Sourajeet Roy is an Assistant Professor in the Department of ECE at the Indian Institute of Technology, Roorkee, India where he leads the Computational Modeling and Simulation (CMAS) research group. He has previously worked at Colorado State University as an Assistant Professor from 2013 to 2018. He is a Senior Member of the IEEE and his areas of research include numerical modeling and simulation of high-speed devices/circuits and computational algorithms for uncertainty quantification.

Chapter 1

Uncertainty quantification in electromagnetic device, circuit, and system simulation: its importance and value

Sourajeet Roy[1]

Numerical simulations, theoretical analyses, and practical experiments are the three different approaches to observe, study, and examine the behavior of any system. However, with the continuous improvement in computing hardware, falling hardware costs, and increasing memory and data processing capabilities in a computer, it is numerical simulations that have emerged as the more suitable approach for system identification and behavioral analysis in the early product design cycles. In this chapter, we will explore the evolution of numerical simulation techniques for electromagnetic devices, circuits, and systems from the classical deterministic approach to the emergent stochastic approaches. In course of this exploration, we will identify the different sources of parametric uncertainty that can arise in a numerical simulation, the different forms these uncertainties can assume, and the various state-of-the-art mathematical and algorithmic techniques to probe these uncertainties in order to gain deeper insights into the behavior of the device, circuit, or system under test.

1.1 Basics of numerical simulation

Numerical simulation involves the development and solution of mathematical models of any system under test. All numerical simulations consist of the following basis steps.

Model Formulation: Within the context of electromagnetic structures, the behavior of the system under test is usually described using partial differential equations (PDE) such as the fundamental Maxwell's equations, wave equation, Poisson or Laplace equation, etc. [1–5]. Next, knowledge pertaining to all the geometrical, physical, and material parameters of the system is collated. At the geometrical boundaries, the boundary conditions to the governing PDEs are imposed. Assuming that the problem being investigated is transient in nature, the

[1]Computational Modeling and Simulation Laboratory, Indian Institute of Technology, Roorkee, India

initial conditions (i.e., conditions at zero time) are either assumed or known *a priori*. If there are any physical constraints in the system, these too are added to the problem definition. All this information together forms the mathematical model of the system under test [6].

Spatial Discretization: Next, assuming no analytic or closed-form solution of the mathematical model of the system exists, the governing PDEs in the model are spatially discretized into a system of coupled ordinary differential equations (ODEs). Of course, the discretization of the governing PDEs must be in the one-, two-, or three-dimensional solution space of the system under test. The discretization schemes adopted may be a forward, backward, and central difference scheme. Furthermore, the discretization can be uniform, non-uniform, or adaptive based on the structure of the system under test. Finally, if nonlinearity is present in the system (e.g., due to the presence of nonlinear devices such as transistors, diodes), the ODEs obtained will include a nonlinear term. For purposes of illustration, let us assume that the discretization of the governing PDEs for the electromagnetic system under test yields a system of coupled first-order ODEs described as [6]

$$\mathbf{GX} + \mathbf{C}\frac{d\mathbf{X}}{dt} + \mathbf{F}(\mathbf{X}) = \mathbf{B}(t) \tag{1.1}$$

In (1.1), **G** and **C** are the coefficient matrices of the linear and differential terms, **F** is the vector of nonlinear terms, **B** is the vector of independent inputs, and **X** is the vector of unknown quantities such as currents/voltages or electromagnetic field quantities.

Model Solution: The mathematical model of (1.1) is solved using numerical integration techniques. These numerical integration techniques can be of the explicit, semi-explicit, or the implicit variety [6]. The key task of numerical integration techniques is to utilize time-stepping algorithms to convert the differential term in (1.1) into a set of linear terms. This converts the model of (1.1) from a system of ODEs into a system of nonlinear equations of the form:

$$\mathbf{AX} + \mathbf{F}(\mathbf{X}) = \mathbf{U} \tag{1.2}$$

The nonlinear system of equations of (1.2) is then converted into a linear system of equations using root finding techniques such as the Newton-Raphson method [6]. Eventually, the linear system of equations is solved to find $\mathbf{X}(t)$ using direct or indirect matrix inversion. This procedure for numerical simulations is visually explained in Figure 1.1.

1.2 Deterministic numerical simulation

A key assumption in the classical numerical simulation procedure of above is that the geometrical, material, and physical parameters of the system, the boundary conditions of the governing PDEs, and the physical constraints of the system under test are perfectly known. This ensures that all the entries in the **G**, **C**, and **B** matrices as well as the nonlinear expressions in **F** of (1.1) are exactly known and a self-consistent solution to the model of (1.1) is possible. Hence, the model of (1.1)

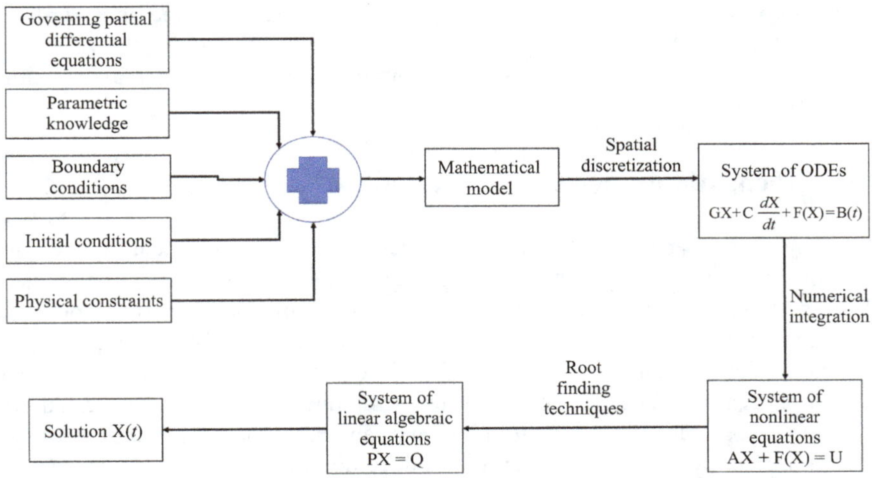

Figure 1.1 Flowchart showing the discrete steps in the deterministic simulation of an electromagnetic device, circuit, or system

is known as a *deterministic model* and the above solution technique as a *deterministic* numerical simulation. The main challenges for deterministic simulation are:

- *Memory Challenge*: Storing the entire model of (1.1) where the number of equations in (1.1), say N, is often significantly large (in the order of thousands) requires large memory resources.
- *Time Challenge*: Let us assume that an iterative root finding algorithm is used to convert the nonlinear system of equations of (1.2) into a linear form and the average number of iterations required is N_{avg}. The linear system of equations obtained is then solved via a direct matrix inversion. Thus, the total time cost of all the matrix inversions involved will scale as $O(N_{\text{avg}}N_tN^3)$ where N_t is the number of non-uniform time steps used in the numerical integration scheme. It is clear that even for a modest number of N, this time cost can be intractably large.
- *Stability Challenge*: A key aspect of deterministic simulations is the numerical integration scheme involved. If an explicit scheme or multistep methods are used, care must be taken such that the time step is small enough to ensure that the solution is stable. This means that N_t must be sufficiently large to ensure stability of the predicted solution of (1.2), which in turn exacerbates the time challenge.
- *Local Truncation Error Challenge*: Even when implicit numerical integration schemes are used, non-zero local truncation errors always arise due to the approximation of the differential operator of (1.1) with a linear operator. Typically, the local truncation error vector ($\mathbf{E}$) is expressed as [6]

$$\mathbf{E} = kh^{m+1}\frac{d^{m+1}\mathbf{X}}{dt^{m+1}} \tag{1.3}$$

where m is the order of the numerical integration scheme and h is the time step. It is observed from (1.3) that in order to reduce the local truncation error, the value of h must decrease (i.e., N_t must increase) which in turn would again increase the time cost of the matrix inversions.

1.3 Stochastic numerical simulation

One major drawback of deterministic numerical simulation is that it does not account for the possibility that the geometrical, material, and physical parameters of the system, the boundary conditions of the governing PDEs, and the physical constraints of the system under test may not always be perfectly known [7–12]. In realistic modeling scenarios, this lack of knowledge about the system can arise due to multiple reasons. The most common reason is that in nanoscale integrated circuits, the lithography, etching, patterning, and fabrication processes of devices and circuits are not perfect and invariably introduce non-negligible variations – variations that have a profound impact on the robustness, reliability, failure probability, and optimal design of the devices and circuits [7–10]. The second reason is that measurement noise, imperfect calibration of equipment, approximation errors and local truncation errors in numerical simulations, and human error can never be truly eliminated and may also introduce additional uncertainty [7–10]. Finally, even if the model of (1.1) is perfectly known, the same may not apply for the input data. These reasons together ensure that there is always some non-zero uncertainty in the geometrical, material, and physical parameters of the system, the boundary conditions of the governing PDEs, and the physical constraints of the system under test. These uncertainties get propagated through the model of (1.1) and hence the response vector $\mathbf{X}(t)$ also becomes an uncertain process. This forward propagation of uncertainty is illustrated in Figure 1.2. An outcome of the forward propagation of uncertainty is that if we want to optimize the different responses of interest of the system under test, it becomes necessary to first quantify the impact of the input parametric uncertainty on these very responses in the early design cycles. Numerical simulations that help circuit designers to account for the effects of input

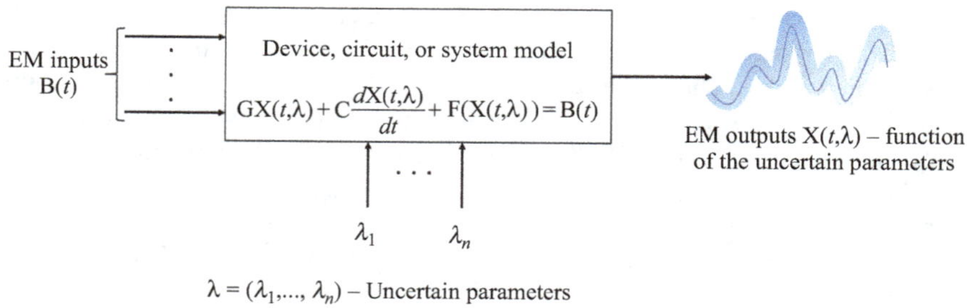

Figure 1.2 Forward propagation of uncertainty through a mathematical model of EM device, circuit, or system

parametric uncertainty on the model responses are called *stochastic* numerical simulations. Thus, it is appreciated that stochastic simulations are the opposite of deterministic simulations. Finally, the task of performing stochastic simulations to quantify the uncertainty in the model responses is referred to as uncertainty quantification [11,12].

1.3.1 Different types of uncertainty

The uncertainty in the input parameters of an electromagnetic device, circuit, or system can generally be of two broad types:

(i) *Aleatory Uncertainty*: Aleatory uncertainty is the type of uncertainty present in any input parameter that is mathematically expressed using a probability density function [11]. This probability density function can be described using a closed-form mathematical equation, for example, the normal distribution function, or simply using data, for example, by the knowledge of the probability of discrete values of the parameter. Typically, this form of uncertainty is considered to be irreducible meaning that it is not possible to reduce this uncertainty by performing more experiments, measurements, or simulations on the system under test. The traditional approach of quantifying this form of uncertainty is using probabilistic approaches. Probabilistic approaches aim to quantify the statistical moments of the responses of interest from a prior knowledge of the probability density function of the input parameters.

(ii) *Epistemic Uncertainty*: Epistemic uncertainty (or systematic uncertainty) refers to the uncertainty in any input parameter where the interval of support over which the parameter varies is known but no knowledge of the probability density function is available [13–17]. This type of uncertainty is reducible in the sense that it is possible to garner additional knowledge, for example, through additional experiments, measurements, and simulations that can reduce the epistemic uncertainty in a parameter. In particular, epistemic uncertainty cannot be quantified using probabilistic approaches. Rather, epistemic uncertainty is quantified using non-probabilistic techniques such as the Dempster-Shafer theory of evidence and fuzzy sets [18,19]. These techniques aim to quantify the interval of support of the responses of interest from a prior knowledge of the interval of support of the input parameters.

(iii) *Mixed Epistemic-Aleatory Uncertainty*: The most general form of uncertainty is the combination of both epistemic and aleatory uncertainty in an input parameter [20–27]. For example, an input parameter may be aleatory in nature with a known probability density function. Now, the mean and standard deviation of the parameter are usually obtained from legacy data, expert opinion, numerical simulations, experimental measurements, or a combination of all these. It is understood that if sufficiently large number of simulations or experiments are performed then the mean and standard deviation values can be inferred from the collected results. However, due to the prohibitive computational costs, manufacturing costs, and labor costs associated with a massive number of simulations and experiments, only a few of these may be realistically possible.

Similarly, approximation errors in the numerical simulations, calibration errors of the instruments used in the experiments, and human errors can also be present. All these causes together introduce epistemic uncertainty into our knowledge of the mean and standard deviation. Therefore, the parameter in question possesses both epistemic and aleatory uncertainty. Another possibility could be that some of the input aleatory parameters behave as design parameters whose mean value can be intentionally tuned within a specific interval of support in order to meet certain performance specifications. In this scenario, again the input parameter will include both aleatory and epistemic uncertainty. It is expected that a combination of probabilistic and non-probabilistic approaches will be required to perform uncertainty quantification for mixed epistemic-aleatory problems [20–27]. In mixed epistemic-aleatory problems, the responses of interest will be defined by statistical moments where each moment will lie within an interval of support (i.e., the exact value of the statistical moment will be unknown). The task of uncertainty quantification involves evaluating the intervals of support for the different statistical moments of the responses of interest.

1.3.2 Value of uncertainty quantification

Given the possibility of parametric uncertainty in electromagnetic device, circuit, and system models, it is important to appreciate the value of uncertainty quantification. Typically, uncertainty quantification enables the circuit designer to perform the following key tasks that are not possible using deterministic models.

(i) *Robust Design Optimization (RDO):* RDO refers to the task of tuning the value of various design parameters in order to minimize the variance of the device, circuit, or system responses of interest. In effect, by performing RDO, the device, circuit, or system responses are made robust to the input parametric uncertainty. In order to perform such an optimization, the input–output mapping between the input design and uncertain parameters and the statistical moments of the responses of interest must be determined via stochastic numerical simulations.

(ii) *Reliability-based Design Optimization (RBDO)*: RBDO involves tuning the values of various design parameters in order to minimize the area bounded by the probability density function curve of the device, circuit, or system response of interest and the failure threshold. In effect, by performing RBDO, the failure (success) probability of the device, circuit, or system under test is minimized (maximized) in the presence of input parametric uncertainty. Once again, in order to perform such an optimization, we need to know the input–output mapping between the input design and uncertain parameters and the probability density functions of the responses of interest via stochastic numerical simulations.

(iii) *Worst Case Analysis*: Worst case analysis refers to the task of quantifying the maximum or minimum bounds of the responses of interest based on prior knowledge of the maximum and minimum bounds (i.e., the intervals of support) of the input parameters of a device, circuit, or system. This type of analysis is only possible for problems involving purely epistemic uncertainty [28–32].

(iv) *Sensitivity Analysis*: Often, it is important to understand how the uncertainty in one input parameter affects the uncertainty of the device, circuit, or system responses of interest and their failure thresholds. Sensitivity analysis is the task of mathematically quantifying such an input–output mapping [33,34].

Within an aleatory context, moment dependent sensitivity analysis involves calculating how the variance of an input parameter affects the total variance of the responses of interest (e.g., Sobol's sensitivity indices) [33,34]. Moment independent sensitivity analysis involves calculating how the uncertainty of an input parameter over its entire interval of support changes the shape of the entire probability density function of the responses of interest (e.g., Borgonovo's sensitivity indices) [35–37]. Within an epistemic context, sensitivity analysis involves calculating how as an input parameter varies over its specific interval of support, the maximum and minimum bounds (i.e., the interval of support) of the responses of interest of the device, circuit, or system changes. Finally, sensitivity analysis can also be performed in a mixed epistemic-aleatory context. For this purpose, let us assume that an input aleatory parameter contains epistemic uncertainty in the mean and/or the standard deviation. In that case, sensitivity analysis involves calculating how the maximum and minimum bounds of any pertinent statistics of the responses of interest vary as the mean, standard deviation, or both of the input parameter varies over their specific intervals of support.

While all the aforementioned analyses are of immense value to circuit designers, within the context of this book, we are mainly focused on the topic of aleatory uncertainty quantification and its value in RDO and reliability-based design optimization (RBDO). When performing these design optimizations, it is understood that the design parameters are tuned iteratively. In each iteration, the full set of input uncertain parameters has to be taken into account to calculate the statistics (e.g., the variance, probability/cumulative density function, and failure probability) of the responses of interest. These statistical results are used to check if the variance of the responses has been minimized (for RDO) or if the failure probability of the responses has been minimized (for RBDO). If not, then the design parameters are updated in the next iteration and the response statistics are recalculated. This process continues till we reach the point of convergence – the point when all response statistics satisfy the target criteria. Thus, it is clear that design optimizations involve the repeated calculation of the response statistics such as the variance, probability/cumulative density function, and failure probability for each new value of the design parameters – in effect, the uncertainty quantification needs to be redone for every new value of the design parameters till convergence is reached. This process of design optimization is illustrated in Figure 1.3. Therefore, it is not enough simply to perform uncertainty quantification, but rather, it needs to be done in a highly efficient and rapid manner so that the process can be repeated over and over again. In here lies the critical challenge of uncertainty quantification – how to perform fast uncertainty quantification when considering a massively large set of input parameters and when the device, circuit, or system model itself is very large and cumbersome to solve.

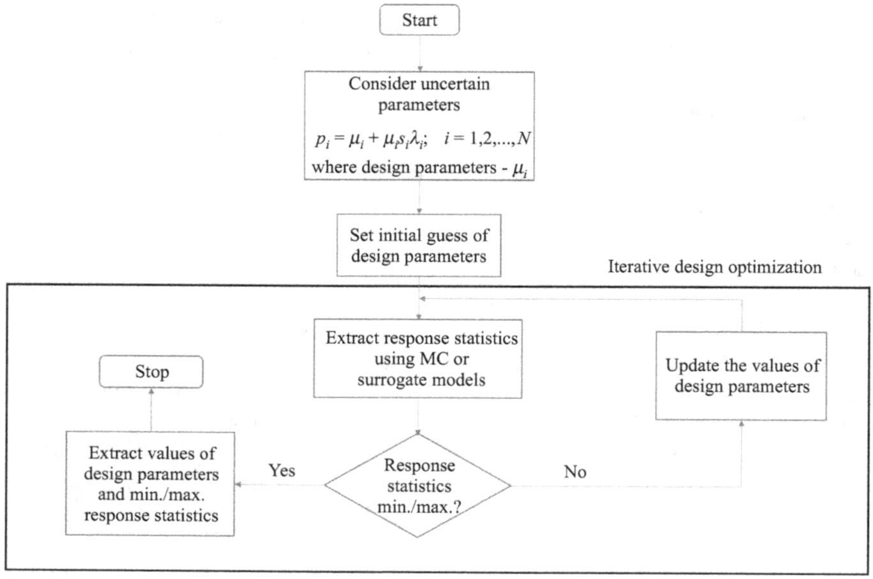

Figure 1.3 Flowchart showing the iterative approach to design optimization with the step where surrogate models will be deployed clearly shown

1.4 State-of-the-art in performing uncertainty quantification

Based on the discussion of the previous section, it is appreciated that design optimization involves the repeated estimation of the statistics of the responses of interest of the device, circuit, or system under test. Hence, design optimization is closely tied to the idea of aleatory uncertainty quantification. This is why, within the context of this book, by the term uncertainty quantification, we will refer to aleatory uncertainty quantification. For more insights into the more general mixed epistemic-aleatory uncertainty quantification, readers are referred to Chapter 9 and the associated references.

Currently, the state-of-the-art in aleatory uncertainty quantification can be broadly divided into two types of numerical approaches – the random sampling-based approaches [38–40] and the surrogate model (or metamodel) based approaches [41–50]. Both of these approaches have their own unique pros and cons as described next.

1.4.1 Random sampling-based approaches

The most basic yet effective of the random sampling-based approaches is Monte Carlo [38,40]. Let us assume an electromagnetic device, circuit, or system possesses N number of input uncertain parameters defined as

$$p_i = \mu_i + \mu_i s_i \lambda_i; \quad i = 1, 2, ..., N \tag{1.4}$$

In (1.4), μ_i refers to the nominal value (mean) of the ith parameter, s_i refers to the relative standard deviation of the ith parameter, and λ_i refers to the random variable describing the uncertainty in the ith parameter. It is assumed that all the random variables are independent and together occupy the interval of support Ω. Now, if the response of interest is $x(\lambda)$ where $\lambda = [\lambda_1, \ldots \lambda_N]$, any arbitrary kth *statistical moment of the output can be described using Monte Carlo as* [38]

$$M_k = \int_\Omega (x(\lambda))^k \rho(\lambda)\, d\lambda \approx \frac{1}{N_s} \sum_{j=1}^{N_s} (x(\lambda_j))^k \tag{1.5}$$

where N_s refers to the number of samples and $\lambda_j = [\lambda_1^{(j)}, \lambda_2^{(j)}, \ldots, \lambda_N^{(j)}]$ is the jth N-dimensional sample point in the interval of support Ω. Thus, Monte Carlo involves the generation of a sequence of samples of the random variables λ. Importantly, the choice of the sample sequence is based on the shape of the joint probability density function $\rho(\lambda)$. In other words, the probability of choosing λ_j as the jth sample point in the sequence depends on the value $\rho(\lambda = \lambda_j)$. This is why Monte Carlo is referred to as a random sampling approach. At this juncture, it is emphasized that if the samples of (1.5) are generated using a computer, it is never truly possible to generate a random sequence of samples. Rather, the sequence produced is pseudo-random in nature. In effect, we understand from (1.5) that Monte Carlo is nothing other than a numerical integration technique based on repeated sampling of the random variables λ [38–40]. Once the sequence of samples is available, for each sample in the sequence $\lambda = \lambda_j$, we need to first put the value of $\lambda_j = [\lambda_1^{(j)}, \lambda_2^{(j)}, \ldots, \lambda_N^{(j)}]$ in (1.4) and find the corresponding values of the input parameters. These values of the input parameters, p_i of (1.4), are then replaced into the device, circuit, or system model of (1.1). Thereafter, the model of (1.1) needs to be solved at this $\lambda = \lambda_j$ sample point in a deterministic manner and the response recorded. Finally, by using the recorded responses at the sample points of (1.5), the response statistics are estimated. Thus, Monte Carlo allows the harnessing of legacy deterministic electromagnetic and circuit solvers to estimate the statistical moments of (1.5).

The key limitation of Monte Carlo is the rate of convergence. In particular, the error between the actual integral of (1.5) and the Monte Carlo approximation scales as [38–40]

$$O(e) = O\left(\frac{1}{\sqrt{N_s}}\right) \tag{1.6}$$

This means that a massive number of samples will be required in order to ensure a small enough approximation error of (1.5). Here lies the challenge of Monte Carlo – that the deterministic solution of the model of (1.1) needs to be performed N_s number of times where typically N_s is massively large. In fact, it is not uncommon for N_s to be in the order of hundreds of thousands (i.e., Monte Carlo requires hundreds of thousands of discrete deterministic simulations). Such a task is, of course, computationally intractable for most devices and structures, especially where the model of (1.1) is large as commonly seen in distributed circuits.

In order to address this convergence issue, low discrepancy sampling sequences such as Sobol's sequence and Halton sequence can be used [39,40]. When such low discrepancy sequences are used, the corresponding method to calculate the statistical moments is referred to as quasi-Monte Carlo [39,40]. In quasi-Monte Carlo, the rate of convergence of (1.6) is modified to [39,40]

$$O(e) = O\left(\frac{(\log N_s)^N}{N_s}\right) \tag{1.7}$$

It is observed that the rate of convergence of (1.7) is smaller than that of (1.6) provided N is small and N_s is large. Unfortunately, this is usually not true for realistic problems. Therefore, irrespective of the sampling sequence adopted, for realistic electromagnetic problems, Monte Carlo methods still require at least a few thousands of deterministic samples to converge reliably.

1.4.2 Surrogate modeling or metamodeling approaches

It is observed from the discussion of the previous subsection that Monte Carlo methods are unable to capture the global nature of the statistical moments of (1.5). Rather, they try to approximate the global nature of the statistical moments by extracting the model responses at a set of local values of the random variables λ. This is an expected feature of all numerical algorithms. Hence, a very large number of local points or samples are required to approximate the global attributes of the statistical moments with a high degree of accuracy. This translates to an equally large number of deterministic simulations at massive time and memory costs. Surrogate models or metamodels attempt to address this challenge by developing alternate closed-form representative models of the response of interest $x(\lambda)$ as $x = F(\lambda)$where F is the input–output nonlinear map of the impact of the random variables on the response. Currently, two well-known approaches to formulate the function F are:

- Using spectral metamodels such as polynomial chaos models [11,12,43].
- Using machine learning regression models [47,48,51–54].

The key attributes of both these approaches are discussed next.

1.4.2.1 Polynomial chaos models

In spectral models, the functional map F is expressed as a summation or linear combination of terms as

$$x(\lambda) = \sum_{k=0}^{\infty} x_k \phi_k(\lambda) \approx \sum_{k=0}^{P} x_k \phi_k(\lambda) \tag{1.8}$$

In (1.8), x_k is the kth coefficient and $\varphi_k(\lambda)$ is the appropriate basis function chosen. Importantly, in the spectral expansion of (1.8), for practical reasons, only the most dominant $P+1$ terms are considered. For the particular case of generalized polynomial chaos theory [11,12], the basis functions are chosen to be orthonormal polynomial functions. Specifically, the condition of orthonormality is described with respect to the inner product integral

$$\langle \phi_p(\lambda)\phi_q(\lambda)\rangle = \int_\Omega \phi_p(\lambda)\phi_q(\lambda)\rho(\lambda)d\lambda = \begin{cases} 0 & \text{if } p \neq q \\ 1 & \text{if } p = q \end{cases} \tag{1.9}$$

It is noted that in (1.9), the weight of the inner product integral is set to the joint PDF of λ. By doing so, we essentially ensure a one-to-one correspondence between the family of the orthonormal basis functions chosen in the spectral expansion of (1.8) and the shape of the joint PDF, also referred to as the Askey-Weiner scheme [55]. The key point in surrogate models is the evaluation of the coefficients of (1.8) for the specific electromagnetic device, circuit, or system at hand. This task is referred to as the training of the surrogate model. The training of the polynomial chaos surrogate model or metamodel of (1.8) can proceed in an intrusive or non-intrusive manner [11,12,56]. Once trained, the closed-form model of (1.8) serves as a surrogate to the larger and more cumbersome deterministic model of (1.1). Now, for any Monte Carlo sample, we no longer have to solve the massive model of (1.1). Rather we simply need to replace $\lambda = \lambda_j$ in the metamodel of (1.8) to obtain the response $x(\lambda = \lambda_j)$ at virtually negligible computational costs. Therefore, even for a prohibitively large number of Monte Carlo samples, the statistical moments of the response can be obtained using (1.5) and (1.8) in a remarkably fast manner. A more detailed examination of polynomial chaos metamodels is left for Chapters 2 and 3.

1.4.2.2 Machine learning models

In machine learning, the functional map $x = F(\lambda)$ is learned from existing data using regression. In particular, regression models such as artificial neural networks (ANNs) [51,52], support vector regression (SVR) [53,54], or Gaussian process regression [47,48] are commonly used. For example, Figure 1.4 shows an ANN where a three-layer perceptron model is used to capture the functional dependency of the random variables λ on the response $x(\lambda)$ [52,57]. In the feedforward process, the weighted outputs from neurons of the previous layer are fed into the neurons in the current layer. The outputs of the first (input) layer are simply the random variables, given as

$$z_i^{(1)} = \lambda_i; \quad i = 1, 2, \ldots, N \tag{1.10}$$

The inputs to the general ith neuron of the second (or hidden layer) are the weighted outputs of the first layer. The sum of the weighted outputs of the first layer are then fed into a nonlinear activation function $f(.)$ such that the output of the ith neuron of the second layer is given as

$$z_i^{(2)} = f\left(\sum_{j=1}^{N} w_{ij}^{(2)} z_j^{(1)} + b_i^{(2)}\right) \tag{1.11}$$

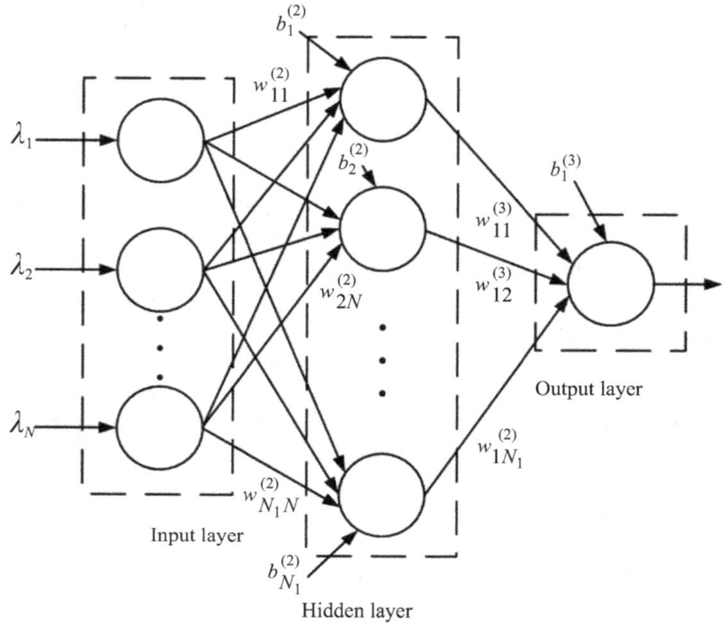

Figure 1.4 Basic architecture of three layer artificial neural network

where $b_i^{(2)}$ is the bias term added to the ith neuron. The outputs of the second layer are again weighted and fed into the third (output) layer. As before, the inputs to the third layer are summed and fed through another activation function to emerge as the output $x(\lambda)$ as

$$x(\lambda) = z_1^{(3)} = f\left(\sum_{j=1}^{N_2} w_{1j}^{(3)} z_j^{(2)} + b_1^{(3)}\right) \tag{1.12}$$

In (1.12), we assume that there are N_2 neurons in the second layer. The activation function in each layer can be set by the user. The weights and bias terms are tuned using various optimization algorithms such as back propagation, stochastic gradient descent, simplex method, etc., to ensure that the norm of the error between the output of the ANN and the response $x(\lambda)$ is minimized. The tuning of the weights and bias terms using data samples of the response $x(\lambda)$ obtained from discrete simulations of the original model of (1.1) is generally referred to as the "training process." Once trained, the ANN of Figure 1.4 will act as a closed-form surrogate to the original model of (1.1) that can be probed in an analytic manner for any value of the random variables λ. Of course, while only a three-layer perceptron in shown in Figure 1.4, the number of layers and the number of hidden neurons in each layer are referred to as hyperparameters and the value of these hyperparameters are decided based on how complex the functional map F is. The deployment of ANN regression models for uncertainty quantification is left for Chapters 7–9.

When using SVR, the functional map $x = F(\lambda)$ is modeled as [53]

$$x(\lambda) = \langle \mathbf{w}, \boldsymbol{\psi}(\lambda) \rangle + b \tag{1.13}$$

where $\mathbf{w} = [w_1, \ldots, w_Q]$ is the vector of weights, $\psi = [\psi_1(\lambda), \ldots, \psi_Q(\lambda)]$ is the vector of multidimensional nonlinear functions, and b is the bias term. The main objective of training is to evaluate the weights and bias term from existing data so as to minimize the error between the predicted value of the response $x(\lambda)$ in (1.13) and the true value over the entire interval of support Ω. It is worthwhile to note that for SVR, training of the weights and bias terms can be done by solving both a nonlinear convex optimization problem or by solving a least-square formulation of the same [54]. The second approach is referred to as a least-square support vector regression (LS-SVR) [54]. Finally, we conclude this section by noting that while ANN and SVR are highly popular machine learning (ML) regression techniques, they are not the only ones. Other techniques such as Gaussian process regression, generalized regression neural networks, relevance vector machines, etc., also exist [47,48,58,59]. Nonetheless, all these techniques share the common feature of using training data to model/emulate the nonlinear input–output data dependency $x = F(\lambda)$ as shown in (1.12) and (1.13).

1.5 Organization of the book

This book consists of nine chapters organized into two parts – Chapters 2–5 and the remaining Chapters 6–9. Chapters 2–5 deal with the topic of aleatory uncertainty quantification of electromagnetic devices, circuits, and systems. Importantly, these chapters focus primarily on the generalized polynomial chaos approach as the key methodology to perform uncertainty quantification. Chapters 6–9 provide a broader examination of parametric uncertainties and deal with epistemic and mixed epistemic-aleatory uncertainty problems. These chapters focus on machine learning techniques as the key approach to tackle different forms of uncertainties.

Chapter 2 deals with the mathematical fundamentals underpinning the generalized polynomial chaos approach. In particular, this chapter explains in detail how polynomial chaos metamodels of any given order can be constructed for any general problem. Thereafter, the intrusive stochastic Galerkin approach [11,12,43] and the non-intrusive stochastic testing [60,61] and linear regression approaches [62,63] for training the coefficients of polynomial chaos metamodels is described. This leads to a deep understanding of the limitations and computational challenges facing the training of such metamodels as applicable for electromagnetic devices and circuits. The application of polynomial chaos metamodels for specifically electrical problems are covered in Chapter 3.

Chapter 4 explores the specific challenge of the curse of dimensionality plaguing surrogate modeling techniques in general and the polynomial chaos approach in particular. This chapter begins with an in-depth look into the notion of statistical sensitivity and the myriad of different sensitivity measures that can be calculated. Among the different types of statistical sensitivity measures possible, the Sobol's sensitivity indices is identified as a particularly effective measure that can be parlayed to identify the least important or impactful dimensions of the input. This allows for a more compact and easily trainable reduced dimensional polynomial chaos metamodel to be used to approximate the random outputs/responses of interest in electromagnetic devices and circuits. Multiple numerical examples are provided to compare the efficacy of existing dimension reduction schemes based on different sensitivity measures.

Chapter 5 presents a look into predictor-corrector type of algorithms for addressing the curse of dimensionality in polynomial chaos metamodels. This approach represents a completely new line of inquiry that is far removed from the dimension reduction approach. In particular, predictor-corrector algorithms intelligently crosscut the numerical efficiency of an approximate but compact low-fidelity model with the accuracy of a rigorous but cumbersome high-fidelity model in order to achieve a better computational time cost versus accuracy tradeoff when evaluating the coefficients of a metamodel. In this chapter, special attention is devoted to identifying the low- and high-fidelity models for multi-walled carbon nanotube interconnect networks.

Chapter 6 deals with constructing polynomial chaos metamodels when the input random parameters exhibit non-Gaussian correlations. Under non-Gaussian correlations, it is no longer possible to treat each random variable individually – in other words, it is no longer possible to construct the orthonormal basis functions for each random variable separately and then use their tensor products to obtain multi-dimensional basis functions. For such scenarios, Chapter 6 explains how a tensor-based approach can be exploited to directly construct the multidimensional orthonormal basis functions without the need of constructing unidimensional basis functions. Additionally, this chapter also details a new sparse solver that can train the polynomial chaos metamodels for such non-Gaussian correlated input variables.

Chapter 7 marks the shift from using PC metamodels to the emerging machine learning metamodels for uncertainty quantification of highly nonlinear systems. This chapter provides a thorough background of how metamodels can be trained using conventional regression techniques, such as LS-SVR, ridge regression, LASSO, etc. Thereafter, this chapter explains how metamodels can be trained using the more flexible machine learning techniques, such as SVR, LS-SVR, and Gaussian process regression. The advantage of machine learning regression techniques over conventional regression techniques is demonstrated using numerous examples.

Chapter 8 delves into the application of another type of machine learning technique – ANNs for the yield optimization of electromagnetic devices and circuits in the presence of uncertainties. Techniques to address the training cost for ANNs using space mapping techniques have been detailed in this chapter. Besides neural networks, the utilization of PC metamodels for yield optimization has also been discussed.

Chapter 9 focuses on order reduction clustering to filter the most influential random variables involved in uncertainty quantification of electromagnetic devices and circuits. Such techniques are very valuable when dealing with a high number of random and/or dependent parameters. The efficacy of the proposed methodology is illustrated through the assessment of electromagnetic composite models assuming stochastic structures within microwave frequency bandwidth.

Chapter 10 focuses on performing the more complicated mixed epistemic-aleatory uncertainty quantification using a combination of PC and machine learning metamodels. In this chapter, epistemic variables are characterized using the Dempster-Schafer theory of evidence while aleatory variables are characterized as conventional random variables. Thereafter, the Bayesian optimization technique have been combined with PC metamodels to quantify the impact of both forms of uncertainty simultaneously on the response in a numerically efficient manner.

1.6 Conclusion

In this first chapter, we lay out the motivation driving the current need for uncertainty quantification of engineering systems in general and electromagnetic devices, circuits, and systems in particular. In addition, we have looked closely at what the term "uncertainty" means, what are the various forms of uncertainty arising in electromagnetic devices and circuits, and what are the different numerical techniques that exists in order to quantify the impact of these different types of uncertainties. In particular, the idea of surrogate modeling has been explored and the spectral and machine learning approaches to construct surrogate models have been discussed. The future chapters will look more deeply into the spectral and machine learning-based surrogate models, their advantages and disadvantages, and what computational algorithms and techniques have emerged to address the disadvantages.

References

[1] S. Ramo, J. R. Whinnery, and T. Van Duzer, *Fields and Waves in Communication Electronics*. New York, NY: Wiley, 1994.

[2] J. M. Jin, *The Finite Element Method in Electromagnetics*, 2nd ed. New York, NY: John Wiley, 2002.

[3] A. Taflove and S. C. Hagness, *Computational Electrodynamics: The Finite-Difference Time-Domain Method*, 3rd ed. Norwood, MA: *Artech House*, 2005.

[4] R. F. Harrington, *Field Computation by Moment Methods*. New York, NY: Macmillan, 1968.

[5] K. S. Yee, "Numerical solution of initial boundary value problems involving Maxwell's equations in isotropic media," *IEEE Transactions on Antennas and Propagation*, vol. AP-14, no. 5, pp. 302–307, 1966.

[6] J. Vlach and K. Singhal, *Computer Methods for Circuit Analysis and Design*. New York, NY: Van Nostrand Reinhold, 1983.

[7] D. S. Boning, K. Balakrishnan, C. Hong, et al., "Variation," *IEEE Transactions on Semiconductor Manufacturing*, vol. 21, no. 1, pp. 63–71, 2008.

[8] R. Shen, S. X.-D. Tan, and H. Yu, *Statistical Performance Analysis and Modeling of Nanometer VLSI*. New York, NY: Springer, 2012.

[9] X. Li, J. Le, and L. T. Pillegi, "Statistical performance modeling and optimization, foundation and trends," *Electronic Design Automation*, vol. 1, no. 4, pp. 331–480, 2006.

[10] J. F. Swidzinski and K. Chang, "Nonlinear statistical modeling and yield estimation technique for use in Monte Carlo simulations [microwave devices and ICs]," *IEEE Transactions on Microwave Theory and Techniques*, vol. 48, no. 12, pp. 2316–2324, 2000.

[11] D. Xiu, *Numerical Methods for Stochastic Computations: A Spectral Method Approach*. Princeton, NJ: Princeton University Press, 2010.

[12] O. P. Le Maitre and O. M. Konio, *Spectral Methods for Uncertainty Quantification*. New York, NY: Springer-Verlag, 2010.

[13] W. L. Oberkampf, S. M. DeLand, B. M. Rutherford, K. V. Diegert, and K. F. Alvin, "Error and uncertainty in modeling and simulation," *Reliability Engineering and System Safety*, vol. 75, no. 3, pp. 333–357, 2002.

[14] R. Ghanem, D. Higdon, and H. Owhadi (Editors), *Handbook of Uncertainty Quantification*. Geneva: Springer International Publishing, 2017.

[15] A. D. Kiureghiana and O. Ditlevsen, "Aleatory or epistemic? Does it matter?," *Structural Safety*, vol. 31, no. 2, pp. 105–112, 2009.

[16] F. O. Hoffman and J. S. Hammonds, "Propagation of uncertainty in risk assessments: the need to distinguish between uncertainty due to lack of knowledge and uncertainty due to variability," *Risk Analysis*, vol. 14, no. 5, pp. 707–712, 1994.

[17] S. Ferson and L. R. Ginzburg, "Different methods are needed to propagate ignorance and variability," *Reliability Engineering and System Safety*, vol. 54, no. 2–3, pp. 133–144, 1996.

[18] A. P. Dempster, "Upper and lower probabilities induced by a multivalued mapping," *The Annals of Mathematical Statistics*, vol. 38, no. 2, pp. 325–339, 1967.

[19] G. Schafer, *A Mathematical Theory of Evidence.* Princeton, NJ: Princeton University Press, 1976.

[20] M. S. Eldred and L. P. Swiler, "Efficient algorithms for mixed aleatory-epistemic uncertainty quantification with applications to radiation hardened electronics," Technical Report SAND2009-5805, Sandia National Laboratories, 2009.

[21] M. S. Eldred, L. P. Swiler, and G. Tang, "Mixed aleatory-epistemic uncertainty quantification with stochastic expansions and optimization-based interval estimation," *Reliability Engineering and System Safety*, vol. 96, no. 9, pp. 1092–1113, 2011.

[22] J. Wu, Z. Luo, H. Li, and N. Zhang, "A new hybrid uncertainty optimization method for structures using orthogonal series expansion," *Applied Mathematical Modeling*, vol. 45, pp. 474–490, 2017.

[23] J. Wu, Z. Luo, N. Zhang, and Y. Zhang, "A new uncertain analysis method and its application in vehicle dynamics," *Mechanical Systems and Signal Processing*, vol. 50–51, pp. 659–675, 2015.

[24] J. L. Wu, Y. Q. Zhang, and Z. Luo, "A new method for hybrid of probability and interval analysis," in *Proceedings of the International Conference on Computational Methods*, 2014, pp. 1–8.

[25] J. Wu, Z. Luo, N. Zhang, and Y. Zhang, "A new interval uncertain optimization method for structures using Chebyshev surrogate models," *Computers and Structures*, vol. 146, pp. 185–196, 2015.

[26] S. Yin, D. Yu, H. Yin, and B. Zai, "Interval and random analysis for structure–acoustic systems with large uncertain-but-bounded parameters," *Computer Methods in Applied Mechanics and Engineering*, vol. 305, pp. 910–935, 2016.

[27] C. Jiang, J. Zheng, and X. Han, "Probability-interval hybrid uncertainty analysis for structures with both aleatory and epistemic uncertainties: a review," *Structural amd Multidisciplinary Optimization*, vol. 864, no. 4, pp. 1–18, 2017.

[28] S. R. Nassif, A. J. Strojwas, and S. W. Director, "A methodology for worst-case analysis of integrated circuits," *IEEE Transactions on Computer-Aided Design of Integrated Circuits Systems*, vol. 5, no. 1, pp. 104–113, 1986.

[29] W. Tian, X.-T. Ling, and R.-W. Liu, "Novel methods for circuit worstcase tolerance analysis," *IEEE Transactions on Circuits Systems I, Fundamental Theory and Applications*, vol. 43, no. 4, pp. 272–278, 1996.

[30] M. Tian and C. J. R. Shi, "Worst case tolerance analysis of linear analog circuits using sensitivity bands," *IEEE Transactions on Circuits Systems I, Fundamental Theory and Applications*, vol. 47, no. 8, pp. 1138–1145, 2000.

[31] L. Kolev, "Worst-case tolerance analysis of linear DC and AC electric circuits," *IEEE Transactions on Circuits Systems I, Fundamental Theory and Applications*, vol. 49, no. 12, pp. 1693–1701, 2002.

[32] M. Sengupta, S. Saxena, L. Daldoss, G. Kramer, S. Minehane, and J. Cheng, "Application-specfic worst case corners using response surface and statistical models," *IEEE Transactions on Computer-Aided Design of Integrated Circuits and Systems*, vol. 24, no. 9, pp. 1372–1380, 2005.

[33] E. H. Sandoval, F. Anstett-Collin, and M. Basset, "Sensitivity study of dynamic systems using polynomial chaos," *Reliability Engineering and System Safety*, vol. 104, pp. 15–26, 2012.

[34] A. Saltelli, S. Tarantola, F. Campolongo, and M. Ratto, *Sensitivity Analysis in Practice: A Guide to Assessing Scientific Models*. Sussex: John Wiley, 2004.

[35] E. Borgonovo, "A new uncertainty importance measure," *Reliability Engineering and System Safety*, vol. 92, no. 6, pp. 771–784, 2007.

[36] E. Borgonovo, "Measuring uncertainty importance: investigation and comparison of alternative approaches," *Risk Analysis*, vol. 26, no. 5, pp. 1349–1361, 2006.

[37] Q. Liu and T. Homma, "A new importance measure for sensitivity analysis," *Journal of Nuclear Science and Technology*, vol. 47, no. 1, pp. 53–61, 2013.

[38] G. S. Fishman, Monte Carlo: Concepts, Algorithms, and Applications. New York, NY: *Springer-Verlag*, 1996.

[39] H. Niederreiter, *Random Number Generation and Quasi-Monte Carlo Methods*. Philadelphia, PA: SIAM, 1992.

[40] H. Niederreiter, P. Hellekalek, G. Larcher, and P. Zinterhof, *Monte Carlo and Quasi-Monte Carlo Methods*. New York, NY: Springer-Verlag, 1998.

[41] D. Xiu, "Fast numerical methods for stochastic computations: A review," *Communications in Computational Physics*, vol. 5, no. 2–4, pp. 242–272, 2009.

[42] D. Xiu and G. E. Karniadakis, "The Wiener-Askey polynomial chaos for stochastic differential equations," *SIAM Journal on Scientific Computing*, vol. 24, no. 2, pp. 619–644, 2002.

[43] R. G. Ghanem and P. Spanos, *Stochastic Finite Elements: A Spectral Approach*. New York, NY: Springer-Verlag, 1991.

[44] H. Fang and M. F. Horstemeyer, "Global response approximation with radial basis functions," *Engineering Optimization,* vol. 38, no. 4, pp. 407–424, 2006.

[45] S. Chen, C. F. N. Cowan, and P. M. Grant, "Orthogonal least squares learning algorithm for radial basis function networks," *IEEE Transaction on Neural Networks*, vol. 2, no. 2, pp. 302–309, 1991.

[46] I. C. Kampolis, E. I. Karangelos, and K. C. Giannakoglu, "Gradient-assisted radial basis function networks: theory and applications," *Applied Mathematical Modeling*, vol. 28, no. 2, pp. 197–209, 2004.

[47] C. E. Rasmussen and C. K. I. Williams, *Gaussian Process for Machine Learning*. Cambridge, MA: MIT Press, 2006.

[48] T. J. Santner, B. J. Williams, and W. I. Notz, *The Design and Analysis of Computer Experiments*. New York, NY: Springer-Verlag, 2003.

[49] J. H. Friedman, "Multivariate adaptive regression splines," *The Annals of Statistics*, vol. 19, no. 1, pp. 1–67, 1991.

[50] M. B. Yelten, T. Zhu, S. Koziel, P. D. Franzon and M. B. Steer, "Demystifying surrogate modeling for circuits and systems," *IEEE Circuits and Systems Magazine*, vol. 12, no. 1, pp. 45–63, First quarter 2012.

[51] C. M. Bishop, *Neural Networks for Pattern Recognition*. New York, NY: Oxford University Press Inc., 1996.

[52] K. Hornik, M. Stinchcombe, and H. White, "Multi-layered feed-forward neural networks are universal approximations," *Neural Networks*, vol. 2, pp. 259–266, 1990.

[53] V. N. Vapnik, *Statistical Learning Theory*. Hoboken, NJ: John Wiley and Sons Inc., 1998.

[54] J. A. K. Suykens, T. V. Gestel. J. D. Brabanter, B. D. Moor, and J. Vandewalle, *Least Squares Support Vector Machines*. Singapore: World Scientific Publishing, 2002.

[55] R. Askey and J. Wilson, "Some basic hypergeometric polynomials that generalize Jacobi polynomials," *Memoirs of the American Mathematical Society*, vol. 54, no. 319, 1985.

[56] M. S. Eldred, "Recent advances in non-intrusive polynomial chaos and stochastic collocation methods for uncertainty analysis and design," in

Proceedings of 50th AIAA Structures, Structural Dynamics, and Materials Conference, Palm Springs, CA, 2009, pp. 1–37.

[57] F. Scarselli and A. C. Tsoi, "Universal approximations using feedforward neural networks: a survey of some existing methods and same new results," *Neural Networks*, vol. 11, pp. 15–37, 1998.

[58] M. E. Tipping, "Sparse Bayesian learning and the relevance vector machine," *Journal of Machine Learning Research*, vol. 1, pp. 211–244, 2001.

[59] D. F. Specht, "A general regression neural network," *IEEE Transactions on Neural Networks*, vol. 2, no. 6, pp. 568–576, 1991.

[60] Z. Zhang, T. A. El-Moselhy, I. M. Elfadel, and L. Daniel, "Stochastic testing method of transistor level uncertainty quantification based on generalized polynomial chaos," *IEEE Transactions on Computer-Aided Design*, vol. 32, no. 10, pp. 1533–1545, 2013.

[61] P. Manfredi and F. Canavero, "Efficient statistical simulation of microwave devices via stochastic testing-based circuit equivalents of nonlinear components," *IEEE Transactions on Microwave Theory and Techniques*, vol. 63, no. 5, pp. 1502–1511, 2015.

[62] M. Ahadi and S. Roy, "Sparse linear regression (SPLINER) approach for efficient multidimensional uncertainty quantification of high-speed circuits," *IEEE Transactions on Computer-Aided Design of Integrated Circuits and Systems*, vol. 35, no. 10, pp. 1640–1652, 2016.

[63] A. K. Prasad, M. Ahadi, and S. Roy, "Multidimensional uncertainty quantification of microwave/RF networks using linear regression and optimal design of experiments," *IEEE Transactions on Microwave Theory and Techniques*, vol. 64, no. 8, pp. 2433–2446, 2016.

Chapter 2

Polynomial chaos based uncertainty quantification in electrical engineering: theory

Paolo Manfredi[1] and Dries Vande Ginste[2]

We start by considering the generic stochastic function

$$y = f(\boldsymbol{\xi}), \tag{2.1}$$

where $\boldsymbol{\xi} = (\xi_1, \ldots, \xi_D) \in \mathbb{R}^D$ is a vector encompassing D uncertain parameters, whose joint probability density function (PDF) is denoted by $w(\boldsymbol{\xi})$. At this stage, y is understood to be an arbitrary output variable of interest that depends on the uncertain parameters $\boldsymbol{\xi}$, which in turn could be, for example, circuit element values, or geometrical dimensions affected by manufacturing tolerances, or material parameters that are known only to a limited precision. In the context of electrical engineering, y could be, for example, a circuit voltage or current, an S-parameter, an electromagnetic field intensity, an absorbed power, etc. The function $f : \mathbb{R}^D \rightarrow \mathbb{R}$ (or $f : \mathbb{R}^D \rightarrow \mathbb{C}$ for frequency-domain analysis) generically indicates the dependency of the output quantity y on the parameters $\boldsymbol{\xi}$, which is rarely analytically known in closed form. Indeed, y is typically the solution of a system of differential algebraic equations.

We aim at quantifying the stochasticity of y in terms, for example, of its mean, standard deviation, and probability distribution. An obvious and naive approach to the abovementioned problem is to apply Monte Carlo (MC) sampling [1], that is, draw a (large) number of random samples of the uncertain parameters $\boldsymbol{\xi}$ and calculate (2.1) for each configuration thereof. However, this strategy rapidly becomes intractable as the complexity of the model grows.

2.1 The polynomial chaos expansion

We introduce here a more efficient method based on polynomial chaos expansion (PCE). In this Section 2.1, we focus on the properties of PCEs. Methods to calculate its coefficients are described in Sections 2.2–2.4.

[1]Department of Electronics and Telecommunications, Politecnico di Torino, Turin, Italy
[2]quest, Department of Information Technology, Ghent University/imec, Ghent, Belgium

Under the assumption that the stochastic quantity y has a finite second-order moment (variance), the goal is to construct an approximation of the form

$$\widehat{y} = \widehat{f}(\boldsymbol{\xi}) = \sum_{k=1}^{K} y_k \varphi_k(\boldsymbol{\xi}) \approx f(\boldsymbol{\xi}) = \sum_{k=1}^{\infty} y_k \varphi_k(\boldsymbol{\xi}), \tag{2.2}$$

where the functions $\{\varphi_k\}_{k=1}^{\infty}$ form a complete basis of multivariate orthogonal polynomials of degree k and y_k are the corresponding coefficients. By taking the full basis, the representation is exact. Yet, it is truncated to a number of K terms for practical reasons, leading to the approximate model $\widehat{y}$.

The orthogonality of the basis functions is defined via the inner product

$$\langle f, g \rangle = \int_{\mathbb{R}^D} f(\boldsymbol{\xi}) g(\boldsymbol{\xi}) w(\boldsymbol{\xi}) d\boldsymbol{\xi}, \tag{2.3}$$

where $w(\boldsymbol{\xi})$ is the joint PDF of $\boldsymbol{\xi}$. Hence,

$$\langle \varphi_k, \varphi_m \rangle = \|\varphi_k\|^2 \delta_{km}, \tag{2.4}$$

where δ_{km} denotes the Kronecker delta and $\|\cdot\|$ is in this case the L^2-norm associated with the inner product (2.3).

In case of independent parameters, the joint PDF of $\boldsymbol{\xi}$ is merely the product of the PDFs of each variable ξ_j ($j = 1, \ldots, D$), that is,

$$w(\boldsymbol{\xi}) = \prod_{j=1}^{D} w_j(\boldsymbol{\xi}_j), \tag{2.5}$$

and in turn the basis functions are conveniently constructed as the product of univariate polynomials, that is,

$$\phi_k(\boldsymbol{\xi}) = \prod_{j=1}^{D} \varphi_{k_j}(\boldsymbol{\xi}_j), \tag{2.6}$$

with ϕ_{k_j} an orthogonal polynomial of degree k_j w.r.t. the distribution $w(\boldsymbol{\xi}_j)$, where orthogonality is again defined according to (2.3) with $D = 1$. Therefore, the orthogonality of such multivariate polynomials stems directly from the orthogonality of the univariate polynomials, since

$$\begin{aligned} \langle \phi_k, \phi_m \rangle &= \int_{\mathbb{R}^D} \prod_{j=1}^{D} \varphi_{k_j}(\boldsymbol{\xi}_j) \prod_{j=1}^{D} \varphi_{m_j}(\boldsymbol{\xi}_j) \prod_{j=1}^{d} w(\boldsymbol{\xi}_j) d\boldsymbol{\xi} \\ &= \prod_{j=1}^{D} \left(\int_{\mathbb{R}} \varphi_{k_j}(\boldsymbol{\xi}_j) \varphi_{m_j}(\boldsymbol{\xi}_j) w(\boldsymbol{\xi}_j) d\boldsymbol{\xi}_j \right) \neq 0 \Leftrightarrow k_j = m_j \quad \forall j. \end{aligned} \tag{2.7}$$

For standard distributions, well-known sets of polynomials exist [2]. For example, Hermite and Legendre polynomials are orthogonal w.r.t. the standard normal distribution and the uniform distribution in the interval $[-1, 1]$, respectively. For this reason, it is always convenient to normalize parameters exhibiting Gaussian or uniform uncertainty by means of proper translation and scaling. In the general case of arbitrary, and possibly correlated distributions, numerical techniques are available to generate suitable orthogonal polynomials starting from a set of multivariate monomials, for example, via Gram-Schmidt orthogonalization [3] or Cholesky decomposition [4].

2.1.1 *Truncation*

Without loss of generality, we consider the case of independent uncertain parameters. Because of (2.6), we can rewrite (2.2) as

$$\widehat{y} = \sum_{\mathbf{k} \in \mathscr{K}} y_{\mathbf{k}} \varphi_{\mathbf{k}}(\boldsymbol{\xi}), \tag{2.8}$$

where $\mathbf{k} = (k_1, \ldots, k_D) \in \mathbb{N}^D$ is a multi-index indicating the degrees of the univariate polynomials in each variable. Several truncation schemes are available for the definition of the set $\mathscr{K}$. The most common one bounds the *total degree* of the multivariate basis functions to a given order P, that is,

$$\mathscr{K} = \left\{ \mathbf{k} : \sum_{j=1}^{D} k_j = \|\mathbf{k}\|_1 \le P \right\}. \tag{2.9}$$

The resulting cardinality of $\mathscr{K}$, that is, the total number of terms in the PCE, is

$$K = \frac{(P+D)!}{P!\, D!}. \tag{2.10}$$

Another possible choice is the tensor-product truncation, retaining all multi-indices such that $\|\mathbf{k}\|_\infty \le P$, which however yields a much larger expansion with $K = (P+1)^D$ terms. A popular alternative, yielding a sparser PCE, is the hyperbolic truncation [5], retaining multi-indices such that $\|\mathbf{k}\|_u \le P$, with $0 < u < 1$. In many applications, order $P = 2$ or $P = 3$ usually provides satisfactory accuracy [6,7].

In the following, the PCE notations of (2.2) and (2.8) are interchanged. It is understood that φ_k, with a scalar (1-based) index k, indicates a multivariate polynomial with a corresponding multi-index $\mathbf{k}$ belonging to set $\mathscr{K}$, which is further assumed to be sorted in graded lexicographic order. Table 2.1 illustrates the set of basis functions for a total-degree expansion of order $P = 2$ and $D = 3$ Gaussian random variables, based on the first three univariate probabilists' Hermite polynomials: $\phi_0 = 1$, $\phi_1 = \xi$ and $\phi_2 = \xi^2 - 1$.

From now on, we further assume that the basis functions are *orthonormal* (i.e., $\|\varphi_k\| = 1\ \forall k$). This is usually not the case for classical polynomials. For example, the norm of the classical probabilists' Hermite polynomials is $\|\phi_k\| = \sqrt{k!}$. A proper normalization is thus required and readily implemented.

2.1.2 *Statistical properties*

The main reason for adopting the PCE (2.2) is that it provides a compact closed-form model which is accurate in statistical sense. Indeed, the error norm obeys [8]

$$\lim_{P \to \infty} \|y - \widehat{y}\| = \left(\int_{\mathbb{R}^D} (f(\boldsymbol{\xi}) - \widehat{f}(\boldsymbol{\xi}))^2 w(\boldsymbol{\xi}) d\boldsymbol{\xi} \right)^{\frac{1}{2}} = 0, \tag{2.11}$$

Table 2.1 Trivariate basis functions for a second-order total-degree truncation based on Hermite polynomials

Scalar index k	Multi-index k	Basis function φ_k	Degree
1	$(0,0,0)$	$\phi_0(\xi_1)\phi_0(\xi_2)\phi_0(\xi_3) = 1$	0
2	$(1,0,0)$	$\phi_1(\xi_1)\phi_0(\xi_2)\phi_0(\xi_3) = \xi_1$	1
3	$(0,1,0)$	$\phi_0(\xi_1)\phi_1(\xi_2)\phi_0(\xi_3) = \xi_2$	1
4	$(0,0,1)$	$\phi_0(\xi_1)\phi_0(\xi_2)\phi_1(\xi_3) = \xi_3$	1
5	$(2,0,0)$	$\phi_2(\xi_1)\phi_0(\xi_2)\phi_0(\xi_3) = \xi_1^2 - 1$	2
6	$(1,1,0)$	$\phi_1(\xi_1)\phi_1(\xi_2)\phi_0(\xi_3) = \xi_1\xi_2$	2
7	$(1,0,1)$	$\phi_1(\xi_1)\phi_0(\xi_2)\phi_1(\xi_3) = \xi_1\xi_3$	2
8	$(0,2,0)$	$\phi_0(\xi_1)\phi_2(\xi_2)\phi_0(\xi_3) = \xi_2^2 - 1$	2
9	$(0,1,1)$	$\phi_0(\xi_1)\phi_1(\xi_2)\phi_1(\xi_3) = \xi_2\xi_3$	2
10	$(0,0,2)$	$\phi_0(\xi_1)\phi_0(\xi_2)\phi_2(\xi_3) = \xi_3^2 - 1$	2

typically with exponential convergence rate. As is seen from (2.11), the quadratic error is weighted by the probability distribution of the uncertain parameters. Briefly speaking, this implies that a larger error is tolerated for unlikely occurrences, thus still guaranteeing an overall good accuracy in statistical sense. In turn, this implies that a small error is obtained when applying MC sampling to the approximate model (2.2) (instead of the original function (2.1)). Hence, via MC sampling, any statistical information is inexpensively extracted from the PCE, as this closed-form expression is rapidly evaluated. Moreover, owing to the orthogonality of the basis functions, the first two statistical moments are obtained analytically from the PCE coefficients as

$$\mathrm{E}\{y\} \approx \mathrm{E}\{\widehat{y}\} = \int_{\mathbb{R}^D} \sum_{\mathbf{k}\in\mathscr{K}} y_{\mathbf{k}}\phi_{\mathbf{k}}(\boldsymbol{\xi})w(\boldsymbol{\xi})d\boldsymbol{\xi} = \sum_{\mathbf{k}\in\mathscr{k}} y_{\mathbf{k}}\langle\phi_{\mathbf{k}}, 1\rangle = y_{\mathbf{0}}, \tag{2.12}$$

and

$$\begin{aligned}\mathrm{Var}\{y\} \approx \mathrm{Var}\{\widehat{y}\} &= \int_{\mathbb{R}^D} \left(\sum_{\mathbf{k}\in\mathscr{K}} y_{\mathbf{k}}\phi_{\mathbf{k}}(\boldsymbol{\xi}) - y_{\mathbf{0}}\right)^2 w(\boldsymbol{\xi})d\boldsymbol{\xi} \\ &= \sum_{\mathbf{k}\in\mathscr{K}\setminus\mathbf{0}} y_{\mathbf{k}}^2 \underbrace{\int_{\mathbb{R}^D} \phi_{\mathbf{k}}^2(\boldsymbol{\xi})w(\boldsymbol{\xi})d\boldsymbol{\xi}}_{\|\phi_{\mathbf{k}}\|^2=1} \\ &\quad + 2\sum_{\mathbf{k}\in\mathscr{K}\setminus\mathbf{0}} \sum_{\mathbf{m}<\mathbf{k}} \underbrace{\int_{\mathbb{R}^D} \phi_{\mathbf{m}}(\boldsymbol{\xi})\phi_{\mathbf{k}}(\boldsymbol{\xi})w(\boldsymbol{\xi})d\boldsymbol{\xi}}_{\langle\phi_{\mathbf{m}},\phi_{\mathbf{k}}\rangle=0} = \sum_{\mathbf{k}\in\mathscr{K}\setminus\mathbf{0}} y_{\mathbf{k}}^2,\end{aligned} \tag{2.13}$$

leveraging the (normalized) basis function $\phi_{\mathbf{0}}(\boldsymbol{\xi}) = 1$, where $\mathbf{0} = (0,\ldots,0)$ is the null element of $\mathbb{N}^D$. In other words, the variance is given by the sum of the squares of all coefficients but the zeroth-order one, which provides instead the mean.*

*For complex-valued outputs, the variance is given by the sum of the squared magnitudes of the coefficients.

It is also worth mentioning that any linear function f of the uncertain parameters $\boldsymbol{\xi}$ follows the same type of distribution as $\boldsymbol{\xi}$, and is represented exactly by a first-order PCE, whereas a deterministic (i.e., constant) quantity can be expressed as a zeroth-order PCE.

2.1.3 Calculation of the expansion coefficients

The problem is now reduced to the determination of the PCE coefficients y_k, $k = 1, \ldots, K$. Following the Hilbert projection theorem, their rigorous computation is by projection of the stochastic function onto the basis, that is,

$$y_k = \langle y, \phi_k \rangle = \int_{\mathbb{R}^D} f(\boldsymbol{\xi}) \phi_k(\boldsymbol{\xi}) w(\boldsymbol{\xi}) d\boldsymbol{\xi}, \tag{2.14}$$

which minimizes the error norm $\|y - \widehat{y}\|$. Obviously, exact analytical integration is possible only for a limited set of closed-form functions, while numerical integration rapidly becomes intractable as the dimensionality D increases and/or when the function (2.1) is expensive to compute, as is often the case in practice. Therefore, several techniques are employed for the effective calculation of the PCE coefficients, the most popular of which are summarized in the following sections.

2.2 Non-intrusive methods

A method is referred to as "non-intrusive" when it relies only on a collection of response samples of f for the determination of the PCE coefficients. Non-intrusive methods are thus sampling-based techniques, similar to the MC method, yet they exploit specific assumptions (such as smoothness of the stochastic function) to improve the accuracy and efficiency of the calculation.

2.2.1 Pseudo-spectral collocation method

The pseudo-spectral collocation (PSC) method approximates the projection integral (2.14) by means of a suitable Gaussian quadrature. We start from the univariate case ($D = 1$), for which such an approximation reads

$$y_k = \int_{\mathbb{R}} f(\xi) \varphi_k(\xi) w(\xi) d\xi \approx \sum_{q=1}^{Q} f(\xi_q) \varphi_k(\xi_q) w_q, \tag{2.15}$$

where $\mathcal{Q} = \left\{ (\xi_q, w_q) \right\}_{q=1}^{Q}$ is a set of quadrature node-weight pairs. The nodes ξ_q coincide with the roots of the Qth-order orthogonal polynomial for $w(\xi)$, that is, $\phi_Q(\xi_q) = 0 \; \forall q = 1, \ldots, Q$. For standard polynomials, closed-form formulae are available for the corresponding weights, while for the general case, nodes and weights are found by solving an eigenvalue problem [9].

The method requires only Q evaluations of the stochastic function $f(\boldsymbol{\xi})$ at the quadrature nodes, which are the same for every coefficient y_k. A useful property of Gaussian quadratures is that they are *exact* for a polynomial integrand of degree $2Q-1$ or less, not including the weighting function $w(\boldsymbol{\xi})$. In (2.15), the integrand is thus the product of $f(\boldsymbol{\xi})$ and $\phi_k(\boldsymbol{\xi})$. While the latter is a polynomial of degree k, the former is implicitly assumed to be well approximated by a polynomial of degree P when using a Pth-order PCE. Therefore, the integrand has approximately a degree of $P+k \leq 2P$. This suggests taking (at least) $Q = P+1$ points for the quadrature. In practice, the accuracy of the calculation will reduce for higher-order coefficients, as a result of the increment of the integrand order. However, this has a marginal impact on the overall accuracy, as the higher-order coefficients are also becoming increasingly negligible.

For multidimensional problems ($D > 1$), a straightforward extension is to adopt a tensor-product rule:

$$y_k \approx \sum_{q_1=1}^{Q_1} \cdots \sum_{q_D=1}^{Q_D} f(\boldsymbol{\xi}_{q_1}, \ldots, \boldsymbol{\xi}_{q_D}) \varphi_k(\boldsymbol{\xi}_{q_1}, \ldots, \boldsymbol{\xi}_{q_D}) w_{q_1} \cdots w_{q_D}$$

$$= \sum_{q=1}^{Q} f(\boldsymbol{\xi}_q) \phi_k(\boldsymbol{\xi}_q) w_q, \tag{2.16}$$

where now

$$\mathcal{Q} = \left\{ (\boldsymbol{\xi}_q, w_q) \right\}_{q=1}^{Q} = \bigotimes_{j=1}^{D} \left\{ (\boldsymbol{\xi}_{q_j}, w_{q_j}) \right\}_{q_j=1}^{Q_j}. \tag{2.17}$$

This leads to a total number of $Q = \prod_{j=1}^{d} Q_j$ function evaluations, or $Q = (P+1)^D$ if the same order is used along each dimension, which rapidly becomes intractable. More advanced solutions make use of sparse grids [10,11], although the sparsity becomes significant only for high-order integrations, which are often unnecessary.

2.2.2 Least-square regression method

Another popular, non-intrusive method is the least-square regression (LSR) method, which calculates the PCE coefficients by fitting the model (2.2) to a set of available data in the least-square sense. We assume that a set of L samples of the uncertain parameters and corresponding function evaluations, $\mathcal{R} = \{(\boldsymbol{\xi}_l, f(\boldsymbol{\xi}_l))\}_{l=1}^{L}$, be available. For each of these samples, the residual between the PCE model and the actual function reads

$$r_l = \widehat{f}(\boldsymbol{\xi}_l) - f(\boldsymbol{\xi}_l) = \sum_{k=1}^{K} y_k \varphi_k(\boldsymbol{\xi}_l) - f(\boldsymbol{\xi}_l), \quad l = 1, \ldots, L. \tag{2.18}$$

In matrix form, $\forall l$, we obtain the system

$$\underbrace{\begin{pmatrix} r_1 \\ \vdots \\ r_L \end{pmatrix}}_{\mathbf{r}} = \underbrace{\begin{pmatrix} \varphi_1(\boldsymbol{\xi}_1) & \cdots & \varphi_K(\boldsymbol{\xi}_1) \\ \vdots & \ddots & \\ \varphi_1(\boldsymbol{\xi}_L) & & \varphi_K(\boldsymbol{\xi}_L) \end{pmatrix}}_{\boldsymbol{\Psi}} \underbrace{\begin{pmatrix} y_1 \\ \vdots \\ y_K \end{pmatrix}}_{\widetilde{\mathbf{y}}} - \underbrace{\begin{pmatrix} f(\boldsymbol{\xi}_1) \\ \vdots \\ f(\boldsymbol{\xi}_L) \end{pmatrix}}_{\mathbf{f}}. \tag{2.19}$$

When the model is exact, then $\mathbf{r} = \mathbf{0}$, and the solution for the PCE coefficients is found by taking any set of $L = K$ distinct samples of the uncertain parameters and inverting the resulting (square) system. However, in practice, the model is never exact. In this case, the solution is found by taking $L > K$ regression samples and computing the set of PCE coefficients that minimize the residual norm, that is,

$$\widetilde{\mathbf{y}}^* = \arg\min_{\widetilde{\mathbf{y}}} \|\boldsymbol{\Psi}\,\widetilde{\mathbf{y}} - \mathbf{f}\| = \boldsymbol{\Psi}^+\mathbf{f}, \tag{2.20}$$

where $\boldsymbol{\Psi}^+ = (\boldsymbol{\Psi}^{\mathrm{T}}\boldsymbol{\Psi})^{-1}\boldsymbol{\Psi}^{\mathrm{T}}$ is the Moore-Penrose pseudo-inverse of matrix $\boldsymbol{\Psi}$. The accuracy improves by letting the regression system (2.19) be sufficiently over-dermined, that is, by using $L \gg K$. Literature recommendations prescribe to use at least $L = 2K$ regression samples [12].

To obtain the correct solution, it is important that the regression samples $\boldsymbol{\xi}_l$ are drawn according to the actual distribution $w(\boldsymbol{\xi})$ of the uncertain parameters. If this is the case, then

$$\begin{aligned} \|\mathbf{r}\|^2 &= \sum_{l=1}^{L} r_l^2 = \sum_{l=1}^{L} \left(\widehat{f}(\boldsymbol{\xi}_l) - f(\boldsymbol{\xi}_l)\right)^2 \approx L \cdot \mathrm{E}\left\{(\widehat{y} - y)^2\right\} \\ &= L\int_{\mathbb{R}^D} \left(\widehat{f}(\boldsymbol{\xi}) - f(\boldsymbol{\xi})\right)^2 w(\boldsymbol{\xi})d\boldsymbol{\xi} = L\|\widehat{y} - y\|^2, \end{aligned} \tag{2.21}$$

which means that the residual norm is a MC estimate of the L^2 norm of the error. Therefore, minimizing the residual norm is roughly equivalent to minimizing the error norm, thus leading to a solution that is consistent with the projection (2.14).

2.2.3 *Stochastic testing method*

Originally presented in an intrusive fashion [13], the stochastic testing (ST) method was later reformulated as a non-intrusive technique [14]. It can be regarded as a sparse Gaussian quadrature, since it amounts to selecting a subset of $L = K$ nodes, called "ST points", from a Pth-order tensor-product Gauss quadrature (2.16). Function values at other points are then found via interpolation. The solution for the PCE coefficients is in fact also equivalent to (2.20) with a square matrix $\boldsymbol{\Psi}$, which now collects the basis functions calculated at the ST points. The vector of PCE coefficients is obtained as

$$\widetilde{\mathbf{y}} = \boldsymbol{\Psi}^{-1}\mathbf{f}, \tag{2.22}$$

and the resulting PCE model interpolates the function values at the ST points.

To illustrate the algorithm for the selection of the testing points, we first assume that the set of node-weight pairs (2.17) of the full tensor-product Gauss quadrature rule is sorted according to the weight in decreasing order, and further, we define a vector-valued function

$$\varphi(\boldsymbol{\xi}) = (\varphi_1(\boldsymbol{\xi}), \ldots, \varphi_K(\boldsymbol{\xi}))^{\mathrm{T}} \in \mathbb{R}^K,$$

that is, a column vector collecting the values of all PCE basis functions for a given $\boldsymbol{\xi}$. The algorithm for the node selection is summarized as follows:

- Take the first quadrature node $\boldsymbol{\xi}_1$ (i.e., the one with largest weight) as the first ST point.
- Initialize column matrix $\mathbf{M}$ as

$$\mathbf{M} = \frac{\varphi(\boldsymbol{\xi}_1)}{\|\varphi(\boldsymbol{\xi}_1)\|}.$$

- Start an iterative selection procedure:
  1. Take the next quadrature point $\boldsymbol{\xi}_q$ and calculate

$$\mathbf{V} = \varphi\,(\boldsymbol{\xi}_q) - \mathbf{M}\mathbf{M}^{\mathrm{T}}\varphi\,(\boldsymbol{\xi}_q).$$

  2. Check whether

$$\frac{\|\mathbf{V}\|}{\|\varphi\,(\boldsymbol{\xi}_q)\|} > \beta,$$

where β is a predefined threshold (e.g., $\beta = 10^{-3}$). If this is not the case, discard $\boldsymbol{\xi}_q$ and return to point 1. Otherwise, select $\boldsymbol{\xi}_q$ as an additional ST point and update matrix $\mathbf{M}$ by appending a column as follows:

$$\mathbf{M} = \left(\mathbf{M} \quad \frac{\mathbf{V}}{\|\mathbf{V}\|}\right).$$

So, briefly speaking, the algorithm selects $\boldsymbol{\xi}_q$ if $\varphi\,(\boldsymbol{\xi}_q)$ has a large enough orthogonal component to $\mathbf{M}$. When $L = K$ testing points are selected, the procedure is terminated.

Other similar techniques use a different strategy for the selection of the subset of Gauss quadrature nodes. For example, the D-optimal criterion is used in [15].

Some important considerations follow:

- The method requires to compute only $L = K$ response samples, thus less than the PSC and LSR methods.
- The univariate case ($D = 1$) is trivial: since $Q = P + 1 = K$, all quadrature nodes are taken as testing points, and the ST method is equivalent to the PSC method.
- For $D > 1$, $Q > K$, and the ST method is in fact a sparse version of a Gaussian quadrature.

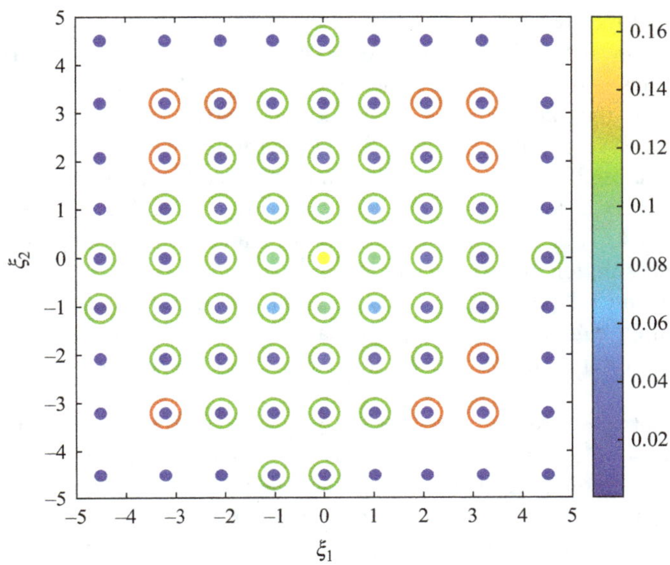

Figure 2.1 Illustration of the ST point selection for $P = 8$, $D = 2$, and Gaussian distribution. The dots are the nodes of the initial tensor-product Gauss quadrature, with color indicating the weight. Green and red circles denote selected and discarded nodes, respectively

- The rejection of quadrature nodes is appreciable for large P and/or D. Otherwise, the algorithm tends to just select the first K quadrature nodes with the largest weights.

Figure 2.1 illustrates the procedure of selecting the ST points for a two-dimensional Gaussian distribution and an eighth-order total-degree PCE expansion (i.e., $P = 8$ and $D = 2$). The dots represent the starting $Q = 81$ nodes of the tensor-product Gauss quadrature rule, with color indicating the corresponding weight. Green and red circles denote selected and discarded nodes, respectively, based on a threshold of $\beta = 10^{-3}$. The algorithm starts from the central node, $\boldsymbol{\xi}_1 = (0, 0)$, and moves outwards, as the weight reduces radially with circular symmetry. The algorithm stops when $K = 45$ points are selected.

2.2.4 Features

First, it is useful to show that (2.16) can be cast in matrix form in analogy with (2.20) and (2.22), that is,

$$\widetilde{\mathbf{y}} = \boldsymbol{\Psi}'\mathbf{f} = \boldsymbol{\Psi}^{\mathrm{T}}\mathbf{W}\mathbf{f}, \tag{2.23}$$

where $\boldsymbol{\Psi}$ and $\boldsymbol{f}$ are defined as in (2.19), but here $L = Q$, and with $\mathbf{W} = \mathrm{diag}(w_1, \ldots, w_Q) \in \mathbb{R}^{Q \times Q}$, such that $\boldsymbol{\Psi}' \in \mathbb{R}^{K \times Q}$ with entries $\Psi'_{kq} = \varphi_k(\boldsymbol{\xi}_q) w_q$. It is possible to prove that if $Q = K$ (this is the case for $D = 1$ or for $D > 1$ with appropriate truncation of the tensor-product), then the PSC method interpolates the function values at the quadrature nodes like the ST method, thus corroborating the statement that the latter can be regarded as a sparse version of the former. Moreover, in this case $\Psi' = \Psi^{\mathrm{T}}\mathbf{W} = \Psi^{-1}$, consistently with (2.22).

Second, a peculiar aspect, which is common to the PSC, LSR, and ST methods, is that the matrix Ψ appearing in (2.20), (2.22), and (2.23) does not depend on the data in vector $\mathbf{f}$. This, at first glance irrelevant feature, has the important implication that the solution of these methods is easily vectorized for multiple output quantities y which depend on the same uncertain parameters, as is often the case in practice (think, e.g., of the transient behavior of a multiplicity of voltages and currents within a circuit, or the scattering parameters of a multiport electromagnetic structure).

Next, let us assume that L samples of a multioutput function $\mathbf{y} = f(t; \boldsymbol{\xi}) : \mathbb{R}^D \to \mathbb{R}^M$, where t is a *continuous* sweep variable such as time, frequency, power, or temperature, are available at N discrete points of the continuous variable. By rearranging such a data *tensor* into a matrix $\mathbf{F} \in \mathbb{R}^{L\times MN}$, the solution of the outlined non-intrusive methods is readily obtained by replacing $\mathbf{f}$ with $\mathbf{F}$ in (2.20), (2.22), or (2.23), which yields a matrix collecting the PCE coefficients of the each output variable and at each discrete point with the same arrangement as the data in matrix $\mathbf{F}$. Hence, the PCE coefficients for each output at each discrete point are computed at once, with a considerable computational efficiency. It is important to point out that this is no longer the case for more advanced non-intrusive techniques, such as the sparse and rational methods briefly discussed in Section 2.4, in which the information on the output data is included in the optimization process.

Lastly, another very important feature of non-intrusive methods is that they readily apply, without modification, also to nonlinear systems. This is because they merely require samples of f, which is in contrast to *intrusive* methods. The most popular intrusive method is the stochastic Galerkin (SG) method, which is discussed in Section 2.3.

2.2.5 Comparison

Since they rely on a collection of data samples, the efficiency of non-intrusive methods is mainly determined by the number of samples they require. Indeed, in many applications, data acquisition is the most time-consuming step, whereas the computational overhead required by the technique itself is usually marginal. An exception could be the ST method, whose algorithm of node selection becomes costly for large dimensions. Nevertheless, the testing points and the corresponding interpolation matrix Ψ can be pre-computed and stored into look-up tables for a predefined set of dimensions, expansion orders, and distributions, thus avoiding the need for an on-the-fly calculation. The time required to manipulate matrix Ψ is also negligible in most cases.

Concerning the overall performance, the PSC method can be regarded as the most accurate non-intrusive method, as it provides a good approximation of the projection (2.14). Nevertheless, it also requires the largest number of samples, rendering it impractical for large dimensions. The LSR method requires, instead, a number of samples that is proportional to the number of PCE coefficients to estimate, and it is therefore attractive as long as this number is significantly lower than the one of a typical MC analysis. However, since it uses random samples, the solution itself is stochastic (as is the case for MC analysis). When the output exhibits a large variability, the solutions $\widetilde{\mathbf{y}}^*$ (2.20) of several runs of the LSR method might deviate considerably from one another, and might therefore become occasionally inaccurate. The use of Latin hypercube sampling designs usually helps to reduce the deviation. Finally, the ST method provides a deterministic solution that uses the lowest number of data samples. Nonetheless, it has been reported to be inaccurate in some applications.

2.3 The SG method

As opposed to non-intrusive approaches, the SG method operates directly on the system of governing equations that relates the uncertain parameters to the outputs of interest [2]. Therefore, assume—to clarify the discussion—that the stochastic output $\mathbf{y}$ collects the M components of a (linear) system of differential equations

$$\dot{\mathbf{y}}(t, \boldsymbol{\xi}) = \mathbf{A}(\boldsymbol{\xi})\mathbf{y}(t, \boldsymbol{\xi}) + \mathbf{u}(t, \boldsymbol{\xi}), \tag{2.24}$$

where $\mathbf{A} \in \mathbb{R}^{M \times M}$ and $\mathbf{u} \in \mathbb{R}^M$ are a generic (stochastic) system matrix and input vector, respectively. In the framework of this work (2.24) could be, for example, the state-space equation governing electrical circuits, in which the output variables are capacitor voltages and inductor currents. The method is readily extended to similar and more general formulations, like modified nodal analysis (MNA) [16], telegrapher's equations [17], or electromagnetic equations [18].

In a first step, all stochastic quantities, including the system matrix $\mathbf{A}$ and input vector $\mathbf{u}$, are approximated by means of PCEs:

$$\mathbf{y}(t, \boldsymbol{\xi}) \approx \widehat{\mathbf{y}}(t, \boldsymbol{\xi}) = \sum_{k=1}^{K} \mathbf{y}_k(t)\varphi_k(\boldsymbol{\xi}), \tag{2.25a}$$

$$\mathbf{A}(\boldsymbol{\xi}) \approx \widehat{\mathbf{A}}(\boldsymbol{\xi}) = \sum_{k=1}^{K} \mathbf{A}_k\varphi_k(\boldsymbol{\xi}), \tag{2.25b}$$

$$\mathbf{u}(t, \boldsymbol{\xi}) \approx \widehat{\mathbf{u}}(t, \boldsymbol{\xi}) = \sum_{k=1}^{K} \mathbf{u}_k(t)\varphi_k(\boldsymbol{\xi}) \tag{2.25c}$$

where, for the sake of simplicity, the same truncation is assumed for all PCEs. It is understood that matrix- and vector-valued quantities are described by corresponding coefficients (indicated with subscript k in (2.25)) of compatible size, and all properties described in Section 2.1.2 apply element-wise.

The determination of the expansion coefficients of $\mathbf{A}$ and $\mathbf{u}$ is trivial for entries that are constant (i.e., deterministic) or linear w.r.t. the uncertain parameters.

Otherwise, it can be carried out using any of the non-intrusive methods described in Section 2.2, or even with the direct calculation of (2.14) via accurate numerical techniques. At this point, it should be noted that the relationship between the system matrix and the uncertain parameters is often analytical, thus making this preliminary step relatively affordable.

Next, substitution of the PCE approximations (2.25) into (2.24), and using the basis function φ_m to *test* the resulting equation leveraging the inner product (2.3), leads to

$$\langle \dot{\widehat{\mathbf{y}}}, \varphi_m \rangle = \langle \widehat{\mathbf{A}}\widehat{\mathbf{y}} + \widehat{\mathbf{u}}, \varphi_m \rangle \tag{2.26}$$

The linearity of the inner product allows us to rewrite (2.26) explicitly as

$$\sum_{k=1}^{K} \dot{\mathbf{y}}_k(t) \ \langle \varphi_k, \varphi_m \rangle = \sum_{k=1}^{K}\sum_{n=1}^{K} \mathbf{A}_k \mathbf{y}_n(t) \ \langle \varphi_k \varphi_n, \varphi_m \rangle + \sum_{k=1}^{K} \mathbf{u}_k(t) \ \langle \varphi_k, \varphi_m \rangle. \tag{2.27}$$

By invoking the orthogonality condition (2.4), this is further reduced to

$$\dot{\mathbf{y}}_m(t) = \sum_{k=1}^{K}\sum_{n=1}^{K} \alpha_{knm} \mathbf{A}_k \mathbf{y}_n(t) + \mathbf{u}_m(t), \tag{2.28}$$

where the (real) scalar coefficients

$$\alpha_{knm} = \langle \varphi_k \varphi_n, \varphi_m \rangle = \int_{\mathbb{R}^D} \varphi_k(\boldsymbol{\xi}) \varphi_n(\boldsymbol{\xi}) \varphi_m(\boldsymbol{\xi}) w(\boldsymbol{\xi}) d\boldsymbol{\xi} \tag{2.29}$$

have been introduced. It is worth noting that (2.28) is a *deterministic* equation in the unknown PCE coefficients $\mathbf{x}_n$ of $\mathbf{x}$, since the projection (2.26) eliminates the dependence on the uncertain parameters $\boldsymbol{\xi}$. In order to obtain a square system, the equations (2.34) are constructed for $m = 1, \ldots, K$. Casting them in matrix form yields

$$\dot{\widetilde{\mathbf{y}}}(t) = \widetilde{\mathbf{A}}\widetilde{\mathbf{y}}(t) + \widetilde{\mathbf{u}}(t), \tag{2.30}$$

where $\widetilde{\mathbf{y}} = (\mathbf{y}_1^{\mathrm{T}}, \ldots, \mathbf{y}_K^{\mathrm{T}})^{\mathrm{T}}$ and $\widetilde{\mathbf{u}} = (\mathbf{u}_1^{\mathrm{T}}, \ldots, \mathbf{u}_K^{\mathrm{T}})^{\mathrm{T}}$ are vectors $\in \mathbb{R}^{KM}$ collecting the PCE coefficients of $\mathbf{y}$ and $\mathbf{u}$, respectively, whereas $\widetilde{\mathbf{A}} \in \mathbb{R}^{KM \times KM}$ is a block matrix which is constructed from the coefficients of $\mathbf{A}$ as

$$\widetilde{\mathbf{A}} = \sum_{k=1}^{K} \mathbf{A}_k \otimes \mathbf{M}_k, \tag{2.31}$$

where $\mathbf{M}_k \in \mathbb{R}^{K \times K}$ is an auxiliary matrix with entries $[\mathbf{M}_k]_{mn} = \alpha_{knm}$, and $\otimes$ denotes the Kronecker product.

At this point, it is important to note that (2.30) is a *coupled* deterministic system of the same type as (2.24), though K times larger, in which $\widetilde{\mathbf{A}}$ and $\widetilde{\mathbf{u}}$ play the role of an *augmented* system matrix and input vector, respectively. The solution of this augmented, deterministic system (2.30) provides the sought-after PCE coefficients of the solution $\mathbf{y}$ of the original, stochastic equation (2.24). The SG method is readily generalized to an arbitrary system of linear differential algebraic equations. Specifically, an augmented equation of the same type is obtained by replacing (i) stochastic vectors with the vector of their PCE coefficients, and (ii) stochastic matrices with an augmented matrix constructed as in (2.31). It is useful to mention that, in the special case of a deterministic vector $\mathbf{u}$ or matrix $\mathbf{A}$, their augmented counterparts reduce to

$$\widetilde{\mathbf{u}}(t) = (\mathbf{u}^{\mathrm{T}}(t), 0, \ldots, 0)^{\mathrm{T}}, \tag{2.32a}$$

or

$$\widetilde{\mathbf{A}} = \begin{pmatrix} \mathbf{A} & & \\ & \ddots & \\ & & \mathbf{A} \end{pmatrix}, \tag{2.32b}$$

respectively.

In the construction of the augmented matrix, the coefficients α_{knm} must be computed by evaluating the triple-product integral (2.29). In the general case, this is readily achieved by numerical integration, which is relatively straightforward as the integrands are polynomials. When dealing with independent uncertain parameters, the calculation is further eased by reducing the integral to the product of univariate integrals, similar as in (2.7). Furthermore, for classical polynomials, there often exist closed-form formulae (see, e.g., [19] for Hermite polynomials and [20] for Legendre polynomials). At least for standard distributions, it is convenient to precompute these coefficients up to a predefined order and make them available via look-up tables.

2.3.1 Discussion

Sampling-based stochastic techniques, such as the MC method and the non-intrusive techniques of Section 2.2, are readily implemented. Indeed, the data (samples) can be collected by means of available (sometimes commercial) software, developed to simulate deterministic problems. For example, in the case of electronic network problems, one often relies on SPICE-type solvers.

The situation is quite different for the SG method. Although the augmented equations (e.g., (2.30)) have the same shape as the original ones (e.g., (2.24)), solving the former might require to develop a dedicated code, as commercial software does not always allow supplying user-defined equations. Sometimes, this limitation can be circumvented, for example in the case of electrical circuits, as described in Section 2.3.2. It is important to mention that the use of ortho*normal* basis function preserves the possible symmetry of the system matrices, which is often helpful in this regard [21].

Besides the possible need for a dedicated solver, an important limitation of the SG method is that it cannot be directly applied to nonlinear systems, since the linearity of the inner product can no longer be used. Let us consider, for simplicity, the system (2.24) with $M = 1$ (hence, scalar) and with an additional nonlinear term $\mathbf{F}(y)$ in the r.h.s., that is,

$$\dot{y}(t, \boldsymbol{\xi}) = A(\boldsymbol{\xi})y(t, \boldsymbol{\xi}) + \mathbf{F}(y(t, \boldsymbol{\xi})) + u(t, \boldsymbol{\xi}). \tag{2.33}$$

After introducing the PCEs into (2.25), the Galerkin projection of the nonlinear term reads

$$\langle \mathbf{F}(\widehat{y}), \varphi_m \rangle = \langle \mathbf{F}\left(\sum_{k=1}^{K} y_k(t)\varphi_k(\boldsymbol{\xi}) \right), \varphi_m \rangle \neq \mathbf{F}(y_m(t)), \tag{2.34}$$

as would be the case if the function $\mathbf{F}$ were linear. A closed-form solution for the SG projection (2.34) exists for polynomial nonlinearities [22], whereas ad hoc solutions are required for the inclusion of generic nonlinear terms [23,24].

The size of the SG equations is smaller than the number of equations to be solved in the case of the PSC and LSR methods, and equal to the one of the ST method. However, the fact that the SG equations are coupled usually leads to superlinear scaling, especially for large D. Nevertheless, this shortcoming is typically offset by the superior accuracy, which makes the SG method one of the most popular PCE-based techniques for linear systems. Indeed, the SG method is considered to offer the most accurate solution with the least number of equations [8].

In this regard, it has been proven that the SG and the PSC methods are equivalent (i.e., they provide the same solution), under the assumption that [25]

1. a tensor-product truncation is considered for the PCEs;
2. the system matrices are linear w.r.t. the uncertain parameters or, in other words, they are represented exactly by means of a first-order PCE (i.e., with $P = 1$).

If this is not the case, then the two methods provide different solutions, although the difference is appreciable only for a very large variability of the uncertain parameters [26].

This notable theoretical result is readily illustrated by considering the following two equations:

$$\dot{y}(t, \xi) = -\gamma(\xi)y(t, \xi) + u, \tag{2.35}$$

and

$$\dot{y}(t, \xi) = -\gamma^2(\xi)y(t, \xi) + u, \tag{2.36}$$

with $\gamma(\xi) = \gamma_0 + \gamma_1 \cdot \xi$. Both equations depend on one uncertain parameter ξ, with linear and quadratic dependence, respectively. Because of the findings in [26], we expect the PSC and SG methods to provide the same PCE coefficients only for (2.35).

In order to make the difference appreciable, we deliberately ascribe a large variability to γ, by taking $\gamma_0 = 2$, $\gamma_1 = 1.6$, and the standard uniform distribution within $[-1, 1]$ for $\boldsymbol{\xi}$. This means that the parameter γ varies uniformly within $\pm 80\%$ around its central value. We calculate the coefficients of a second-order PCE of the solution y. To apply the SG method, we first need to express the system parameter $A = -\gamma$ or $A = -\gamma^2$ as the PCE $A_0\phi_0 + A_1\phi_1 + A_2\phi_2$, with $\phi_0 = 1$, $\phi_1 = \sqrt{3}\cdot\xi$, and $\phi_2 = \frac{3\sqrt{5}}{2}\cdot\xi^2 - \frac{\sqrt{5}}{2}$ being the first three orthonormal Legendre polynomials. The coefficients for the linear case are $A_0 = -\gamma_0$, $A_1 = -\gamma_1/\sqrt{3}$, and $A_2 = 0$, whereas for the quadratic parameter they become

$$A_0 = \gamma_0^2 + \frac{1}{3}\gamma_1^2,$$

$$A_1 = \frac{2}{\sqrt{3}}\gamma_0\gamma_1,$$

$$A_2 = \frac{2}{3\sqrt{5}}\gamma_1^2.$$

It should be noted that the PCE representation is *exact* in both cases. From the above coefficients, the SG-augmented matrix is readily constructed using (2.31).

Figure 2.2 shows the absolute value of the PCE coefficients of the solution y computed for the two cases with both the SG (solid blue lines) and PSC (dashed yellow lines) methods, and forcing term $u = 3$. The left and right panels refer to (2.35) and (2.36), respectively. The difference between the SG and PSC solution is highlighted using dotted green lines, which represent the absolute error between the two solutions. It is confirmed that the two methods provide the same result only for the first equation, whereas they exhibit a substantial difference for the second one. However, the difference becomes negligible if a higher expansion order or a smaller variability of the parameter γ (e.g., 30% or less) are considered.

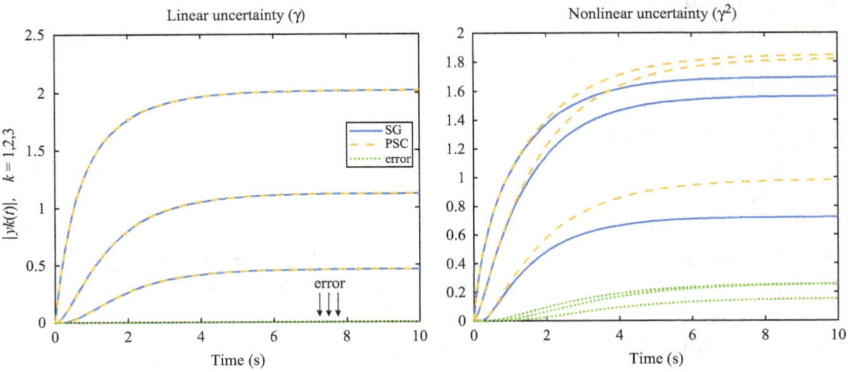

Figure 2.2 Comparison between the SG and PSC methods in the calculation of the PCE coefficients of the solution of (2.35) and (2.36)

2.3.2 Application to electrical circuits

The SG method is readily applied to the governing equations of (linear) electrical circuits. Besides the state-space representation, analogous to (24), a more general and versatile formulation that is implemented in most commercial simulators is the MNA equation

$$-\mathbf{C}(\boldsymbol{\xi})\dot{\mathbf{y}}(t,\boldsymbol{\xi}) = \mathbf{G}(\boldsymbol{\xi})\mathbf{y}(t,\boldsymbol{\xi}) - \mathbf{u}(t), \tag{2.37}$$

where the outputs $\mathbf{y}$ are the nodal voltages and some of the branch currents (e.g., those flowing into inductors or independent voltage sources), whereas $\mathbf{C}$ and $\mathbf{G}$ are the matrices accounting for the effect of memory (inductive or capacitive) and memoryless (resistive) elements, respectively [16]. The application of the SG method to the stochastic MNA equation (40) leads to a K times larger deterministic MNA equation

$$-\widetilde{\mathbf{C}}\dot{\widetilde{\mathbf{y}}}(t) = \widetilde{\mathbf{G}}\widetilde{\mathbf{y}}(t) - \widetilde{\mathbf{u}}(t), \tag{2.38}$$

characterized by the pertinent augmented matrices $\widetilde{\mathbf{C}}$ and $\widetilde{\mathbf{G}}$, constructed as in (2.31). Two application examples are found in [27] and [28]. A solution for the inclusion of general nonlinear elements was proposed in [24].

In [29], the SG method was applied to the governing equations of individual circuit elements, resulting in deterministic companion models that were implemented as library elements of a SPICE-type solver, which allowed the stochastic analysis of arbitrary circuit configurations in a modular fashion. The overall network model for the nodal analysis-based simulation was obtained through direct construction, that is, by connecting the terminals of the companion models in accordance with the original circuit topology. The underlying idea is that all nodal voltages and branch currents are expressed by PCEs, and each voltage and current coefficient corresponds, respectively, to a new node or branch in an equivalent augmented network. Ad hoc solutions were also provided for certain types of nonlinear elements, such as diodes and bipolar transistors.

2.3.3 SPICE-compatible equivalent models for linear elements

Following the same rationale, fully SPICE-compatible equivalent circuit interpretations were derived for the SG equations of stochastic resistors, inductors, and capacitors [30]. Consequently, deterministic equivalent networks can be simulated with commercial software, and thus, the voltage and current PCE coefficients are computed without the need for a dedicated solver. By construction, the overall augmented network is described by a system of equations that is equivalent to the SG equation (2.38).

2.3.3.1 Model for stochastic resistors

Let us first consider the case of a stochastic resistor, whose governing equation reads

$$v(t,\boldsymbol{\xi}) = R(\boldsymbol{\xi})i(t,\boldsymbol{\xi}) \tag{2.39}$$

where v is the voltage across the resistor, i is the current flowing into it, and R is the pertinent (stochastic) resistance. Introducing PCEs for all quantities and applying SG projection, we obtain

$$\widetilde{\mathbf{v}}(t) = \widetilde{\mathbf{R}}\widetilde{\mathbf{i}}(t) \tag{2.40}$$

where, conceptually similar as in (2.30), $\widetilde{\mathbf{v}}$ and $\widetilde{\mathbf{i}}$ collect the voltage and current PC coefficients, and $\widetilde{\mathbf{R}}$ is an augmented system (resistor) matrix. Each row of (2.40) is an equation of the form

$$v_m(t) = \sum_{k=1}^{K} \widetilde{R}_{mk} i_k(t) = \widetilde{R}_{mm} i_m(t) + \sum_{k \neq m} \widetilde{R}_{mk} i_k(t), \quad m = 1, \ldots, K, \tag{2.41}$$

which corresponds to the series connection of a resistor with resistance $\widetilde{R}_{mm}$ and current-controlled voltage sources with transresistance $\widetilde{R}_{mk}$. As illustrated in Figure 2.3(a), the controlling currents i_k $(k \neq m)$ flow into similar branches which represent the other rows of (2.40). The dependent sources provide the coupling between the various equations (branches) as per the SG equation.

However, in an MNA-based formulation, the model of Figure 2.3(a) introduces additional unknowns for the voltage of the inner nodes. Therefore, a more convenient model is obtained by inverting the relation (2.40) and rewriting it as

$$\widetilde{\mathbf{i}}(t) = \widetilde{\mathbf{R}}^{-1}\widetilde{\mathbf{v}}(t) = \widetilde{\mathbf{G}}\widetilde{\mathbf{v}}(t), \tag{2.42}$$

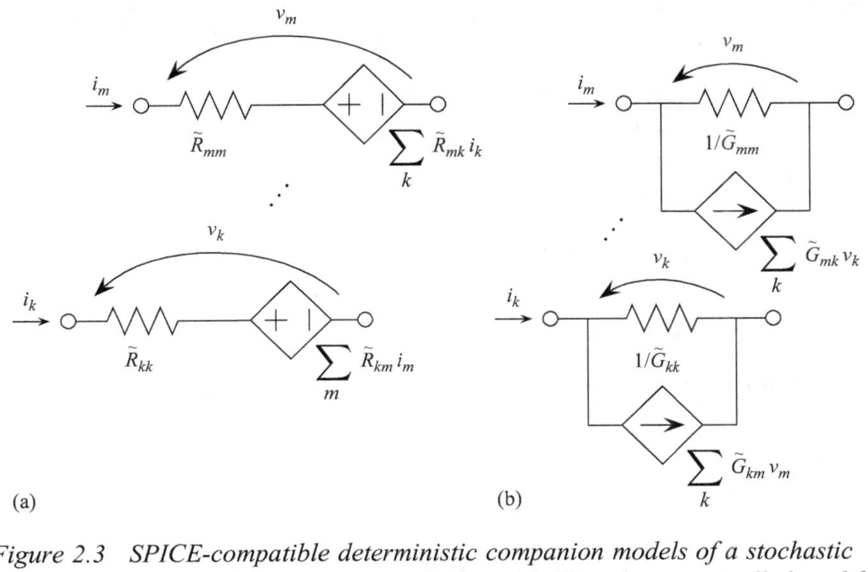

Figure 2.3 SPICE-compatible deterministic companion models of a stochastic resistor: (a) current-controlled model; (b) voltage-controlled model

whose rows have now the form of

$$i_m(t) = \sum_{k=1}^{K} \widetilde{G}_{mk} v_k(t) = \widetilde{G}_{mm} v_m(t) + \sum_{k \neq m}^{K} \widetilde{G}_{mk} v_k(t), \quad m = 1, \ldots, K, \tag{2.43}$$

and correspond to the parallel connection of a resistor with conductance $\widetilde{G}_{mm}$ (or, equivalently, resistance $1/\widetilde{G}_{mm}$) and voltage-controlled current sources with transconductance $\widetilde{G}_{mk}$. These depend on the voltages v_k of the other branches, as illustrated in Figure 2.3(b).

2.3.3.2 Model for stochastic inductors

We consider now the equation of a stochastic inductor, that is,

$$v(t, \boldsymbol{\xi}) = L(\boldsymbol{\xi}) \frac{d}{dt} i(t, \boldsymbol{\xi}). \tag{2.44}$$

Similar steps as for the resistor lead to

$$\widetilde{\mathbf{v}}(t) = \widetilde{\mathbf{L}} \frac{d}{dt} \widetilde{\mathbf{i}}(t). \tag{2.45}$$

Each row is an equation of the form

$$v_m(t) = \sum_{k=1}^{K} \widetilde{L}_{mk} \frac{d}{dt} i_k(t) = \widetilde{L}_{mm} \frac{d}{dt} i_m(t) + \sum_{k \neq m} \widetilde{L}_{mk} \frac{d}{dt} i_k(t), \quad m = 1, \ldots, K, \tag{2.46}$$

which corresponds to an inductor with inductance $\widetilde{L}_{mm}$ that is coupled, with mutual inductance $\widetilde{L}_{mk}$ (or coupling coefficient $\widetilde{K}_{mk} = \widetilde{L}_{mk} / \sqrt{\widetilde{L}_{mm} \widetilde{L}_{kk}}$), to the inductances of the other branches, as illustrated in Figure 2.4(a).

2.3.3.3 Model for stochastic capacitors

The equation of a stochastic capacitor

$$i(t, \boldsymbol{\xi}) = C(\boldsymbol{\xi}) \frac{d}{dt} v(t, \boldsymbol{\xi}) \tag{2.47}$$

is converted, by means of the SG method, into the augmented equation

$$\widetilde{\mathbf{i}}(t) = \widetilde{\mathbf{C}} \frac{d}{dt} \widetilde{\mathbf{v}}(t). \tag{2.48}$$

Each row is an equation of the form

$$\begin{aligned} i_m(t) &= \sum_{k=1}^{K} \widetilde{C}_{mk} \frac{d}{dt} v_k(t) = \widetilde{C}_{mm} \frac{d}{dt} v_m(t) + \sum_{k \neq m} \widetilde{C}_{mk} \frac{d}{dt} v_k(t) \\ &= \widetilde{C}_{mm} \frac{d}{dt} v_m(t) + \sum_{k \neq m} \frac{\widetilde{C}_{mk}}{\widetilde{C}_{kk}} i_{\widetilde{C}_{kk}}(t), \quad m = 1, \ldots, K, \end{aligned} \tag{2.49}$$

method falls under the umbrella of global sensitivity analyses. However, it is pointed out that the key challenge is how to efficiently determine the variances of the HDMR terms because computing the variances of (11)–(13) requires solving many multidimensional integrals.

4.2.2.2 Moment independent sensitivity indices

Instead of calculating the S_i only with respect to the variance of the QoI $y(\lambda)$, a far more accurate and holistic set of S_i can be calculated with respect to the probability or cumulative density measure of $y(\lambda)$ [3–5]. Such a set of S_i is referred to as moment independent sensitivity indices because they measure the sensitivity not with respect to a particular statistical moment (e.g., the variance) but with respect to the entire distribution of the response $y(\lambda)$. For example, in [3], the area enclosed by the unconditional and conditional PDFs of $y(\lambda)$ is expressed as

$$A_i = \int \left(\rho_y(y) - \rho_{y/\lambda_i}(y)\right)dy \tag{14}$$

where $\rho_y(y)$ is the unconditional PDF and $\rho_{y/\lambda_i}(y)$ is the conditional PDF when the value of λ_i is fixed. Then, the moment independent S_i of the dimension λ_i is expressed as the expectation

$$S_i = \frac{1}{2}\int A_i\rho(\lambda_i)d\lambda_i = \frac{1}{2}E(A_i) \tag{15}$$

Similarly, the work of [5] expressed the moment independent sensitivity index of the dimension λ_i as

$$S_i = \frac{1}{2}\int\int (C_y(y) - C_{y/\lambda_i}(y))\rho(\lambda_i)d\lambda_i dy \tag{16}$$

where $C_y(y)$ is the unconditional cumulative density function (CDF) and $C_{y/\lambda_i}(y)$ is the conditional CDF.

4.2.3 *Other approaches to perform dimension reduction*

At this juncture, it is important to emphasize that there exist other approaches to perform dimension reduction. For example, the Karhunen–Loève (KL) transformation which is based on the eigenvalue decomposition of correlated input random variables is very well-known [23–25]. However, this transformation does not utilize any information of the QoI $y(\lambda)$ during the eigenvalue decomposition process. As a result, the KL transformation may fail to sufficiently reduce the number of original dimensions of a problem. To address this issue, other approaches such as those based on active subspaces [6, 7], sliced inverse regression [8–10], basis adaptations [11–13], and partial least squares [26] have also been developed for PC metamodels. These approaches are considered to be more effective than KL because they employ statistical information of the QoI $y(\lambda)$ to ensure that the reduced number of dimensions achieved is as small as possible. In fact, these approaches identify critical input–output information such as the covariance matrix of the gradient of the QoIs [6, 7], the covariance matrix of the sliced conditional expectations of the inputs inferred from the QoIs [8–10], or the matrix of the partial least square weights of the QoIs [26]. Thereafter, by taking the dominant eigenvalues of these matrices and stacking the corresponding eigenvectors together allows the construction of a linear transformation matrix. This transformation matrix maps the original set of random dimensions to a reduced set of random dimensions. Constructing a PC metamodel within this reduced dimensional subspace leads to a more compact set of terms and fewer training simulations to evaluate the coefficients.

Effective though these transformation-based dimension reduction techniques maybe, the focus of this chapter will be on how to perform dimension reduction using S_is. It is understood from the discussion of Section 4.2.1 that the key challenge in S_i based dimension reduction techniques is how to efficiently quantify the S_i of (11)–(13) or (15)–(16) in the very first place. In the next section, a methodology based on HDMR is shown to efficiently quantify the S_is, especially for time-varying or frequency-varying QoIs as commonly encountered in electronic and electromagnetic devices and circuits.

4.3 Sensitivity analysis via HDMR

4.3.1 S_i *computations*

Any S_i based algorithm to perform dimension reduction consists of three principal tasks:

1. Quantify the S_is of each dimension using a suitable sensitivity analysis technique.
2. Based on the S_is, separate the important dimensions from the unimportant dimensions.
3. Train the PC metamodel in the reduced dimensional subspace.

Thus, the computational cost incurred to perform the first task can be thought of as the overheads for the third task. This means that an effective dimension reduction strategy must be able to perform the first task in a very efficient manner. In here lies the key challenge to dimension reduction – how can the S_i of each dimension be computed efficiently?

In order to answer this question, a HDMR based strategy is adopted [21]. According to HDMR, the impact of each dimension on the QoI $y(\lambda)$ taken singly is expressed using the 0th and first-order functions of (7). Now, calculating $y(\lambda^{(0)})$ requires a single deterministic simulation of the network under test where the uncertain parameter values p_i are set to their mean values u_i in (1). For calculating the first-order functions of (7), 1D PC metamodels can be used as

$$y_i(\lambda_i) \approx \sum_{k=1}^{m} y_{i,k}\phi_k(\lambda_i) = y(t,\lambda)\Big|_{\lambda^{(0)}/\lambda_i} - y_0(t,\lambda) \quad \forall i = 1, 2, ..., N \tag{17}$$

Now, by definition, the coefficients of the 1D PC metamodel satisfy the inner product with respect to the marginal PDF of λ_i (i.e., $\rho(\lambda_i)$) as

$$y_{i,k} = \langle y_i(\lambda_i), \phi_k(\lambda_i) \rangle = \int y_i(\lambda_i)\phi_k(\lambda_i)\rho(\lambda_i)d\lambda_i \tag{18}$$

The inner product integral of (18) is evaluated using 1D Gaussian quadrature [27] as

$$y_k \approx \sum_{p=0}^{m} w_p y_i(\lambda_i^{(p)})\phi_k(\lambda_i^{(p)}) = \sum_{p=0}^{m} w_p \phi_k(\lambda_i^{(p)})\left(y(\lambda_i^{(p)}) - y_0(t,\lambda)\right) \tag{19}$$

where w_p represents the pth 1D Gaussian quadrature weight corresponding to the pth quadrature node $\lambda_i^{(p)}$. Note that the quadrature nodes are all roots of the $m+1$th order basis function chosen from the Wiener-Askey scheme. In (19), the value of the QoI at the quadrature node $\lambda_i = \lambda_i^{(p)}$ can be obtained via a deterministic simulation of the system under test at the multidimensional node $\lambda^{(p)} = [0,\ldots,0, \lambda_i^{(p)}, 0,\ldots\ 0]$ and the prior knowledge of $y_0(t,\lambda)$ as per (17). From this discussion, it is concluded that, in order to compute the PC coefficients of (17) only a mere $(m+1)N+1$ deterministic simulations are required to find the requisite network responses.

Once the PC representations of the first-order functions of (7) are determined, a crude approximation of the QoI can be constructed using only these first-order functions and neglecting all higher order terms in the HDMR of (7). Consequently, the total variance of the QoI $y(t,\lambda)$ is approximated as the sum of the variances of the first-order contributions as

$$\mathrm{Var}(y) \approx \sum_{i=1}^{N} Var(y_i(\lambda_i)) \tag{20}$$

Thus, based on the definition of Sobol's S_i for each dimension, the impact of each dimension on the variance of $y(\lambda)$ is calculated as [1, 2, 20]

$$S_i = \frac{\mathrm{Var}(y_i(t,\lambda_i))}{\mathrm{Var}(y)} \tag{21}$$

The S_i of (21) is calculated directly from the 1D PC coefficient of (17) as

$$\mathrm{Var}(y_i(\lambda_i)) = \sum_{q=1}^{m} (y_{i,q})^2 \tag{22}$$

In other words,

$$S_i = \sum_{q=1}^{m} (y_{i,q})^2 \Big/ \sum_{i=1}^{N}\sum_{q=1}^{m} (y_{i,q})^2 \tag{23}$$

The S_i of (23) is an approximate measure of how the variance of each dimension λ_i contributes to the variance of the entire QoI $y(\lambda)$. Thus, using this S_i, the N dimensions of the system can be ranked based on their importance. Only the first R

dimensions with the largest S_i are considered to be important while the remaining N–R dimensions with lower S_i are considered to be unimportant. As a rule of thumb, the sum of S_i of the unimportant N–R dimensions adds up to a prescribed threshold percentage, say ε, usually set to 5%.

In the work of [28], the approximation of $y(t,\lambda)$ using only the 0th and first-order contributions is done for the sole purpose of extracting the Sobol's S_i of each dimension via (23) in a highly efficient manner. It is understood that if the higher order interaction terms of the HDMR was included to model $y(t,\lambda)$, this would have led to a more accurate calculation of the S_i but at the cost of many more deterministic training simulations. In fact, it is pointed out here that we are not interested in calculating the correct value of the S_i of each dimension or even in the relative ranking of the dimensions so long as we can just extract the correct subset of unimportant N–R dimension. Thus, under these circumstances, the approximation of $y(t,\lambda)$ using only the 0th and first-order HDMR terms is considered to be reasonable.

4.3.2 S_i *for dynamic systems*

Often, when dealing with electromagnetic and electronic devices or circuits, the QoI $y(\lambda)$ is time-varying or frequency-varying in nature. For example, step responses of linear circuits are time-varying while **S**-parameters of any microwave circuit are frequency-varying. In such cases, the S_i of (23) too becomes time-varying or frequency-varying. As a result, at different time points, different dimensions may become important or unimportant, thereby making the identification of a global set of unimportant dimensions very difficult. In order to address this issue, the work of [28] suggested integrating the S_i of each dimension over the entire time (or frequency) window of observation – in other words, to measure the total area under each S_i curve given that these curves are always going to have a value greater than or equal to zero. Therefore, a new set of integrated S_i are proposed as

$$S_i(t) = \sum_{q=1}^{m} \left(y_{i,q}(t)\right)^2 \Big/ \sum_{i=1}^{N} \sum_{q=1}^{m} \left(y_{i,q}(t)\right)^2; \quad \widehat{S}_i = \frac{1}{T_2 - T_1} \int_{T_1}^{T_2} S_i(t)dt \tag{24}$$

where $[T_1 - T_2]$ is the time window of interest. The integration of (24) can be performed using any numerical integration scheme [29]. In (24), the integration with respect to time is shown. A similar integration with respect to frequency can also be performed if the real and imaginary part of the QoI is frequency-varying. In effect, the new integrated S_i of (24) measures the average value of the original S_i of (23) and this averaged S_i is considered as a holistic measure of the importance of a dimension. Thus, now the ranking of the dimensions is done based on the averaged S_i of (24) instead of the time-varying or frequency-varying S_i of (23).

4.3.3 *Constructing the reduced dimensional PC metamodel*

Let the vector λ_r contain all R the important dimensions for a given problem. The PC metamodel for the QoI $y(t,\lambda)$ can now be expressed as

$$y(t, \lambda_r) \approx \sum_{k=0}^{Q} \hat{y}_k(t)\psi_k(\lambda_r) \tag{25}$$

where $\psi_k(\lambda_r)$ is the kth reduced dimensional basis function, $\hat{y}_k(t)$ is the corresponding coefficient and the number of terms in the expansion of (25) is truncated to [29, 30]

$$Q + 1 = \frac{(R + m)!}{R!m!}; \text{where } Q + 1 << P + 1 \tag{26}$$

Next, in order to intelligently evaluate the coefficients of (25), any non-intrusive approach can be directly applied. However, as per the work of [28], a more accurate strategy would be to separate the metamodel of (25) into the mean, 1D terms, and the multidimensional interactions terms as

$$y(t, \lambda_r) \approx y_0(t, \lambda) + \sum_{i=1}^{N} y_i(t, \lambda_i) + \sum_{k=1}^{Q-mN} \hat{y}_k(t)\psi_k(\lambda_r) \tag{27}$$

In (26), the multidimensional interaction terms are collectively expressed as a single PC metamodel while the 1D terms are already expressed as 1D PC metamodels in (17) when evaluating the S_i. Note that in the 1D terms even the unimportant dimensions are included for better accuracy. Therefore, $mN+1$ fewer coefficient has to be evaluated in (26) using any non-intrusive approach where

$$\sum_{k=1}^{Q-mN} \hat{y}_k(t)\psi_k(\lambda_r) = y(t, \lambda_r) - y_0(t, \lambda) - \sum_{i=1}^{N} y_i(t, \lambda_i) \tag{28}$$

If linear regression scheme is used to evaluate the coefficients of (27), only $2(Q-mN)$ deterministic simulations are required. Furthermore, the $(m+1)R$ nodes used to identify the 1D coefficients in (19) for the important R nodes can be reused in (27) to further depress the number of deterministic simulations required from $2(Q-mN)$ to $2(Q-mN) - (m+1)R$.

4.3.4 If multiple QoIs exist

Let us assume that there are L number of QoIs $y_l(t,\lambda)$ where $1 \leq l \leq L$ in a device or circuit. For this device or circuit, the S_i for each dimension corresponding to each QoI is calculated separately. The important set of dimensions for each QoI is $\tilde{\lambda}_l$. Thus, the global set of important dimensions is the union set $\tilde{\lambda}_U = \tilde{\lambda}_1 \cup \tilde{\lambda}_2 \cdots \cup \tilde{\lambda}_L$. This means that a PC metamodel can now be constructed for each QoI using only the dimensions stored in $\tilde{\lambda}_U$. Notably, given that the cost of the deterministic simulations required to evaluate the coefficients in (18) dwarfs the cost of calculating the S_is of (24) for each dimension corresponding to each QoI, the overall cost of constructing the PC metamodels for the different QoIs will remain small.

4.3.5 Complexity analysis of the dimension reduction algorithm

The total number of deterministic simulations for training the conventional PC metamodel is $2(P+1)$ when using the least square linear regression scheme [31, 32]. Thus, the resultant time cost for training is given as

$$C_{\text{conv}} = 2(P+1)C_0 \tag{29}$$

where C_0 is the average time cost of each simulation. For the dimension reduction strategy adopted in Sections 4.3.1–4.3.2, the number of deterministic simulations required are $(m+1)N+1$ to calculate the coefficients of the 0th and 1D terms in (17) plus the $2(Q-mn)$ nodes needed to train the multidimensional terms of (27) minus the reused $(m+1)R$ nodes. The corresponding time cost is therefore equal to

$$\begin{aligned} C_{\text{red}} &= ((m+1)N + 2(Q-mN) - (m+1)R)C_0 \\ &= (2(Q-mN) + (m+1)\Delta_N)C_0; \quad \Delta_N = N - R \end{aligned} \tag{30}$$

where Δ_N is the number of unimportant dimensions. Based on (28) and (29), the speedup achieved using the reduced dimensional PC metamodel is quantified as

$$\eta = \frac{C_{\text{conv}}}{C_{\text{red}}} = \frac{2(P+1)}{2(Q-mN) + (m+1)\Delta_N} \tag{31}$$

Note that in the denominator of (30), usually the first term dominates over the second and so the upper limit of the speedup is

$$\eta_{\max} = \frac{(P+1)}{(Q-mN)} \tag{32}$$

Given that the terms $P+1$ and $Q+1$ scales as $O(N^m)$ and $O(R^m)$, respectively [31, 32], the speedup of (30) can be approximated as

$$\eta_{\max} \approx \frac{N^m}{(R^m - mN - 1)} \approx \left(\frac{N}{R}\right)^m \approx \left(1 + \frac{\Delta_N}{R}\right)^m \tag{33}$$

which is always greater than 1 for any $\Delta_N > 0$. The examples of the next section show that this speedup can range anywhere from one to even two orders of magnitude depending on the value of Δ_N.

4.4 Illustrative examples

In this section, three examples are taken from the work of [28] to illustrate the benefits of the reduced dimensional PC approach described in Sections 4.3.1 and 4.3.2 over full-dimensional PC approaches and competing sparse PC approaches [33, 34]. These examples involve multiconductor transmission line (MTL) networks whose behavior is described using Telegrapher's partial differential equations (PDEs) [35, 36]. These PDEs are converted to ordinary differential equations using the delayed rational function macromodels otherwise referred to as the W-element model available in the HSPICE solver [36]. All remaining computations relevant to the PC are performed using MATLAB® 2013b. For all the examples shown in this section, the threshold for deciding which dimensions are unimportant using the S_i of (24) is set to $\varepsilon = 0.01$ (refer to Section 4.3.1).

4.4.1 Illustrative Example 1

In this example, the performance of the reduced dimensional PC approach of Section 4.3 is compared with that of existing PC approaches, both sparse and

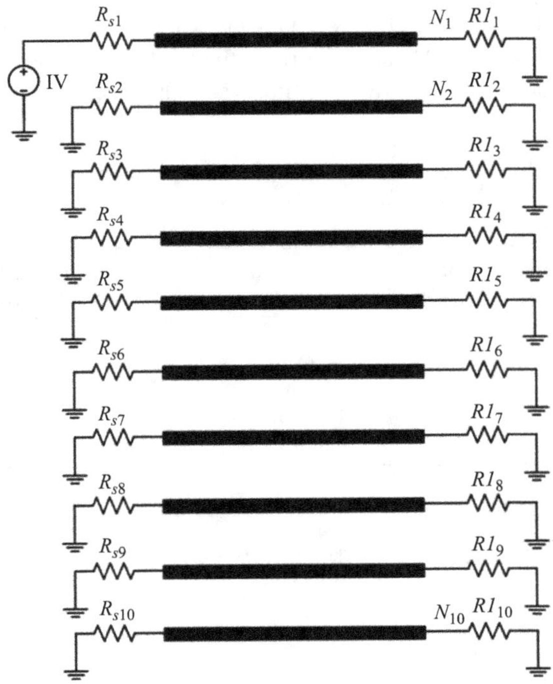

Figure 4.1 Network of Example 1 consisting of 10 coupled microstrip lines

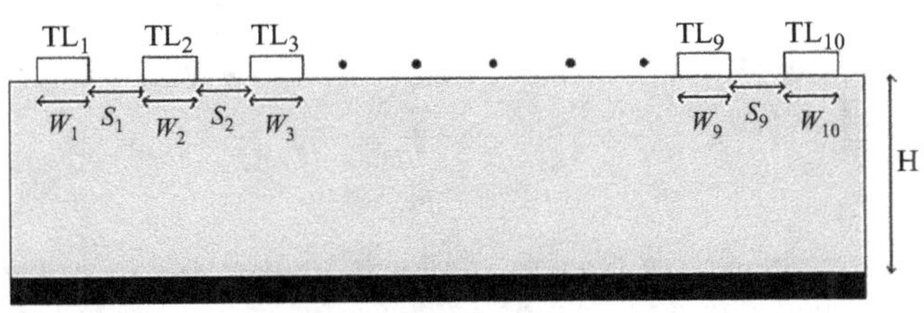

Figure 4.2 Cross-sectional view of the MTL network of Example 1 (Figure 4.1)

full-blown. To that end, a 10 conductor microstrip MTL network as depicted in Figure 4.1 is considered. The cross-section of the MTL network is shown in Figure 4.2. For this example, the uncertainty in the network is introduced via 40 random variables ($N = 40$) with Gaussian distributions as listed in Table 4.1. Thus, typically a Hermite PC expansion of degree $m = 3$ is required.

In the first part of this example, the accuracy of the reduced dimensional PC metamodel is validated. For this purpose, the stochastic modeling of the structure of Figures 4.1 and 4.2 is performed using four techniques – the reduced dimensional PC approach of Section 4.3, the hierarchical sparse PC approach reported in [34], the hyperbolic truncated PC approach of [33], and the brute-force MC approach using 25,000 samples. The structure is excited by a voltage source of unit amplitude whose frequency is swept from 100 MHz to 20 GHz. The frequency domain response at the far-end of line 1 (node N_1 in Figure 4.1) is the output of interest. Note that using the Sobol's integrated S_i of (24) and $\varepsilon = 0.01$, the total number of original dimensions $N = 40$ is reduced to only $R = 7$ important dimensions (i.e., more than 80%

Table 4.1 Uncertain network parameters with normal distribution for Example 1

Random variables	Mean values	% Standard deviation (normal distribution)
Rs_{1-10}	50 Ω	20%
Rl_{1-10}	50 Ω	
w_{1-10}	130 μm	
s_{1-9}	100 μm	
H	100 μm	

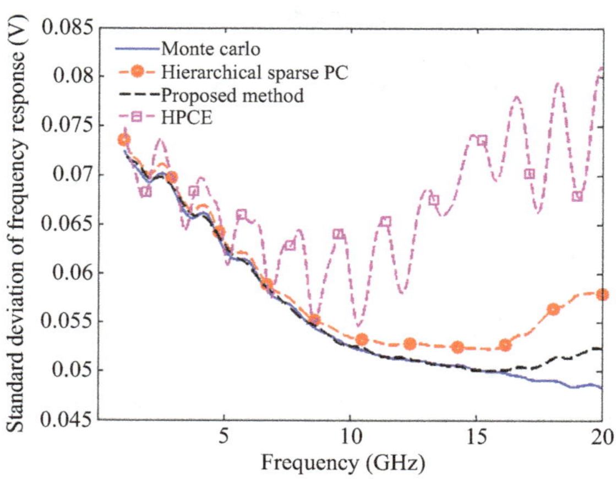

Figure 4.3a (a) Magnitude of standard deviation of the frequency response at N_1 calculated using the proposed dimension reduction approach, the work of [34] and [33], all compared against MC approach (25,000 samples)

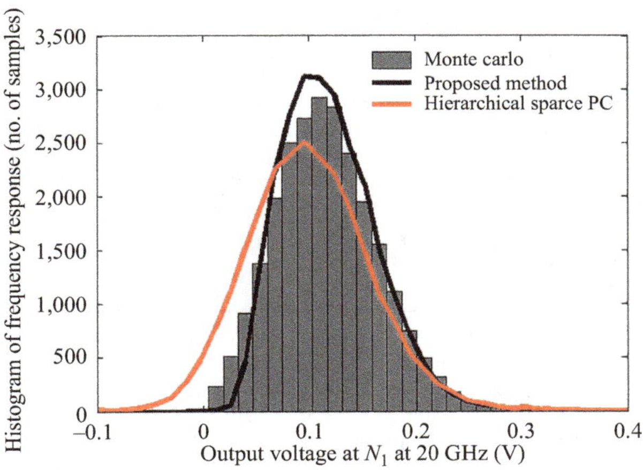

Figure 4.3b (b) PDF of frequency response at N_1 calculated using the proposed dimension reduction approach, the work of [34] and [33], all compared against MC approach (25,000 samples)

compression in number of dimensions). These important dimensions are (Rs_1, Rl_1, w_1, w_2, s_1, s_2, and H) from Table 4.1. The standard deviation of the responses of interest are quantified using the aforementioned techniques and compared in Figure 4.3. It is observed from Figure 4.3(a) that the reduced dimensional PC is able to closely match the MC results up to 15 GHz while the hierarchical sparse PC of [34] can only do so till 5 GHz. This is despite the fact that the hierarchical sparse PC metamodel contains roughly the same number of terms as the reduced dimensional PC metamodel - 209 to 219. Moreover, the hyperbolic truncation PC gives inaccurate results across the entire bandwidth of interest.

To further investigate the accuracy of the reduced dimensional PC over the hierarchical sparse PC approach of [34], the PDF of the response at N_1 at 20 GHz is extracted using these two methods as shown in Figure 4.3(b). As seen from Figure 4.3(b), the reduced dimensional PC is able to capture the PDF reasonably well while the hierarchical sparse PC metamodel of [34] shown distinct errors. These errors can be highly misleading when the circuit designer is performing design analysis and optimization, say for example, yield analysis. The yield of the network is defined to be the probability that the magnitude of the response at N_1 is less than 0.1 V at 20 GHz. From Figure 4.3(b) it is observed that the proposed approach and MC results provide similar yields of roughly 38% while the hierarchical sparse PC metamodel of [30] quantifies the yield as 51.8%.

In the final part of this example, the CPU cost comparison between the different uncertainty quantification approaches is performed. Table 4.2 lists the total number of SPICE simulations required by all the techniques, including the number of simulations required for extracting the Sobol's S_i. The corresponding number of PC bases in each of the competing metamodels is also noted. Furthermore, for

Table 4.2 L_2 *Error norm of standard deviation for Example 1 (MC with 25,000 samples used as reference)*

	L_2 **Error norm**	**# SPICE simulations**	**# Bases**	**Speedup w.r.t full-blown PC(24,682 Simulations)**
Hierarchical sparse PC approach [34]	0.1097	333	207	74.12
Proposed approach	0.0356	331	219	74.57
HPCE approach [33]	0.4679	901	901	27.39

completeness of analysis, the L_2 norm of the standard deviation error incurred by the various PC techniques with respect to the MC results is also listed in Table 4.2. Based on this data, it is clearly shown that the reduced dimensional PC approach of Section 4.3 is more accurate than its existing sparse PC counterparts while still offering the highest speedup (e.g., more than 74).

4.4.2 Illustrative Example 2

In this example, the enhanced performance of the reduced dimensional PC approach of Section 4.3 is examined when considering multiple QoIs. For this purpose, the same MTL network of Figures 4.1 and 4.2 is considered where the outputs of interest are the frequency domain responses at the nodes $N_1 - N_{10}$ in Figure 4.1. By considering the responses at all the output nodes, the global set of important dimensions become (Rs_1, Rl_1, Rl_2, w_1, w_2, s_1, s_2, and H) (i.e., now $R = 8$ instead of the original $N = 40$).

For validating the accuracy of the reduced dimensional PC approach, three techniques for uncertainty quantification are chosen. They are the reduced dimensional PC approach of Section 4.3, the hierarchical sparse PC approach reported in [34], and the brute-force MC approach using 25,000 samples. Next, the standard deviation of all the frequency domain responses is calculated using these four techniques. The results for the responses at nodes N_2 and N_{10} are illustrated in Figure 4.4(a) and (b), respectively. In Table 4.3, the L_2 error norm incurred by the PC approaches with respect to the MC results for the responses at nodes N_1, N_2, and N_{10} is reported. From the results of Table 4.3, it is clear that reduced dimensional PC approach of Section 4.3 is more accurate than the hierarchical sparse PC approach of [34]. Moreover, the error norm for the response at node N_1 for both approaches has improved from the results of the example of Section 4.4.1 (i.e., Table 4.2). This is to be expected because a new dimension Rl_2 has been taken into account to ensure good accuracy for the results of node N_2. Thus, by increasing the number of reduced dimensions, additional bases representing the interaction of the new dimension with the exiting $R = 7$ reduced dimensions of Section 4.4.1 is included leading to more accurate results.

Next, a comparison of the CPU cost incurred by the reduced dimensional PC approach of Section 4.3 and the hierarchical sparse PC approach of [34] is performed and the results included in Table 4.3. As before, the total number of SPICE

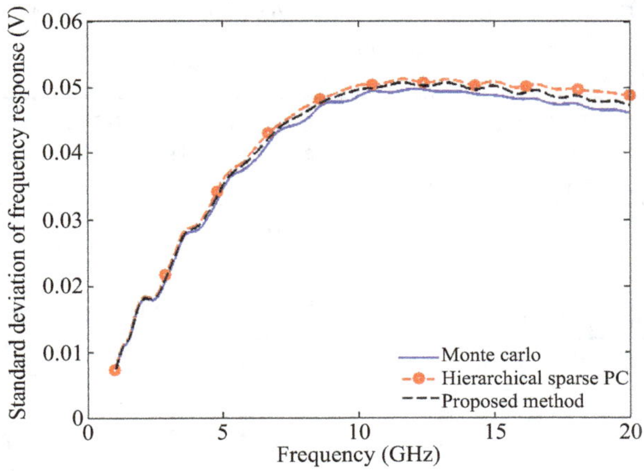

Figure 4.4a *(a) Standard deviation of the frequency response at N_2 calculated using the proposed dimension reduction approach, the work of [34] and MC approach (25,000 samples)*

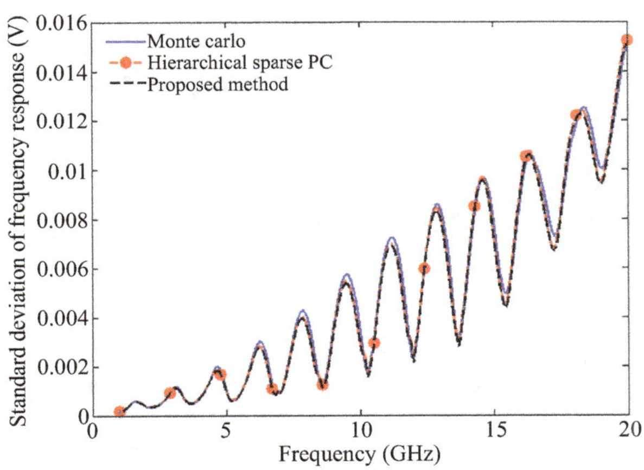

Figure 4.4b *(b) Standard deviation of the frequency response at N_{10} calculated using the proposed dimension reduction approach, the work of [34] and MC approach (25,000 samples)*

simulations includes those required to quantify the Sobol's S_i for all dimensions using (24). It is observed from the data of Table 4.3 that the reduced dimensional PC approach needs slightly more bases but fewer SPICE simulations leading to better speedup. This is consistent with what we saw in the example of Section 4.4.1. Moreover, the reduced dimensional PC approach is now more accurate as well for the responses at node N_2 and N_{10}.

Table 4.3 L_2 *Error norm of standard deviation for Example 2 (MC with 25,000 samples used as reference)*

	L_2 Error norm at N_1	L_2 Error norm at N_2	L_2 Error norm at N_{10}	# SPICE simulations	# Bases	Speedup w.r.t full-blown PC
Hierarchical sparse PC approach [34]	0.0694	0.0520	0.0106	429	255	57.53
Proposed approach	0.0334	0.0295	0.0097	411	261	60.05

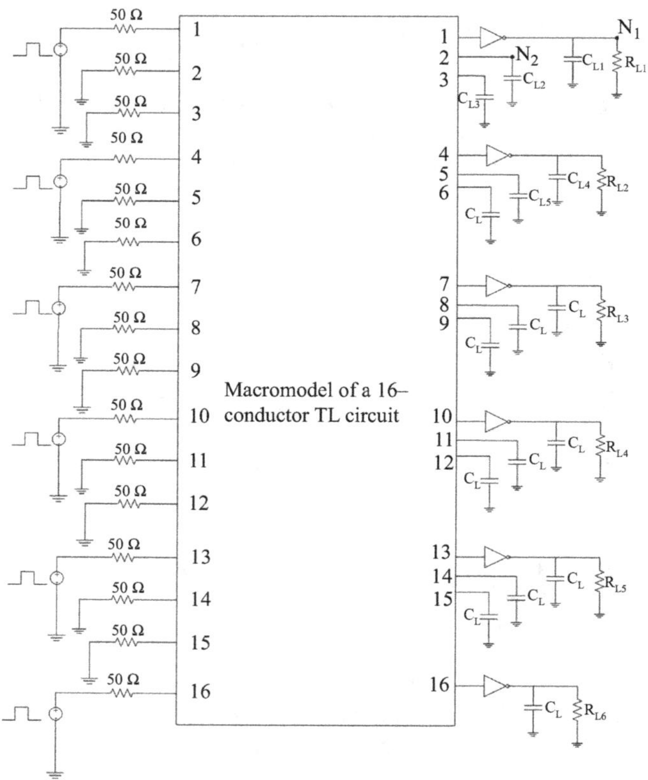

Figure 4.5a (a) Circuit layout of the MTL network for Example 3 with 16 coupled lines terminated by inverters

4.4.3 Illustrative Example 3

In this example, the performance of the reduced dimensional PC approach of Section 4.3 is explored for transient analysis of mutually coupled high-speed interconnect structures. To that end, an MTL stripline network consisting of 16

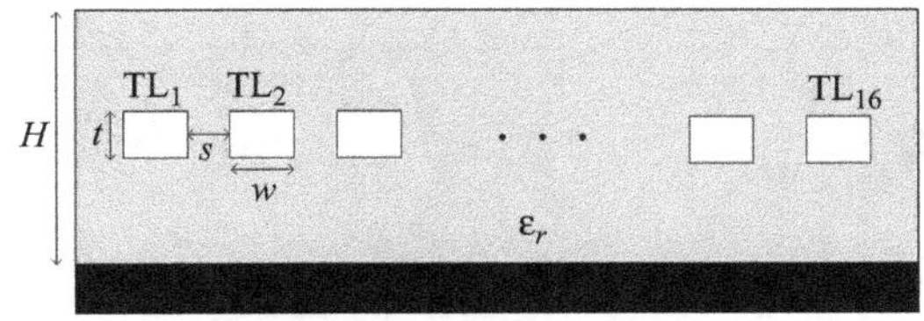

Figure 4.5b (b) Cross-sectional view of the MTL network of Example 3

Table 4.4 Characteristics of random variables of Example 3

Random variables	Mean values	% Standard deviation (uniform distribution)
w (TL width)	150 μm	20%
t (TL height)	30 μm	
H (Dielectric thickness)	450 μm	
s (TL separation)	150 μm	
ε_r (Relative permittivity of dielectric)	100 μm	
L (TL length)	6 cm	
σ (TL conductance)	5.8e7	
PL (PMOS length)	0.1 μm	10%
PW (PMOS width)	10 μm	
NL (NMOS length)	0.1 μm	
NW (NMOS width)	10 μm	
R_{L1-6}	1.5 kΩ	20%
C_{L1-5}	1 pF	

coupled conductors as shown in Figure 4.5 is considered. The network is driven and loaded with complimentary metal oxide semiconductor (CMOS) inverters comprising of SPICE level-49 transistor models [37]. The voltage sources exhibit a trapezoidal waveform with rise/fall time $T_r = 0.1$ ns, pulse width $T_w = 1$ ns, and amplitude of 5 V. The uncertainty in this network is introduced via $N = 22$ uniformly distributed random parameters listed in Table 4.4. The uncertainty quantification in the transient responses of this network is performed using the same three approached adopted for the illustrative example in Section 4.4.2 with the only exception being that the MC results required 35,000 simulations instead of 25,000. In this example, the original $N = 22$ dimension are reduced to simply $R = 4$ dimensions (i.e., (w, s, ε_r, and L) from Table 4.4). This corresponds to more than 80% compression in the number of problem dimensions.

As in the previous illustrative examples of Sections 4.4.1 and 4.4.2, first the accuracy of the reduced dimensional PC is demonstrated. For this purpose, the transient responses at the output nodes N_1 and N_2 are regarded as the QoIs. As shown in Figure 4.6(a) and (b), the standard deviation of these QoIs are probed using the aforementioned three approaches. As per Figure 4.6, both the proposed

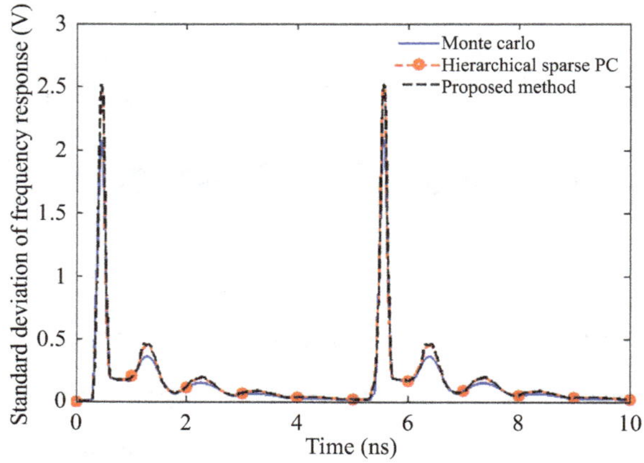

Figure 4.6a (a) Standard deviation of the frequency response at N_1 calculated using the proposed dimension reduction approach, the work of [34], and MC approach (35,000 samples)

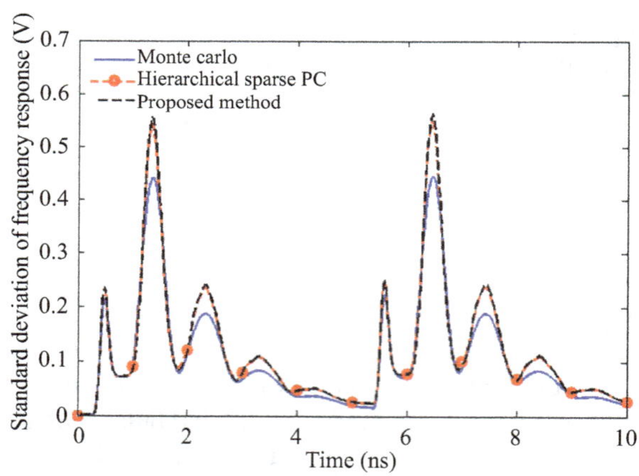

Figure 4.6b (b) Standard deviation of the frequency response at N_2 calculated using the proposed dimension reduction approach, the work of [34], and MC approach (35,000 samples)

approach and that of [34] show good agreement with the MC results for all QoIs. This is corroborated by observing the L_2 error norm for the PC approaches relative to the MC results as reported in Table 4.5.

In the last part of this example, a CPU cost comparison of the various uncertainty quantification techniques is performed. The results of this cost comparison are included in Table 4.5. From these results, it is noted that the reduced dimensional PC approach of Section 4.3 uses slightly more bases than the hierarchical

Table 4.5 L_2 *Error norm of standard deviation for Example 2 (MC with 25,000 samples used as reference)*

	L_2 Error norm at N_1	L_2 Error norm at N_2	# SPICE simulations	# Bases	Speedup w.r.t full-blown PC
Hierarchical sparse PC approach [32]	0.0017	7.53e−4	199	133	150.70
Proposed approach	0.0018	8.41e−4	202	142	148.47

sparse PC but fewer SPICE simulations in keeping with similar results in the illustrative examples of Sections 4.4.1 and 4.4.2. Thus, both approaches offer excellent and similar speedup for this example.

4.5 Conclusion

In this chapter, the strategy of dimension reduction has been explored for addressing the curse of dimensionality facing PC metamodel training. In such strategies, the mapping from the input uncertain parameters of a problem to the output responses of interest have to be measured whether using Sobol's sensitivity indices, covariance of output gradient, inverse regression curves, etc. This mapping information guides the engineer in properly identifying those dimensions that have the most impact on the responses of interest and distinguishing them from those that have the least impact. This allows for the original set of input dimensions to be reduced to a smaller set of dimensions where a PC metamodel can be trained more expediently. Of course, as expected, the main challenge lies in extracting this mapping information because such information is not available *a priori* and the extraction process may add significant CPU cost above and beyond training the reduced dimensional PC metamodel. In particular, the ANOVA based dimension reduction strategy of [22] has been explored in details. For this approach, more than two orders of magnitude in speedup when training PC metamodels is possible while maintaining virtually the same accuracy standard.

References

[1] E. H. Sandoval, F. Anstett-Collin, and M. Basset, "Sensitivity study of dynamic systems using polynomial chaos," *Reliability Engineering and System Safety,* vol. 104, pp. 15–26, 2012.

[2] A. Saltelli, S. Tarantola, F. Campolongo, and M. Ratto, *Sensitivity Analysis in Practice: A Guide to Assessing Scientific Models.* Sussex: J Wiley, 2004.

[3] E. Borgonovo, "A new uncertainty importance measure," *Reliability Engineering and System Safety*, vol. 92, no. 6, pp. 771–784, 2007.

[4] E. Borgonovo, “Measuring uncertainty importance: investigation and comparison of alternative approaches,” *Risk Analysis*, vol. 26, no. 5, pp. 1349–1361, 2006.

[5] Q. Liu and T. Homma, “A new importance measure for sensitivity analysis,” *Journal of Nuclear Science and Technology*, vol. 47, no. 1, pp. 53–61, 2013.

[6] P. G. Constantine, E. Dow, and Q. Wang, “Active subspace methods in theory and practice: applications to kriging surfaces,” *SIAM Journal of Scientific Computation*, vol. 36, no. 4, pp. A1500–A1524, 2014.

[7] P. G. Constantine, *Active Subspaces: Emerging Ideas for Dimension Reduction in Parameter Studies.* Philadelphia, PA: SIAM, 2015.

[8] W. Li, G. Lin, and B. Li, “Inverse regression-based uncertainty quantification algorithms for high-dimensional models: theory and practice,” *Journal of Computational Physics*, vol. 321, no. 3, pp. 259–278, 2016.

[9] K.-C. Li, “Sliced inverse regression,” *Journal of the American Statistical Association*, vol. 86, no. 414, pp. 316–327, 1991.

[10] Q. Pan and D. Dias, “Sliced inverse regression-based sparse polynomial chaos for reliability analysis in high dimensions,” *Reliability Engineering and System Safety*, vol. 167, pp. 484–493, 2017.

[11] R. Tipireddy and R. G. Ghanem, “Basis adaptation in homogeneous chaos spaces,” *Journal of Computational Physics*, vol. 259, no. 2, pp. 304–317, 2014.

[12] P. Tsilifis and R. Ghanem. “Reduced Wiener chaos representation of random fields via basis adaptation and projection,” *Journal of Computational Physics*, vol. 341, pp. 102–120, 2017.

[13] J. Jiang, Y. Chen, and A. Narayan, “A goal-oriented reduced basis methods-accelerated generalized polynomial chaos algorithm,” *SIAM Journal of Uncertainty Quantification*, vol. 4, no. 1, pp. 1398–1420, 2016.

[14] M. Morris, “Factorial sampling plans for preliminary computational experiments,” *Technometrics*, vol. 33, no. 2, pp. 161–174, 1991.

[15] R. Cropp and R. Braddock, “The new Morris method: an efficient second-order screening method,” *Reliability Engineering and System Safety*, vol. 78, no. 1, pp. 77–83, 2002.

[16] J. M. Fedou and M. J. Rendas, “Extending Morris method: identification of the interaction graph using cycle-equitable designs,” *Journal of Statistical Computation and Simulation*, vol. 85, pp. 1398–1419, 2015.

[17] I. Sobol and S. Kucherenko, “Derivative based global sensitivity measures and their links with global sensitivity indices,” *Mathematics and Computers in Simulation*, vol. 79, no. 10, pp. 3009–3017, 2009.

[18] I. Sobol and S. Kucherenko, “A new derivative based importance criterion for groups of variables and its link with the global sensitivity indices,” *Computer Physics Communications*, vol. 181, no. 7, pp. 1212–1217, 2010.

[19] G.S. Fishman, *Monte Carlo: Concepts, Algorithms, and Applications*. New York, NY: Springer-Verlag, 1996.

[20] C. Prieur and S. Tarantola, Variance-Based Sensitivity Analysis: Theory and Estimation Algorithms. in R. Ghanem, D. Higdon, H. Owhadi (Editors)

Handbook of Uncertainty Quantification. Switzerland: Springer International Publishing, 2017.

[21] H. Rabitz and O. F. Alis, "General formulation of high-dimensional model representations," *Journal of Mathematical Chemistry*, vol. 50, no. 2–3, pp. 197–233, 1999.

[22] G. Hardy, J. Littlewood, and G. Polya, *Inequalities* (2nd edition). Cambridge: Cambridge University Press.

[23] D. Xiu and G. E. Karniadakis, "Modeling uncertainty in steady sate diffusion problems via generalized polynomial chaos," *Computer Methods in Applied Mechanics and Engineering*, vol. 191, no. 43, pp. 4297–4248, 2002.

[24] Y. M. Marzouk and H. N. Najm, "Dimensionality reduction and polynomial chaos acceleration of Bayesian inference in inverse problems," *Journal of Computational Physics*, vol. 228, no. 6, pp. 1862–1902, 2009.

[25] A. Doostan, R. G. Ghanem, and J. Red-Gorse, "Stochastic model reduction for chaos representations," *Computer Methods in Applied Mechanics and Engineering*, vol. 196, no. 37–40, pp. 3951–3966, 2007.

[26] L. E. Frank and J. H. Friedman, "A statistical view of some chemometrics regression tools," *Technometrics*, vol. 35, no. 2, pp. 109–135, 1993.

[27] G. H. Golub and J. H. Welsch, "Calculation of Gauss quadrature rules," *Mathematics of Computation*, vol. 23, pp. 221–230, 1969.

[28] A. K. Prasad and S. Roy, "Accurate reduced dimensional polynomial chaos for efficient uncertainty quantification of microwave/RF networks," *IEEE Transactions on Microwave Theory and Techniques.*, vol. 65, no. 10, pp. 3697–3708, 2017.

[29] D. Xiu, *Numerical Methods for Stochastic Computations: A Spectral Method Approach*. Princeton, NJ: Princeton University Press, 2010.

[30] D. Xiu, "Fast numerical methods for stochastic computations: A review," *Communications in Computational Physics*, vol. 5, no. 2–4, pp. 242–272, 2009.

[31] M. Ahadi and S. Roy, "Sparse linear regression (SPLINER) approach for efficient multidimensional uncertainty quantification of high-speed circuits," *IEEE Transactions on Computer-Aided Design of Integrated Circuits and Systems*, vol. 35, no. 10, pp. 1640–1652, 2016.

[32] A. K. Prasad, M. Ahadi, and S. Roy, "Multidimensional uncertainty quantification of microwave/RF networks using linear regression and optimal design of experiments," *IEEE Transactions on Microwave Theory and Techniques*, vol. 64, no. 8, pp. 2433–2446, 2016.

[33] G. Blatman and B. Sudret, "Adaptive sparse polynomial chaos expansion based on least angle regression," *Journal of Computational Physics*, vol. 230, no. 6, pp. 2345–2367, 2011.

[34] Z. Zhang, X. Yang, I. V. Oseledets, G. E. Karniadakis, and L. Daniel, "Enabling high-dimensional hierarchical uncertainty quantification by ANOVA and tensor-train decomposition," *IEEE Transactions on Computer-Aided Design of Integrated Circuits and Systems.*, vol. 34, no. 1, pp. 63–76, 2015.

[35] R. Achar and M. Nakhla, "Simulation of high-speed interconnects," *Proceedings of IEEE*, vol. 89, no. 5, pp. 693–728, 2001.
[36] C. R. Paul, *Analysis of Multiconductor Transmission Line.* New York, NY: Wiley-Interscience, 2008.
[37] HSPICE *Reference Manual Version E 2010.12*, Synopsis Inc., Santa Clara, CA, 2010.

Chapter 5

A predictor–corrector algorithm for fast polynomial chaos based statistical modeling of carbon nanotube interconnects

Surila Guglani[1] and Sourajeet Roy[1]

This chapter introduces a new algorithm to address the curse of dimensionality facing the training of polynomial chaos (PC) metamodels, specifically for on-chip high-speed multi-walled carbon nanotube (MWCNT) interconnect structures. This algorithm is referred to as a predictor–corrector algorithm. In this chapter, the mathematics underpinning this predictor–corrector algorithm and the benefits emerging from it are described in details. Moreover, analysis leading to the identification of an upper bound of the numerical efficiency possible through this algorithm is presented. The chapter concludes with multiple numerical examples of MWCNT networks solved using the predictor–corrector algorithm as well as conventional full-blown PC metamodels.

5.1 Introduction

The accurate characterization, modeling, and simulation of high-speed interconnects at all hierarchical levels (i.e., chip, package, back planes, and printed circuit boards) are a very important task for assessing the performance of modern integrated circuits and systems. This is because of two key reasons. First, with signal speeds and channel data rates reaching 100 Gbps and device operating frequencies crossing the terahertz mark, distributed interconnect effects such as crosstalk noise, signal propagation delay, ringing, attenuation, skin effect and dispersion, etc., have a profound impact on the integrity of communication signals [1–7]. Second, with the sustained miniaturization of device features sizes in advanced technology nodes, the cross-sectional area of interconnects is continually shrinking. From the Fuchs-Sondheimer (F-S) surface scattering theory, it is known that the resistivity of a copper interconnect has an inverse relationship with the line thickness and width, particularly when these are comparable to the mean free path (MFP), that is, 40 nm for copper at room temperature [8–13]. When the interconnect thickness and width shrink below 40 nm, the resistivity of copper

[1]Computational Modeling and Simulation Laboratory, Indian Institute of Technology, Roorkee, India

interconnects increases dramatically beyond its bulk value due to the surface and grain boundary scattering [8–13]. Surface roughness further aggravates the surface and grain boundary scattering leading to even higher resistivity and the associated problems of higher signal attenuation, power losses, propagation delay, thermal breakdown, and electromigration [13]. Moreover, due to surface roughness, the effective surface area between the conductors themselves and between the conductor and ground increases. This leads to greater-than-expected parasitic capacitances and in turn, greater RC signal delay and crosstalk noise [13].

The above scenario in interconnect technology means that copper interconnects are reaching the limits of their performance at the 22 nm technology node. This has motivated the current investigations into newer 2D interconnect materials such as graphene [14,15]. Graphene interconnects typically possess much longer MFPs in the range of several micrometers as opposed to 40 nm for copper at room temperature. By virtue of this property, graphene nanointerconnects exhibit closer-to-ballistic transport compared to copper interconnects, thereby offering smaller scattering resistance. Moreover, they also possess much higher current carrying capacity, higher thermal conductivity, and higher mechanical strength than copper interconnects [16–25]. As a result, graphene nanointerconnects can support extremely high signal speeds (in the range of THz as opposed to GHz), exhibit much smaller propagation delay (in the order of picoseconds instead of nanoseconds), and are far less susceptible to thermal and structural breakdown compared to conventional copper interconnects [14,26]. Among the various types of graphene nanointerconnect structures possible, bundles of single-walled carbon nanotubes (SWCNTs) and standalone MWCNTs are highly common (see Figure 5.1). Of these, MWCNT structures are more viable than SWCNT bundles given the easier control of the growth process involved [18, 27]. Moreover, statistically speaking, only a third of SWCNT shells in a bundle exhibits metallic conductor properties due to their chirality while the rest exhibit semiconductor properties [27]. This is in contrast to MWCNTs where due to the increasing diameters of the outer shells, almost all shells exhibit metallic conduction even if they are of semiconductor chirality [27]. Thus, MWCNT structures are considered to exhibit superior conducting properties for their deployment as on-chip interconnect networks.

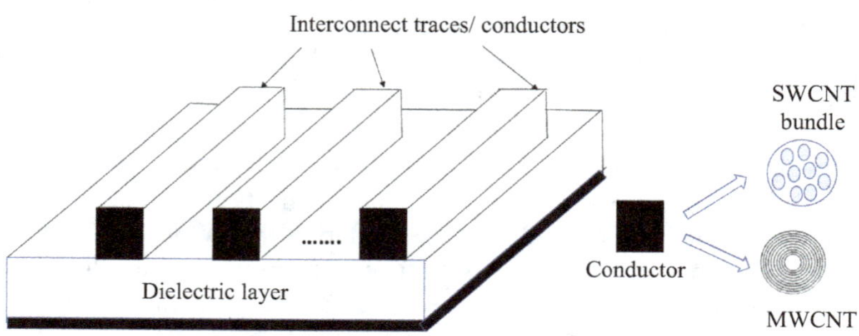

Figure 5.1 On-chip interconnect network consisting of coplanar traces/ conductors. Each conductor can be a SWCNT bundle or a standalone MWCNT structure

Despite the above potential benefits of MWCNT networks, it is worthwhile to appreciate that the growth, patterning, and fabrication processes of MWCNTs are nowhere near as mature as that of copper interconnect networks. Therefore, the fabrication and layout of MWCNT on-chip interconnects invariably introduces non-negligible process variations – variations that have a profound impact on the robustness, reliability, failure probability, and optimal design of the interconnect structures. Thus, for circuit designers to ensure that MWCNT on-chip interconnect networks meet the desired signal integrity standards of sub-22 nm technology nodes while still being fabricated at extremely high yield rates, they must account for the effects of fabrication process variations and manufacturing tolerances in the earliest design cycles. Therefore, uncertainty quantification (UQ) and stochastic modeling of MWCNT structures is a critical aspect of very large scale integration (VLSI) design automation [28–30].

5.2 State-of-the-art in modeling MWCNT networks

5.2.1 Extracting the PUL parameters

Let us begin by considering a general on-chip MWCNT interconnect network as shown in Figure 5.1. This network consists of N parasitically coupled conductors each with n number of conducting shells. For the sake of simplicity, the nonlinear CMOS driver/load of each conductor is replaced by a linear equivalent circuit. Assuming that for any general conductor, the ith shell has a diameter of D_i, the number of conducting channels in that shell is given by [31]

$$N_i = \begin{cases} 0.116D_j + 0.2, & D_i < 4.3 \text{ nm} \\ 0.667, & D_i \geq 4.3 \text{ nm} \end{cases} \tag{5.1}$$

The per-unit-length (PUL) scattering resistance, kinetic inductance, and quantum capacitance of the same shell are expressed as a function of the number of conducting channels N_i as [19]

$$R_i = \frac{h}{2e^2\lambda N_i}, \quad L_i = \frac{h}{4e^2 v N_i}, \quad C_i = \frac{4e^2 N_i}{hv} \tag{5.2}$$

Two other PUL capacitances need to be accounted for. These are the inter-shell capacitances (C_s) and the shell-to-ground capacitance (C_E) of the outermost shell of a conductor, expressed as

$$C_s = \frac{2\pi\varepsilon}{\ln(D_i/D_{i-1})}, \qquad C_E = \frac{2\pi\varepsilon}{\operatorname{arcosh}(D_n/H)} \tag{5.3}$$

where H is the height of the conductor above the ground plane in Figure 5.1. Finally, the inter-shell tunneling effect is represented as a PUL conductive parameter expressed as [19]

$$G = \frac{\alpha\pi D_i}{\delta} \tag{5.4}$$

where δ is the Van der Waal's gap and α is the tunneling conductivity. Based on the above PUL parameters, each shell of a conductor can be modeled as a transmission line using lumped resistance-inductance-capacitance-conductance (RLGC) circuit elements [1,32]. Similar to multiconductor transmission line networks, each conductor is also magnetically and electrically coupled to other conductors. This coupling is captured using the PUL magnetic inductance parameters and the electric field capacitance parameters [19]. However, the magnetic inductance of each line is often much smaller than the kinetic inductance and can be ignored [19,27]. On the other hand, the PUL coupling capacitance between two conductors are given by the well-known formula

$$C_{ij} = \frac{2\pi\varepsilon}{\text{arcosh}\left(S_{ij}/2D_i\right)} \tag{5.5}$$

where S_{ij} is the distance of separation between the ith and jth conductors.

5.2.2 Deterministic simulation of MWCNT networks

For the problem of deterministic simulation, a single conductor MWCNT can be represented using a rigorous cascaded lumped RLGC multiconductor circuit (MCC) model as shown in Figure 5.2. Obviously, the size of the modified nodal analysis matrices representing such an MCC model will be massive and consequently the SPICE simulation time costs of this model will be significantly large, even possibly to the point of being computationally prohibitive. As a result, the work of [33] introduced the idea of compressing the equivalent circuit representation of each conductor into a single line model, also referred to as the equivalent single conductor (ESC) model. The ESC model is predicated on the assumption that the potential at any point along the length of the conductor is equal across all shells in the conductor.

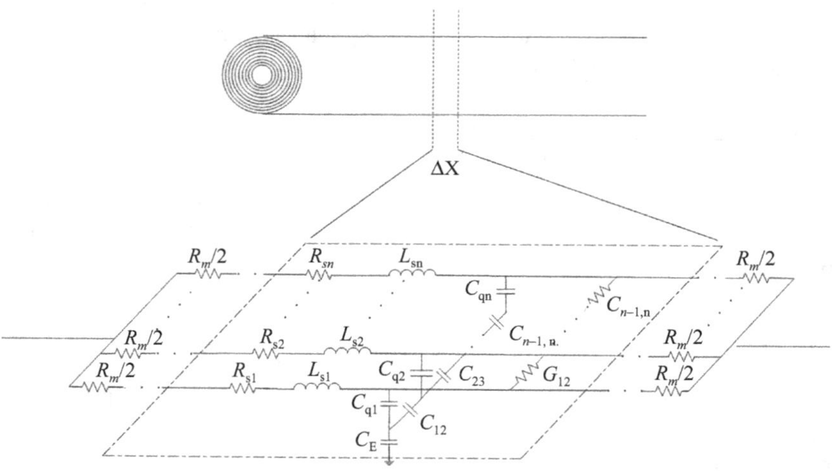

Figure 5.2 Cascaded lumped RLGC model of a single MWCNT conductor. This model is referred to as the MCC model

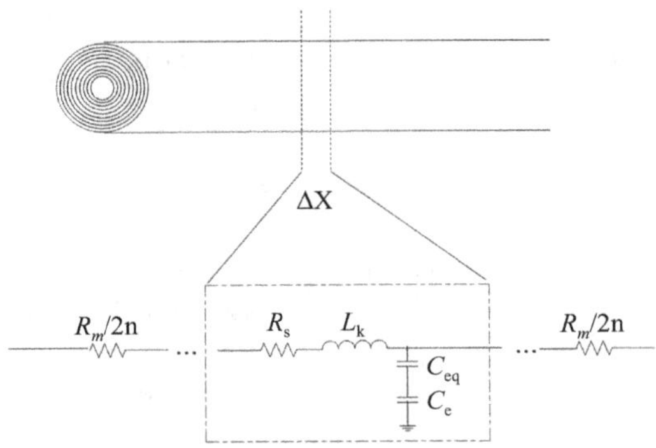

Figure 5.3 Cascaded lumped ESC model of a single MWCNT conductor

In effect, the tunneling conductance is assumed to be infinitely large [33]. Under this equipotential assumption, the lumped RLGC model of Figure 5.2 shrinks to the far more compact RLGC model of Figure 5.3 leading to significantly shorter CPU time cost for simulation. At this juncture, it is important to emphasize that the equipotential assumption is simply an approximation used to curb the simulation cost of MWCNT networks because realistically, the value of the tunneling conductance parameter cannot be infinitely large. Nevertheless, the errors resulting from the equipotential assumption, for many problems has been shown to be small enough to be reasonable [33].

5.2.3 Stochastic simulation and uncertainty quantification

Unfortunately, the deterministic simulation techniques are unable to consider probabilistic variations in the PUL parameters of (5.2)–(5.5) of the MWCNT network. To remedy this issue, probabilistic simulations are required. For this purpose, the fast converging generalized PC theory has been commonly used in the literature [28–30,34]. To better explain this theory, we assume that the impact of the fabrication process variations and manufacturing tolerances in the network of Figure 5.1 is modeled using N mutually uncorrelated random variables $\lambda = [\lambda_1, \lambda_2, \ldots, \lambda_N]$ located within the multidimensional domain of support Ω. The random variables λ will introduce uncertainty into the PUL parameters of the network which will eventually propagate to the transient responses of the network, usually the voltage responses at the far end of the MWCNT network. Under the PC formalism, any specific far-end transient response $y(t,\lambda)$ is cast as a metamodel taking the form of a linear combination of polynomial bases as [30,35–38]

$$y(t,\lambda) \approx \sum_{k=0}^{P} y_k(t)\phi_k(\lambda); \quad P + 1 = \frac{(N+m)!}{N!m!} \tag{5.6}$$

where $\phi_k(\lambda)$ is the kth N-dimensional basis function, $y_k(t)$ is the corresponding coefficient, and m is the maximum order of the expansion of (5.6). In order to ensure optimal convergence of the PC approximation of (5.6), the bases $\phi_k(\lambda)$ are orthonormal with respect to the joint PDF of λ described using the inner product integral [35,38]

$$\langle \phi_p(\lambda), \phi_q(\lambda) \rangle = \int_{\Omega} \phi_p(\lambda)\phi_q(\lambda)\rho(\lambda)d\lambda = \begin{cases} 0 & if \quad p \neq q \\ 1 & if \quad p = q \end{cases} \tag{5.7}$$

where $\rho(\lambda)$ is the joint PDF of λ. This orthonormality condition points to a one-to-one correspondence between the joint PDF $\rho(\lambda)$ and the basis functions $\phi_k(\lambda)$, referred to as the Weiner-Askey scheme [37]. In effect, from the prior knowledge of the marginal PDFs of the d random variables of λ, the optimal choice of the basis functions can be derived.

Once the basis functions of (5.6) is known, the PC coefficients can be determined or 'trained' using a variety of intrusive or non-intrusive techniques [34,39–44]. Interested readers may please refer to Chapter 2 for a detailed description of these intrusive or non-intrusive techniques. After training the PC coefficients, the right hand side of (5.6) will serve as a surrogate expression of the transient response $y(t,\lambda)$. Due to the analytic nature of the metamodel of (5.6), this surrogate expression can be easily inserted and repeatedly probed in a Monte Carlo framework in lieu of the MCC model of the MWCNT network in Figure 5.2 for fast extraction of the statistics of the response.

The major challenge in constructing the PC expansion of the transient response of (5.6) lies in the fact that irrespective of the approach adopted to train the coefficients, the 'curse of dimensionality' dictates that we need $O(N^m)$ deterministic simulations to do so. Typically, these training simulations are the rigorous MCC models of the MWCNT networks solved using SPICE [30]. In essence, we need to perform $O(N^m)$ MCC training simulations in order to determine the PC coefficients of (5.6) – a computationally intractable task. One way to mitigate this computational expense is by using the more compact ESC model, of course, at the expense of the training accuracy [34]. Thus, the accuracy-versus-training time cost tradeoff remains a clear challenge for MWCNT networks. In this chapter, we propose an effective predictor–corrector algorithm to resolve this challenge.

5.3 Predictor–corrector algorithm for MWCNT networks

The basic premise of the predictor–corrector algorithm is closely related to the concept of multi-fidelity training [45–48]. In multi-fidelity training, it is assumed that different models of distinct levels of accuracy or fidelity exist for the system under investigation. Some models are referred to as low-fidelity models while others are referred to as high-fidelity models. The low-fidelity models involve a variety of approaches such as simplifying assumptions in the governing physics, low-order approximation of complex mathematical terms, coarse grid spatial discretization of the partial differential equations involved, and coarse time stepping grids in time-domain analysis to name a few. These approaches allow the low-fidelity models to tradeoff its simulation accuracy in exchange of fast simulation time costs. On the other hand, the high-fidelity models are more rigorous models which yield much more accurate simulation results but at the cost of much greater simulation time costs. Now, if we intelligently cross-cut the numerical efficiency of the low-fidelity model simulations with the accuracy of the high-fidelity model simulations, it is possible to achieve a far more favorable accuracy-versus-time cost tradeoff as opposed to when we exclusively use high-fidelity model simulations [45–48].

5.3.1 Predictor metamodel for MWCNT networks

In our predictor–corrector algorithm, we posit that the ESC model of MWCNT networks will serve as a suitable low-fidelity model and the MCC model as a suitable high-fidelity model. In other words, for the problem of MWCNT networks, the low- and high-fidelity models are very easy to identify and do not require any prior experiments and analysis. This is a major reason why the predictor–corrector algorithm is highly attractive for MWCNT networks.

Next, we propose our predictor metamodel will take the form of the PC metamodel of transient response $y(t,\lambda)$ as shown below

$$y_p(t,\lambda) \approx \sum_{k=0}^{P} y_{pk}(t)\phi_k(\lambda); \quad P+1 = \frac{(N+m)!}{N!m!} \tag{5.8}$$

The subscript 'p' in the coefficients is introduced to denote that these coefficients belong to the predictor metamodel. These coefficients in (5.8) will be trained very efficiently using data extracted from the ESC model simulations of the MWCNT network. At this point, we emphasize that although we will require a massive $O(P+1) = O(N^m)$ ESC model simulations to train all the coefficients of (5.8), by virtue of the compact nature of the model, the cumulative time costs for these training simulations will not be very large. Once the predictor coefficients are trained, the expression of (5.8) will capture the coarse statistical features of the transient response $y(t,\lambda)$ – in other words, it will be a reasonable facsimile or a good 'initial guess' of the transient response obtained fairly cheaply. Of course, it is pointed out that the coefficients of (5.8) will not be able to precisely capture the finer statistical features of the response because their accuracy is limited by the simplifying equipotential assumption of the ESC model. Therefore, the question before us is this: how do we improve the accuracy of the predictor coefficients in (5.8) without incurring significant time costs, that is, how do we preserve the benefit of numerical efficiency provided by the ESC model and yet guarantee better accuracy for the predictor coefficients?

5.3.2 Developing a corrector function

As the name suggests, predictor–corrector algorithms employ a corrector function to compensate for the errors introduced into the predictor metamodel by the ESC model. The corrector function is defined to be the error between the true transient response $y(t,\lambda)$ and the predictor metamodel as

$$F_c(t,\lambda) = y(t,\lambda) - y_p(t,\lambda) = y(t,\lambda) - \sum_{k=0}^{P} y_{pk}(t)\phi_k(\lambda) \tag{5.9}$$

It is observed that the variance of the corrector function is given as

$$\mathrm{Var}(F_c(t,\lambda)) = \mathrm{Var}(y(t,\lambda)) + \mathrm{Var}(y_p(t,\lambda)) - 2\mathrm{Cov}(y_p(t,\lambda), y(t,\lambda)) \tag{5.10}$$

From (5.10), it is appreciated that as the variance of the predictor metamodel approaches the variance of the true transient response $y(t,\lambda)$, the variance of the corrector approaches 0. For the limiting case where the predictor coefficients are trained using MCC simulations, the predictor metamodel will accurately represent the true transient response and so the variance of the corrector will be 0. This will indicate that the predictor metamodel is itself sufficient to capture the uncertainty of the transient response and no corrector function is needed. For the general case, the predictor is a reasonable facsimile of the true transient response and so the variance of the corrector function is typically small. This means that a very sparse set of PC terms will be sufficient to represent the corrector. In this way, we will avoid incurring large computational time costs to extract the corrector (i.e., correct the errors in the predictor).

5.3.3 Training the corrector function

The corrector is represented as a PC metamodel whose form is given as

$$F_c(t,\lambda) = \sum_{k=0}^{Q} y_{ck}(t)\phi_k(\lambda) = y(t,\lambda) - \sum_{k=0}^{P} y_{pk}(t)\phi_k(\lambda) \tag{5.11}$$

where it is expected that $Q << P$. It is important to carefully choose the value of Q because this variable controls the accuracy-versus-time cost tradeoff of the algorithm. For example, a conservative choice of Q will ensure that the corrector is very accurate but, in the process, will incur significant training time costs. Conversely, an optimistic choice of Q may not be sufficient to correct for all the errors in the predictor metamodel although the corrector may be very efficient to train.

We propose to utilize the hyperbolic PC (HPC) truncation scheme introduced in [49] as a methodology to guide us in our best choice of Q. As per the HPC truncation scheme, the pth multidimensional PC basis of order r is expressed as a tensor product of unidimensional bases as

$$\phi_p(\lambda) = \prod_{i=1}^{N} \phi_{p_i}(\lambda_i); \quad \sum_{i=1}^{N} p_i = r \tag{5.12}$$

where p_i represents the order of the basis function corresponding to the ith random variable λ_i. If the orders of the univariate basis functions are stored in a vector

$\boldsymbol{p} = [p_1, \ldots, p_N]$, then the truncation criteria for a mth order PC expansion as in (5.6) or (5.8) is expressed using the L1 vector norm as

$$\|p\|_1 = \sum_{i=1}^{N} p_i \leq m \tag{5.13}$$

Now, using a new hyperbolic criterion in place of the linear criteria of (5.12), a more compact PC expansion can be obtained [49]. This new hyperbolic criterion is mathematically expressed as

$$\sum_{i=1}^{N} p_i^u \leq m^u; \quad 0 < u \leq 1 \tag{5.14}$$

Interested readers are referred to the work of [30,49] for a more in-depth discussion of how the user-define parameter u, also referred to as the hyperbolic shape parameter, controls the number of terms in a PC expansion. Suffice to say, as u increases from 0 to 1, the hyperbolic criterion of (5.14) slowly relaxes till at $u = 1$, it becomes same as the well-known linear truncation criterion of (5.13). In other words, the shape parameter u can be used as a knob to tune the number of terms in the PC expansion of the transient response $y(t,\lambda)$.

We propose an iterative methodology to identify the optimal value of u that will give us a sufficiently large number of terms in our corrector function of (5.11) to correct for the errors in the predictor metamodel while simultaneously limiting the time costs of training the corrector. To explain this methodology, let us assume that the discrete values the shape parameter can take to satisfy the inequality of (5.13) are $U = \{u_1, \ldots, u_R\}$ and the corresponding number of terms in the corrector metamodel are $Q = \{Q_1, \ldots, Q_R\}$. We start with the shape parameter value set to $u = u_1$ and evaluate the corresponding $Q = Q_1$ PC coefficients using the linear regression approach (or any other approach of our choice). Let the value of the coefficients trained be stored as

$$C^{(1)} = \left\{y_{c0}^{(1)}, y_{c1}^{(1)}, \ldots, y_{cQ_1}^{(1)}\right\} \tag{5.15}$$

Note that in (5.15), the superscript denotes the iteration counter. Next, we will calculate the variance of the corrector metamodel and store the result. The above cycle will be repeated for $u = u_2$ and $Q = Q_2$ to evaluate the new set of coefficients stored in $C^{(2)}$. From the knowledge of the new corrector coefficients in $C^{(2)}$, the new variance of the corrector is calculated. The improvement in the variance of the corrector achieved by increasing $u = u_1$ to $u = u_2$ (i.e., by increasing the number of terms in the corrector metamodel from $Q = Q_1$ to $Q = Q_2$) is quantified as the error term

$$e_1 = \left(\sum_{k=0}^{Q_2} (y_{ck}^{(2)})^2\right) - \left(\sum_{k=0}^{Q_1} (y_{ck}^{(1)})^2\right) / \left(\sum_{k=0}^{Q_2} (y_{ck}^{(1)})^2\right) \tag{5.16}$$

The L_2 norm of the error term of (5.15) reliably predicts how relaxing the shape parameter will enrich the accuracy of the corrector. Thus, our iterative method will increase the value of u from $u = u_1$ to $u = u_R$ and keep on recording the L_2 norm of the error term for successive values of u. Once the L_2 norm reaches a prescribed tolerance, then the corrector will be deemed to be accurate enough and the iterations will be stopped. Assume that the iterations are stopped when $u = u_l$ where $1 \leq l \leq R$. Therefore, the optimal value of the shape parameter will be $u_{\text{opt}} = u_l$ and the corresponding corrector function will be expressed as

$$F_c(t, \lambda) = \sum_{k=0}^{Q_l} y_{ck}^{(l)}(t)\phi_k(\lambda) = y(t, \lambda) - \sum_{k=0}^{P} y_{pk}(t)\phi_k(\lambda) \tag{5.17}$$

It is obvious that $Q_l << P$ indicates that only a small subset of the actual PC coefficients is sufficient to model the error between the predictor and the true transient response.

5.3.4 Recovering the PC metamodels of the transient response

Finally, we have to combine the predictor and corrector to recover the PC metamodel of the transient response $y(t,\lambda)$ of the MWCNT network. It is observed from (5.17) that the transient response $y(t,\lambda)$ can be expressed as the sum of the predictor and corrector as

$$\begin{aligned} y(t, \lambda) &= \sum_{k=0}^{Q_l} y_{ck}^{(l)}(t)\phi_k(\lambda) + \sum_{k=0}^{P} y_{pk}(t)\phi_k(\lambda) \\ &= \sum_{k=0}^{Q_l} \left(y_{ck}^{(l)} + y_{pk}(t)\right)\phi_k(\lambda) + \sum_{k=Q_l+1}^{P} y_{pk}(t)\phi_k(\lambda) \end{aligned} \tag{5.18}$$

Thus, the most important first Q_l+1 coefficients of the predictor metamodel are corrected by the corrector coefficients while the remaining lesser important $P - Q_l$ coefficients need no correction.

5.3.5 Numerical cost analysis

From the above subsection it is understood that we need to train $P + 1$ coefficients using the ESC model and $Q_l + 1$ coefficients using the MCC model. Assuming that the time cost of an ESC model is C_{ESC} seconds and that of the MCC model is C_{MCC} seconds, the total cost of implementing the predictor–corrector algorithm will be

$$C_{pc} = (P + 1)C_{\text{ESC}} + (Q_l + 1)C_{\text{MCC}} \tag{5.19}$$

In contrast, training the conventional PC metamodel of (5.6) involves evaluating $P + 1$ coefficients using only the MCC model for a total cost of

$$C_{\text{conv}} = (P + 1)C_{\text{MCC}} \tag{5.20}$$

Consequently, the speedup achieved by the predictor–corrector algorithm is given as

$$\eta = \frac{C_{\text{conv}}}{C_{pc}} = \frac{(P+1)C_{\text{MCC}}}{(P+1)C_{\text{ESC}} + (Q_l+1)C_{\text{MCC}}} \tag{5.21}$$

If the cost of the MCC model simulation is β times greater than the cost of an ESC model simulation, then (5.18) can be compactly expressed as

$$\eta = \frac{C_{\text{conv}}}{C_{pc}} = \frac{\beta(P+1)}{P+1+\beta(Q_l+1)} \tag{5.22}$$

Now, the value β scales with the number of shells (n) in each MWCNT conductor as $O(n^{\xi})$ where $3 \leq \xi \leq 4$ if a lumped RLGC model is used for the ESC and MCC simulations as shown in Figures 5.2 and 5.3 [50]. This means that even for a modest number of shells in each MWCNT conductor (i.e., $n \sim 20$), the value of β will be very high to the point that the speedup achieved can be expressed as

$$\eta_{\max} = \lim_{\beta \to \infty} \frac{C_{\text{conv}}}{C_{pc}} = \frac{P+1}{Q_l+1} \tag{5.23}$$

In effect, the maximum speedup possible saturates at the value of the speedup achieved using the HPC truncation with $u = u_l$. However, it is emphasized that the hyperbolic truncation does not exploit any predictor metamodel and hence always utilizes $u > u_l$in order to achieve the same level of accuracy as the predictor–corrector algorithm. In other words, the predictor–corrector algorithm is expected to be even faster than the HPC truncation scheme of [49] as well as the conventional PC expansion.

5.4 Illustrative examples

In this section, we present multiple numerical examples to highlight the benefits of the predictor–corrector algorithm over conventional PC metamodels. For all ESC and MCC model representations of the MWCNT networks in this section, the lumped RLGC model of Figures 5.2 and 5.3 are utilized and solved in SPICE.

5.4.1 Illustrative Example 1

In this example, we consider a single conductor MWCNT network (i.e., $M = 1$ in Figure 5.4). The number of shells in this conductor is kept varied in order to probe the characteristics of the predictor–corrector algorithm. The network is excited by a saturated ramp voltage waveform of amplitude 1V and rise time $T_r = 0.1$ ps. The relevant response of interest is the far-end transient voltage signal $y(t,\lambda)$. The random parameters of this network are listed in Table 5.1.

We begin by first establishing the accuracy of the predictor–corrector algorithm. For this purpose, the number of shells in the conductor is set to $n = 50$ and the result obtained by the predictor–corrector algorithm is compared against those obtained using the brute-force Monte Carlo approach with 40,000 samples. These

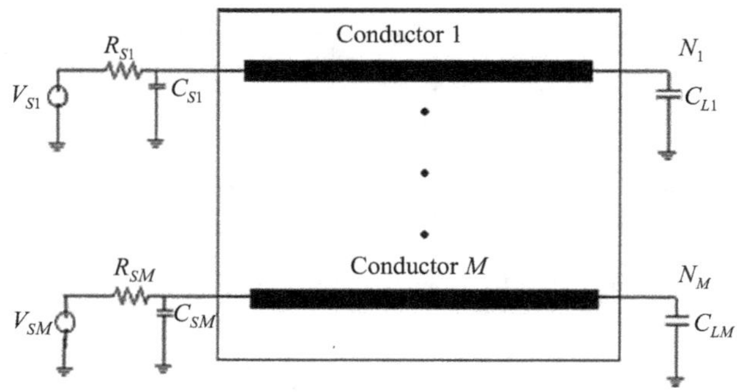

Figure 5.4 Network of Example 1 consisting of a single MWCNT conductor

Table 5.1 Uncertain network parameters with normal distribution for Example 1

No.	Uncertain network parameters	Mean	Relative standard deviation
1	D_{in} (Inner diameter of CNT)	2.28 nm	20%
2	d (Inter-shell distance)	0.34 nm	20%
3	σ (Tunneling conductivity)	20	20%
4	C_{in} (Driver capacitance)	0.14 fF	20%
5	C_{out} (Load capacitance)	0.049 fF	20%
6	H (Height of dielectric)	50 nm	20%
7	ε_r (Dielectric constant)	2	20%
8	R_m (Contact resistance)	1,000 Ω	20%
9	l (length of conductor)	100 μm	5%

results are shown in Figures 5.5 and 5.6 and clearly underscore the good accuracy of the predictor–corrector algorithm. In particular, Figure 5.6 illustrates the significant errors that become embedded in a PC metamodel trained exclusively using the ESC model as has been done in [33].

Next, the numerical efficiency of the predictor–corrector algorithm is established. For this purpose, the number of shells is varied from $n = 30$ to $n = 50$. Irrespective of the value of n, the value of $u_{\text{opt}} = 0.69$ so that the L_2 norm of the error of (5.16) satisfies the following condition

$$\|e\|_2 \leq 10^{-3} \tag{5.24}$$

Under such circumstances, the corresponding number of terms in the corrector is $Q_l + 1 = 145$. From this, we note that only 21% of the full set of PC coefficients in the corrector need to be trained in order to capture the finer statistical features of $y(t,\lambda)$, thereby underscoring the effectiveness of the predictor–corrector algorithm. For the different values of n, the L_2 norm of the standard deviation error of the predictor–corrector algorithm with respect to the Monte Carlo results and the corresponding speedup achieved is reported in Table 5.2. In addition, similar data for

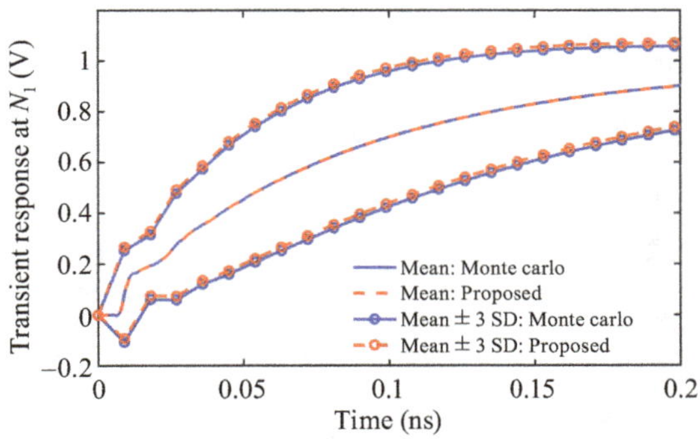

Figure 5.5 Statistics of the transient response at node N_1 of Example 1 computed using the Monte Carlo method (40,000 samples) and the proposed predictor–corrector algorithm

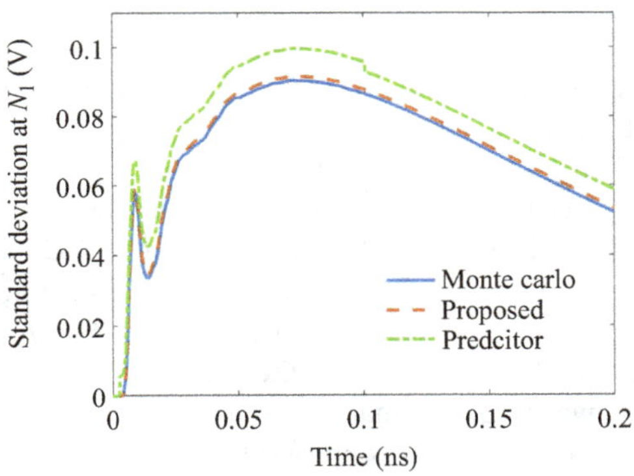

Figure 5.6 Illustrating the errors in the PC metamodel of Example 1 trained exclusively using ESC simulations

Table 5.2 Scaling of numerical efficiency achieved by the predictor–corrector algorithm with respect to the number of shells per conductor of Example 1

# of Shells (*n*)	**Predictor–corrector algorithm**		**Standard PC metamodel**		**Speedup**
	L_2 norm of SD error	**CPU time (ms)**	**L_2 norm of SD error**	**CPU time (ms)**	
30	1.08×10^{-4}	1,904.40	1.19×10^{-4}	8,136.70	4.27
35	1.03×10^{-4}	1,987.70	1.19×10^{-4}	9,237.80	4.64
40	1.06×10^{-4}	2,098.80	1.18×10^{-4}	10,081.50	4.80
45	1.01×10^{-4}	2,148.70	1.21×10^{-4}	10,524.80	4.89
50	1.04×10^{-4}	2,290.80	1.20×10^{-4}	11,225.50	4.90

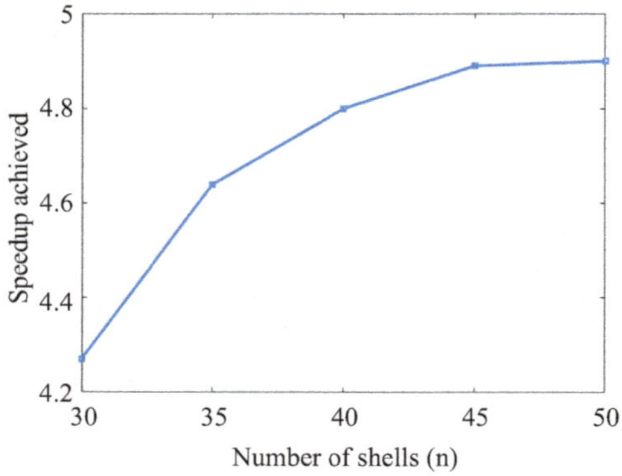

Figure 5.7 Scaling of speedup achieved by the proposed predictor–corrector algorithm with respect to the number of shells in each MWCNT conductor (n)

the conventional PC algorithm is also listed in Table 5.2. Thus, the results of Table 5.2 demonstrate the excellent accuracy-versus-time cost tradeoff that is afforded by the predictor–corrector algorithm over the conventional PC algorithm. From Table 5.2, it is further observed that as the number of shells (n) increases from 30 to 50, the efficiency of the ESC model over the MCC model (i.e., the value of β in (5.22)) increases as well till the speedup achieved saturates at the value of 4.9. This scaling of the speedup with n is illustrated in Figure 5.7. The saturated value of the speedup of 4.9 is very close to the expected maximum speedup of 4.93 from (5.23). Finally, we point out that $u_{\text{opt}} = 0.79$ if the well-known hyperbolic truncation scheme of [49] is used as opposed to the predictor–corrector algorithm where $u_{\text{opt}} = 0.69$. Thus, the predictor–corrector algorithm is much faster than the HPC technique reported in [49]. This is, again, to be expected because the hyperbolic truncation scheme does not exploit any predictor metamodel as an initial guess of the response $y(t,\lambda)$.

5.4.2 Illustrative Example 2

In this example, we assess the performance of the predictor–corrector algorithm when applied to a multiconductor MWCNT network. For this purpose, the number of conductors in the network of Figure 5.4 is set to $M = 3$ and the first line is active while lines 2 and 3 are quiet. The same excitation sources used in Example 1 is also used for this example. This time the responses of interest are the far-end transient voltage for all three conductors. The random parameters of this network are listed in Table 5.3.

In the first part of this example, we investigate the impact of the tunneling conductance on the convergence of the predictor–corrector algorithm. Theoretically, it is expected that as the value of the tunneling conductivity (α) decreases, then the ESC model (where implicitly α is assumed to be infinitely

Table 5.3 Uncertain network parameters with normal distribution for Example 2

No.	Uncertain network parameters	Mean	Relative SD
1–3	D_{in1}–D_{in3} (Inner diameter of conductors 1–3)	2.28 nm	20%
4–6	d_1–d_3 (Inter-shell distance of conductors 1–3)	0.34 nm	20%
7–9	C_{s1}–C_{s3} (Driver capacitance of conductors 1–3)	0.14 fF	20%
10–12	C_{L1}–C_{L3} (Load capacitance of conductors 1–3)	0.49 fF	20%
13–15	R_{m1}–R_{m3} (Contact resistance of conductors 1–3)	1,000 Ω	20%
16	S (Spacing between conductors)	22 nm	20%
17	ε_r (Dielectric constant)	2	20%
18	H (Height of dielectric)	50 nm	20%

Table 5.4 Scaling of numerical efficiency achieved by predictor–corrector algorithm with respect to the tunneling conductance of Example 2

Value of tunneling conductance (σ)	Predictor–corrector algorithm		Standard PC metamodel		Speedup
	L_2 norm of SD error	CPU time	L_2 norm of SD error	CPU time	
0	7.40×10^{-4}	3 h 10 min	8.10×10^{-4}	42 h 55 min	13.52
10	6.22×10^{-4}	3 h 10 min	9.65×10^{-4}	42 h 55 min	13.52
100	6.26×10^{-4}	3 h 10 min	7.80×10^{-4}	42 h 55 min	13.52

large) should introduce more error into the predictor. Consequently, more corrector terms need to be considered to correct this error, leading to larger MCC costs. To check if this is true, in this example, we sweep the value of the tunneling conductivity from the best-case scenario of $\alpha = 100$ to the worst-case scenario $\alpha = 0$. For the different values of the tunneling conductivity, the L_2 norm of the standard deviation error achieved by the predictor–corrector algorithm with respect to the Monte Carlo results is recorded using $u_{opt} = 0.69$ and reported in Table 5.4. In addition, the same data for the conventional PC algorithm is also listed in Table 5.4. As seen from Table 5.4, as the value of the tunneling conductivity decreases, the same corrector yields larger error norms, or in other words, the error in the predictor increases. However, remarkably, this increase in error in the predictor coefficients is very small even for the worst-case scenario of $\alpha = 0$. Thus, for $u_{opt} = 0.69$ (same as in Example 1) we are still able to recover a very accurate PC metamodel for all test cases. This high accuracy of the PC metamodel for $\alpha = 0$ and $u_{opt} = 0.69$ is illustrated in the transient response statistics of Figures 5.8 and 5.9.

In the next part of this example, we study how the CPU time cost for the predictor–corrector algorithm scales with the value of the tunneling conductivity α. In this study, we note that because $u_{opt} = 0.69$ remains sufficient for even the worst-case scenario of $\alpha = 0$, expectedly the numerical efficiency provided by the predictor–corrector algorithm remains unaffected by the value of α. This is

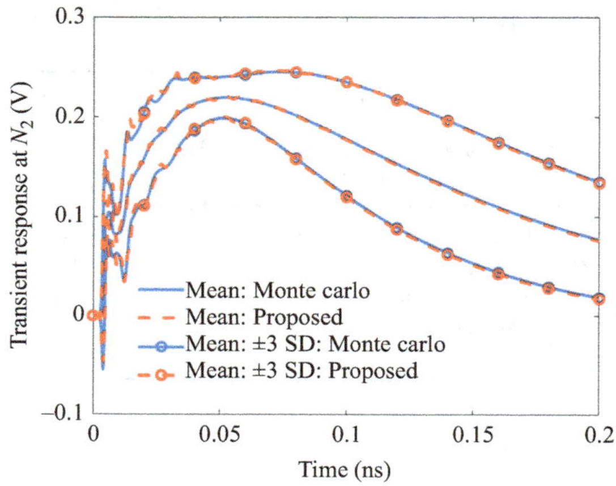

Figure 5.8 Statistics of the transient response at node N_2 of Example 2 computed using the Monte Carlo method (40,000 samples) and the proposed predictor–corrector algorithm

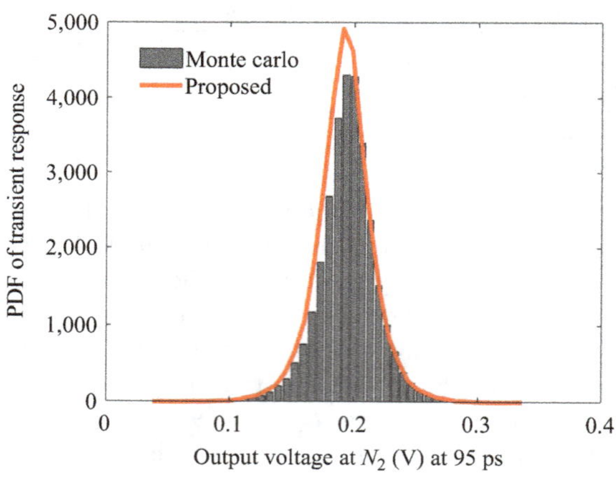

Figure 5.9 PDF of the transient response at node N_2 of Example 2 computed using the Monte Carlo method (40,000 samples) and the proposed predictor–corrector algorithm

reflected in the fact that for the number of terms in the corrector set to $Q_l+1 = 532$ (i.e., less than 8% of the total $P+1 = 7{,}315$ number of terms), sufficient accuracy is reached using the predictor–corrector algorithm irrespective of the value of the tunneling conductivity α (see Table 5.4). This translates to more than 13 times speedup provided by the predictor–corrector algorithm over conventional PC metamodels.

5.4.3 Illustrative Example 3

In this example, a $M = 5$ conductor MWCNT network is explored. The first line is the active line while all other lines are quiet. The first line is excited by a voltage source with exactly the same waveform as seen for Examples 1 and 2. The relevant responses of interest for this example are the far-end transient voltage signals of all five conductors and the random parameters of this network are listed in Table 5.5.

In the first part of this example, the accuracy of the predictor–corrector algorithm with respect to the brute-force Monte Carlo approach is investigated. For this purpose, the PDF of the transient responses of lines 1 and 3 are evaluated using the Monte Carlo method and the predictor–corrector algorithm at the time point when each result exhibits maximum mean value (i.e., at $t = 200$ ps and $t = 73.2$ ps, respectively). The predictor–corrector algorithm displays good agreement with the Monte Caro results as shown in Figure 5.10. The reason the PDF is used in this

Table 5.5 Uncertain network parameters with normal distribution for Example 3

No.	Uncertain network parameters	Mean	Relative SD
1–5	$D_{\text{in}1}$–$D_{\text{in}5}$ (Inner diameter of conductor 1–5)	2.28 nm	15%
6–10	d_1–d_5 (Inter-shell distance of conductor 1–5)	0.34 nm	15%
11–15	C_{s1}–C_{s3} (Driver capacitance of conductor 1–5)	0.14 fF	15%
16–20	C_{L1}–C_{L3} (Load capacitance of conductor 1–5)	0.49 fF	15%
21–25	R_{m1}–R_{m3} (Contact resistance of conductor 1–5)	1,000 Ω	15%
26–30	σ_1–σ_5 (Tunneling conductivity of conductors 1–5)	22 nm	15%
31	ε_r (Dielectric constant)	2	15%
32	H (Height of dielectric)	50 nm	15%

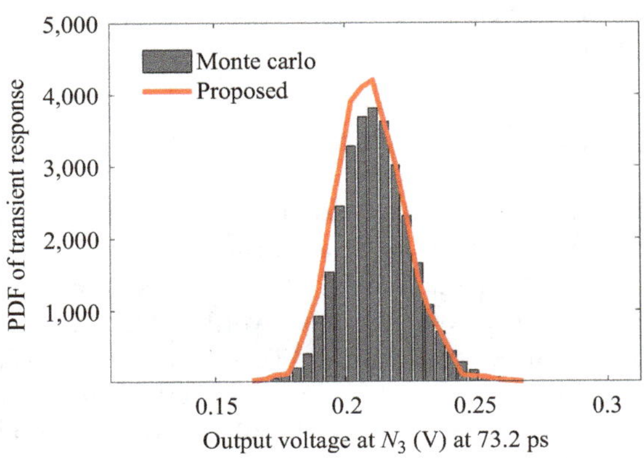

Figure 5.10 PDF of the transient response at node N_3 of Example 3 computed using the Monte Carlo method (40,000 samples) and the proposed predictor–corrector algorithm at time point when the mean has maximum value

Table 5.6 Scaling of numerical efficiency achieved by predictor–corrector algorithm with respect to the tunneling conductivity of Example 3

Value of tunneling conductivity parameter (σ)	Predictor–corrector algorithm		Standard PC metamodel		Speedup
	L_2 norm of SD error	CPU time	L_2 norm of SD error	CPU time	
0	8.51×10^{-4}	5 h 6 min	1.00×10^{-3}	178 h 48 min	35.05
100	7.90×10^{-4}	5 h 6 min	5.43×10^{-4}	(>7 days)	35.05

example is to demonstrate the accuracy of the predictor–corrector algorithm across all statistical moments instead of simply mean or standard deviation.

In the next section of this example, the impact of the tunneling conductance on the convergence of the predictor–corrector algorithm is investigated, similar to Example 2. Here, we only consider the extreme cases where $\alpha = 100$ and $\alpha = 0$. For either case, the L_2 norm of the standard deviation error incurred by the predictor–corrector algorithm with respect to the Monte Carlo results is recorded using $u_{\text{opt}} = 0.69$ and reported in Table 5.6. As expect, for $\alpha = 0$ the error norm is larger than for $\alpha = 100$. However, this increase in error is relatively small and hence in either case, $u_{\text{opt}} = 0.69$ provides a sufficiently accurate PC metamodel.

Given that $u_{\text{opt}} = 0.69$, for either choice of the tunneling conductance (α), the number of terms required in the corrector is $Q_l + 1 = 1{,}617$. This is less than 3% of the total number of terms in a conventional PC metamodel (i.e., $P + 1 = 58{,}905$). Thus, the predictor–corrector algorithm is able to consistently provide a speedup of roughly 35 times over the conventional PC metamodel of (5.2). In other words, the tunneling conductance parameter (α) had little to no impact on the numerical efficiency offered by the predictor–corrector algorithm.

5.5 Improved predictor–corrector algorithms

It is understood from the illustrative examples of the previous section that the predictor–corrector algorithm is an effective approach to address the curse of dimensionality facing PC metamodels, especially as it pertains to MWCNT interconnect networks. However, it is also worth asking if the benefits of this algorithm can be enhanced even further. To that end, we posit that once the predictor metamodel of the network response is derived in (5.8), this predictor can be exploited to quickly extract a wealth of statistical information regarding the network response. For example, the predictor metamodel of (5.8) can be probed to quickly extract the Sobol's sensitivity indices and the integrated sensitivity indices (SIs) of each random dimension λ_i as shown in (4.23)–(4.24) of Chapter 4. We acknowledge the fact that the integrated SIs calculated using (4.24) of Chapter 4 will not be the precisely correct due to the errors arising from the equipotential assumption in the ESC model. However, this disadvantage will be diminished by the following other advantages.

(i) The predictor metamodel contains all the terms in the high dimensional model representation of the network response instead of only the 0th and first orders terms. Hence, when extracting the SIs from the predictor metamodel, all the higher order interactions among the dimensions will be taken into account instead of only considering each random dimension in isolation as proposed in Chapter 4. This is expected to lead to more accurate estimation of the SIs.

(ii) When using the predictor coefficients to calculate the SIs of (4.23)–(4.24) (Chapter 4), it is not necessary to precisely calculate the SIs as long as we can obtain a reasonably approximate value of the same. Given the fact that the ESC model is a reasonable facsimile of the rigorous MCC model, this condition is expected to be met, thus allowing us to be able to distinguish the important R dimensions from the unimportant N–R dimensions in a MWCNT interconnect network.

Now, once the SIs of (4.23)–(4.24) are estimated, they will be used to reduce the original set of random dimensions $\lambda \in \Re^N$ to a new set of the most important random dimensions $\xi \in \Re^R$. Thereafter, the corrector function will be constructed similar to (5.17), with the difference being that this corrector function will contain only the important dimensions ξ instead of λ as

$$F_c(t,\xi) = \sum_{k=0}^{R_l} y_{ck}^{(l)}(t)\phi_k(\xi) = y(t,\lambda) - \sum_{k=0}^{P} y_{pk}(t)\phi_k(\lambda) \tag{5.25}$$

It is obvious that the number of terms in the corrector metamodel of (5.25) will be even smaller than that used in the naïve corrector of (5.17). Thus, even fewer MCC simulations will be required to recover the coefficients (5.25) compared to that of (5.17). Consequently, the maximum speedup achieved will now increase to

$$\eta_{\max} = \lim_{\beta\to\infty} \frac{C_{conv}}{C_{pc}} = \frac{P+1}{R_l+1} \tag{5.26}$$

Of course, it is pointed out that the predictor metamodel of (5.8) can also be leveraged using any other dimension reduction scheme described in Chapter 4.

5.6 Conclusion

In this chapter, we have focused on the stochastic simulation and UQ of MWCNT interconnect networks. In particular, in this chapter, a new predictor–corrector approach to accelerate the training of the PC metamodels of MWCNT interconnect networks is introduced. This algorithm leverages the physics-based approximate and compact ESC model of the MWCNT conductors to extract the coarse statistical features of the transient responses of the interconnect networks. Thereafter, the finer statistical features of the transient responses are captured using a sparse number of simulations of the rigorous MCC model representation of each MWCNT conductor of the network. Thus, the predictor–corrector algorithm intelligently lifts the computational burden of evaluating the PC coefficients from the shoulders of the time-intensive MCC model to that of the more efficient ESC model. Importantly, this algorithm seeks to address the massive computational cost of training PC coefficients without resorting to reducing the number of original dimensions or bases included in the metamodel – in effect, without reducing the size of the PC metamodels involved.

References

[1] R. Achar and M. Nakhla, "Simulation of high-speed interconnects," *Proc. IEEE*, vol. 89, no. 5, pp. 693–728, May 2001.

[2] A. Deutsch, "Electrical characteristics of interconnections for high-performance systems," *Proc. IEEE*, vol. 86, pp. 315–355, Feb. 1998.

[3] Y. I. Ismail and E. G. Friedman, "Effects of inductance on the propoagation delay and repeater insertion in VLSI circuits," *IEEE Trans. Very large Scale Integrated Systems*, vol. 8, no. 2, pp. 195–206, Apr. 2000.

[4] Y. I. Ismail and E. G. Friedman, *On-Chip Inductance in High Speed Integrated Circuits*. Norwell, MA: Kluwer Academic Publishers, 2001.

[5] M. Celik, L. Pileggi, and A. Odabasioglu, *IC Interconnect Analysis*. Norwell, MA: Kluwer Academic Publishers, 2002.

[6] B. Young, *Digital Signal Integrity: Modeling and Simulation with Interconnects and Packages*. Englewood Cliffs, NJ: Prentice Hall, 2001.

[7] M. Swaminathan and A. E. Engin, *Power Integrity Modeling and design for Semiconductors and Systems*. Boston, MA: Prentice Hall, 2007.

[8] J. A. Davis, *et al.*, "Interconnect limits on gigascale integration," *Proc. IEEE*, vol. 89, no. 3, pp. 305–324, March 2001.

[9] W. Steinhögl, G. Schindler, G. Steinlesberger, M. Traving, and M. Engelhardt, "Comprehensive study of the resistivity of copper wires with lateral dimensions of 100 nm and smaller," *J. Appl. Phys*., vol. 97, no. 2, pp. 023706-1–023706-7, 2005.

[10] International Technology Roadmap for Semiconductors, 2015. [Online]. Available: http://www.itrs2.net/2015-itrs.html

[11] S. Kumar and R. Sharma, "Analytical model for resistivity and mean free path in on-chip interconnects with rough surfaces," *IEEE Trans. Emerg. Top. Comp*., vol. 6, no. 2, pp. 233–243, 2018.

[12] A. Sain and K. L. Melde, "Broadband characterization of coplanar wave-guide interconnects with rough conductor surfaces," *IEEE Trans. Comp., Packag. and Manuf. Technol*., vol. 3, no. 6, pp. 1038–1046, 2013.

[13] S. Kumar and R. Sharma, "Analytical modeling and performance bench-marking of on-chip interconnects with rough surfaces," *IEEE Trans. Multi-Scale Comp. Syst*., vol. 4, no. 3, pp. 272–284, 2018.

[14] A. Todri-Sanial, J. Dijon, and A. Maffucci (Editors), *Carbon Nanotubes for Interconnects*. Switzerland: Springer International, 2017.

[15] K. Banerjee and N. Srivastava, *"Are carbon nanotubes the future of VLSI interconnections?" 2006 43rd ACM/IEEE Design Automation Conference*, San Francisco, CA, 2006, pp. 809–814

[16] A. Naeemi and J. D. Meindl, "Compact physical models for multiwall carbon-nanotube interconnects," *IEEE Electron Device Lett.*, vol. 27, no. 5, pp. 338–340, May 2006.

[17] A. Naeemi and J. D. Meindl, "Compact physics-based circuit models for graphene nanoribbon interconnects," *IEEE Trans. Electron. Devices*, vol. 56, no. 9, pp. 1822–1833, Sept. 2009.

[18] H. Li, W. Y. Yin, K. Banerjee, and J. F. Mao, "Circuit modeling and performance analysis of multi-walled carbon nanotube interconnects," *IEEE Trans. Electron Devices*, vol. 55, no. 6, pp. 1328–1337, June 2008.

[19] A. G. Chiariello, A. Maffucci, and G. Miano, "Circuit models of carbon based interconnects for nanopackaging," *IEEE Trans. Compon. Packaging Manuf. Technol.*, vol. 3, no. 11, pp. 1926–1937, Nov. 2013.

[20] A. Nieuwoudt and Y. Massoud, "On the optimal design, performance, and reliability of future carbon nanotube-based interconnect solutions," *IEEE Trans. Electron Devices*, vol. 55, no. 8, pp. 2097–2010, Aug. 2008.

[21] N. Srivastava, H. Li, F. Kreupl, and K. Banerjee, "On the applicability of single-walled carbon nanotubes as VLSI interconnects,"*IEEE Trans. Nanotechnol.*, vol. 8, no. 4, pp. 542–559, July 2009.

[22] H. Li, C. Xu, N. Srivastava, and K. Banerjee, "Carbon nanomaterials for next-generation interconnects and passives: Physics, status, and prospects," *IEEE Trans. Electron Devices*, vol. 56, no. 9, pp. 1799–1821, Sept. 2009.

[23] J. Hone, M. Whitney, C. Piskoti, and A. Zettl, "Thermal conductivity of single-walled carbon nanotubes," *Phys. Rev. B, Condens. Matter*, vol. 59, no. 4, pp. R2514–R2516, 1999.

[24] B. Q. Wei, R. Vajtai, and P. M. Ajayan, "Reliability and current carrying capacity of carbon nanotubes," *Appl. Phys. Lett.*, vol. 79, no. 8, pp. 1172–1174, Aug. 2001.

[25] M. K. Majumder, N. D. Pandya, B. K. Kaushik, and S. K. Manhas, "Analysis of MWCNT and bundled SWCNT interconnects: Impact on crosstalk and area," *IEEE Electron Device Lett.*, vol. 33, no. 8, pp. 1180–1182, Aug. 2012.

[26] R. Mehta, S. Chugh, and Z. Chen, "Enhanced electrical and thermal conduction in graphene-encapsulated copper nanowires," *Nano Letters*, vol. 15, no. 3, pp. 2024–2030, Feb. 2015.

[27] M. Tang and J. Mao, "Modeling and fast simulation of multiwalled carbon nanotube interconnects," *IEEE Trans. Electromagn. Compat.*, vol. 57, no. 2, pp. 232–240, Apr. 2015.

[28] A. K. Prasad, D. Zhou, and S. Roy, "Reduced dimension polynomial chaos approach for efficient uncertainty analysis of multi-walled carbon nanotube interconnects," in *Proc. IEEE MTT-S 64th International Microwave Symposium (IMS)*, San Francisco, CA, May 2016, pp. 1–3.

[29] S. Bhatnagar, A. Merkley, R. Berdine, Y. Li, and S. Roy, "Variability-aware performance assessment of multi-walled carbon nanotube interconnects using a predictor–corrector polynomial chaos scheme," in *Proc. IEEE Conference on Electrical Design of Advanced Packaging and Systems (EDAPS)*, Chandigarh, Dec. 2018, pp. 1–3.

[30] Y. Li, S. Bhatnagar, A. Merkley, D. Weber, and S. Roy, "A predictor–corrector algorithm for fast polynomial chaos-based uncertainty quantification of multi-walled carbon nanotube interconnects," *IEEE Trans. Compon. Packaging Manuf. Technol.*, vol. 9, no. 10, pp. 1963–1975, Oct. 2019.

[31] A. Naeemi and J. D. Meindl, "Performance modelling for single- and multiwall carbon nanotubes as signal and power interconnects in gigascale systems," *IEEE Trans. Electron Devices*, vol. 55, no. 10, pp. 2574–2582, Oct. 2008.

[32] C. R. Paul, *Analysis of Multiconductor Transmission Line*. New York, NY: Wiley-Interscience, 2008.

[33] M. S. Sarto and A. Tamburrano, "Single-conductor transmission-line model of multiwall carbon nanotubes," *IEEE Trans. Nanotechnol.*, vol. 9, no. 1, pp. 82–92, Jan. 2010.

[34] I. S. Stievano, P. Manfredi, and F. G. Canavero, "Carbon nanotube interconnects: process variation via polynomial chaos," *IEEE Trans. Electromagn. Compat.*, vol. 54, no. 1, pp. 140–148, Feb. 2012.

[35] D. Xiu, *Numerical Methods for Stochastic Computations: A Spectral Method Approach*. Princeton, NJ: Princeton University Press, 2010.

[36] D. Xiu, "Fast numerical methods for stochastic computations: a review," *Commun. Comput. Phys.*, vol. 5, no. 2–4, pp. 242–272, Feb. 2009.

[37] D. Xiu and G. E. Karniadakis, "The Wiener-Askey polynomial chaos for stochastic differential equations," *SIAM Journal Sci. Comput.*, vol. 24, no. 2, pp. 619–644, July 2002.

[38] R. G. Ghanem and P. Spanos, *Stochastic Finite Elements: A Spectral Approach*. New York, NY: Springer-Verlag, 1991.

[39] A. Biondi, D. Vande Ginste, D. De Zutter, P. Manfredi, and F. Canavero, "Variability analysis of interconnects terminated by general nonlinear loads," *IEEE Trans. Compon. Packaging Manuf. Technol.*, vol. 2, no. 7, pp. 1244–1251, July 2013.

[40] P. Manfredi, D. Vande Ginste, D. De Zutter, and F. Canavero, "Uncertainty assessment in lossy and dispersive lines in SPICE-type environments," *IEEE Trans. Compon. Packaging Manuf. Technol.*, vol. 2, no. 7, pp. 1252–1258, July 2013.

[41] P. Manfredi, D. Vande Ginste, D. De Zutter, and F. Canavero, "Generalized decoupled polynomial chaos for nonlinear circuits with many random parameters," *IEEE Microw. Wirel. Compon. Lett.*, vol. 25, no. 8, pp. 505–507, Aug. 2015.

[42] A. K. Prasad, M. Ahadi, and S. Roy, "Multidimensional uncertainty quantification of microwave/RF networks using linear regression and optimal

design of experiments," *IEEE Trans. Microw. Theory Tech.*, vol. 64, no. 8, pp. 2433–2446, Aug. 2016.

[43] M. Ahadi and S. Roy, "Sparse linear regression (SPLINER) approach for efficient multidimensional uncertainty quantification of High-speed circuits," *IEEE Trans. Comput.-Aided Design*, vol. 35, no. 10, pp. 1640–1652, Oct. 2016.

[44] M. S. Eldred, "Recent advances in non-intrusive polynomial chaos and stochastic collocation methods for uncertainty analysis and design," in *Proc. 50th AIAA Structures, Structural Dynamics, and Materials Conference*, Palm Springs, CA, May 2009, pp. 1–37.

[45] L. W. T. Ng and M. S. Eldred, "Multifidelity uncertainty quantification using non-intrusive polynomial chaos and stochastic collocation," in *Proc. 53rd AIAA Structures, Structural Dynamics, and Materials Conference*, Apr. 2012, pp. 1–17.

[46] P. S. Palar, T. Tsuchiya, and G. T. Parks, "Multi-fidelity non-intrusive polynomial chaos based on regression," *Compt. Methods Appl. Mech. Eng.*, vol. 315, pp. 579–606, June 2016.

[47] A. Narayan, C. Gittelson, and D. Xiu, "A stochastic collocation algorithm with multifidleity models," *SIAM J. Sci. Comput.*, vol. 36, no. 2, pp. A495–A521, Mar. 2014.

[48] X. Zhu, A. Narayan, and D. Xiu, "Computational aspects of stochastic collocation with multi-fidelity models," *SIAM/ASA J. Uncertain. Quantif.*, vol. 2, no.1, pp. 444–463, Sept. 2014.

[49] G. Blatman and B. Sudret, "Adaptive sparse polynomial chaos expansion based on least angle regression," *J. Comput. Phys.*, vol. 230, no. 6, pp. 2345–2367, Mar. 2011.

[50] N. Nakhla, A. E. Ruehli, M. S. Nakhla, R. Achar, and C. Chen, "Waveform relaxation techniques for simulation of coupled interconnects with frequency-dependent parameters," *IEEE Trans. Adv. Packag.*, vol. 30, no. 2, pp. 257–269, May 2007.

Chapter 6

Uncertainty quantification and design optimization with non-Gaussian correlated process variations

Zheng Zhang[1]

Process variations have been a major challenge in the modeling, simulation and optimization of nano-scale chip design. In the past decade, various stochastic spectral methods [1–3] based on generalized polynomial chaos (gPC) have been applied or developed in the design automation community [4–9], demonstrating orders-of-magnitude speedup over Monte Carlo techniques in many design cases (especially on design cases with a small or medium number of random parameters). Advanced numerical solvers have also been developed to solve uncertainty quantification problems with many random parameters [10–15] and for problems with design hierarchy [16,17].

Classical gPC-based approaches are based on an important assumption: the random parameters are assumed to be mutually independent. This assumption is too strong in many practical chip design: the fabrication process variations are often non-Gaussian and correlated due to the dependence on fabrication steps and spatial correlations. In recent years, we have developed a set of numerical methods to handle non-Gaussian correlated process variations in forward uncertainty quantification [18–20] and design optimization [20,21]. This chapter will summarize our efforts in this direction, and the technical contents are organized as follows:

- Firstly, we will introduce some necessary background, including the classical gPC expansion [22] and a modification proposed by Soize [23].
- Secondly, we will explain how to construct stochastic basis functions to handle the impact of non-Gaussian correlated uncertainty.
- Thirdly, we will explain how to develop a stochastic collocation framework based on the new stochastic basis functions.
- Finally, we will show how the developed surrogate models can be efficiently used in yield-aware circuit design optimization by leveraging chance-constraint optimization.

[1]Department of Electrical and Computer Engineering, University of California, Santa Barbara, CA, USA

6.1 Background: gPC-type methods

Let $\xi = [\xi_1, \ldots, \xi_d] \in \mathbb{R}^d$ denote d random parameters that describe the process variations in chip fabrication. We assume that the joint probability density function of ξ is $\rho(\xi)$. We further assume that $y(\xi) \in \mathbb{R}$ is a parameter-dependent performance metric (e.g., the power consumption or frequency of a chip) that we want to model. When $y(\boldsymbol{\xi})$ is smooth and has a bounded variance, stochastic spectral methods approximate $y(\boldsymbol{\xi})$ via linear combination of some stochastic basis functions:

$$y(\xi) \approx \sum_{|\alpha|=0}^{p} c_\alpha \Psi_\alpha(\xi). \tag{6.1}$$

Here $c_{\boldsymbol{\alpha}}$ is the coefficient of an associated stochastic basis function, and $\{\Psi_{\boldsymbol{\alpha}}(\boldsymbol{\xi})\}$ are a set of properly chosen orthonormal polynomials satisfying

$$E\left[\Psi_\alpha(\xi)\Psi_\beta(\xi)\right] = \begin{cases} 1, & \text{if } \alpha = \beta; \\ 0, & \text{otherwise.} \end{cases} \tag{6.2}$$

Here the operator $\mathbb{E}$ denotes an expectation over ξ; $\alpha = [\alpha_1, \ldots, \alpha_d] \in \mathbb{N}^d$ is an index vector, with each element α_i indicating the highest polynomial order in terms of $\boldsymbol{\xi}_i$. The total polynomial order $|\alpha| = |\alpha_1| + \ldots + |\alpha_d|$ is bounded by p, and thus the total number of basis functions is $N = (p+d)!/(p!d!)$. Once $c_{\boldsymbol{\alpha}}$'s are computed, the mean value, variance and density function of $y(\boldsymbol{\xi})$ can be easily obtained. There exist numerous techniques for computing $c_{\boldsymbol{\alpha}}$ when the elements of $\boldsymbol{\xi}$ are mutually independent. However, this is a non-trivial task when the random parameters are non-Gaussian correlated.

6.1.1 Classical gPC methods

In stochastic spectral methods, the basis functions $\{\Psi_{\boldsymbol{\alpha}}(\boldsymbol{\xi})\}$ are chosen according to the density function $\rho(\boldsymbol{\xi})$, which further decides the numerical solvers to compute $c_{\boldsymbol{\alpha}}$. When all random parameters are mutually independent, the popular gPC theory [22] can be used. In this case, if $\rho_k(\xi_k)$ denotes the marginal density of ξ_k, then the joint density is $\rho(\xi) = \prod_{k=1}^{d} \rho_k(\xi_k)$. Under this assumption, a multivariate basis function has a product form:

$$\Psi_{\boldsymbol{\alpha}}(\xi) = \prod_{k=1}^{d} \varphi_{k,\alpha_k}(\xi_k). \tag{6.3}$$

Here $\varphi_{k,\alpha_k}(\xi_k)$ is a univariate degree α_k orthonormal polynomial of parameter ξ_k, and it is adaptively chosen based on $\rho_k(\xi_k)$ via the three-term recurrence relation [24].

In this case, the unknown coefficients $c_{\boldsymbol{\alpha}}$'s can be computed via various numerical solvers such as stochastic Galerkin [25], stochastic testing [3] and stochastic collocation [1]. In these solvers, a key step is to compute the numerical quadrature points in the random parameter space. Since all random parameters are mutually independent, one can firstly compute the quadrature samples/weights for each parameter (e.g., via the techniques in [26]), then get the multidimensional quadrature samples/weights via tensor products or sparse grids [27].

It is trivial to handle Gaussian correlated random parameters. One just need to de-correlate all random variables and make them independent via a principle component analysis, then apply gPC expansion to the newly independent variables. In this case, the basis functions reduce to the Hermite polynomial chaos.

6.1.2 Soize's method [23] with non-Gaussian correlation

Unfortunately, the random parameters ξ are rarely independent in realistic cases. In this case, principal or independent component analysis [28,29] can be error-prone. In [23], Soize and Ghanem proposed the following modified basis function:

$$\Psi_{\boldsymbol{\alpha}}(\xi) = \left(\frac{\prod_{k=1}^{d} \rho_k(\xi_k)}{\rho(\xi)} \right)^{\frac{1}{2}} \prod_{k=1}^{d} \varphi_{k,\alpha_k}(\xi_k). \tag{6.4}$$

The modified basis functions are guaranteed to be orthonormal even if $\rho(\xi) \neq \prod_{k=1}^{d} \rho_k(\xi_k)$. However, this set of basis functions have two limitations [30]. Firstly, the basis functions are very non-smooth and numerically unstable due to the term in the square root on the right hand side of (6.4). As a result, the modified basis functions have a much slower convergence rate compared with the standard gPC method [22]. Secondly, the basis functions in (6.4) do not allow a closed-form expression for computing the expectation and variance of $y(\xi)$. This is caused by the fact the basis function indexed by $\alpha = 0$ is not a constant any more.

This has motivated us to rethink the choice of basis functions and the associated numerical solvers for the more generic non-Gaussian correlated cases.

6.2 Basis functions with non-Gaussian correlations

In order to accurately capture the impact of non-Gaussian correlated random parameters, a new set of orthonormal basis functions were proposed in [18,31]. Similar to the gPC basis functions that are designed for independent random parameters, our new basis functions allow one to approximate a smooth $y(\xi)$ with a high accuracy and to extract its statistical moments analytically or semi-analytically.

6.2.1 Multivariate orthonormal polynomial basis functions

In the following, we describe two methods that can be used to generate orthogonal and normalized polynomials as the basis functions when the elements of ξ are non-Gaussian correlated.

6.2.1.1 Gram–Schmidt method

The Gram–Schmidt method was widely used for vector orthogonalization in the Euclidean space. In [31], we have used it to construct some orthogonal polynomial functions by replacing the vector inner product with the expectation of the product of functions. Specifically, we can first reorder the monomials $\xi^{\alpha} = \xi_1^{\alpha_1} \ldots \xi_d^{\alpha_d}$ in the graded lexicographic order, and denote them as $\{p_j(\xi)\}_{j=1}^{N_p}$. We can set the first basis function as $\Psi_1(\xi) = 1$. Then we can generate other orthonormal polynomials $\{\Psi_j(\xi)\}_{j=1}^{N_p}$ in the correlated parameter space recursively with the following formula:

$$\widehat{\Psi}_j(\xi) = p_j(\xi) - \sum_{i=1}^{j-1} \mathbb{E}[p_j(\xi)\Psi_i(\xi)]\Psi_i(\xi), \tag{6.5}$$

$$\Psi_j(\xi) = \frac{\widehat{\Psi}_j(\xi)}{\sqrt{\mathbb{E}[\widehat{\Psi}_j^2(\xi)]}}, \quad for\ j = 2, \ldots, N_p. \tag{6.6}$$

The basis functions defined by this approach are unique under the specific order of monomials. Since the basis functions are orthonormal polynomials, the mean value and statistical moment of an approximated stochastic solution can be easily obtained just as in the classical gPC expansion.

6.2.1.2 Cholesky decomposition method

We can also contruct the basis function via a Cholesky decomposition [18], which is easy to implement and suitable for high-dimensional cases. Firstly, we resort all monomials in the graded lexicographic order, denoted as $\mathbf{p}(\xi) = [p_1(\xi), \ldots, p_{N_p}(\xi)]^T$. We further denote the multivariate moment matrix as $\mathbf{M}$

$$\mathbf{M} = \mathbb{E}\big(\mathbf{p}(\xi)\mathbf{p}^T(\xi)\big) \Leftrightarrow m_{ij} = \mathbb{E}[p_i(\xi)p_j(\xi)] \ . \tag{6.7}$$

The resulting $\mathbf{M}$ is a symmetric positive definite matrix, and we can obtain a lower-triangular matrix $\mathbf{L}$ via the Cholesky factorization $\mathbf{M} = \mathbf{L}\mathbf{L}^T$. Finally, the desired basis functions can be computed as as

$$\boldsymbol{\Psi}(\xi) = \mathbf{L}^{-1}\mathbf{b}(\xi). \tag{6.8}$$

Here the N_p-by-1 functional vector $\boldsymbol{\Psi}(\xi)$ stores all basis functions $\{\Psi_{\boldsymbol{\alpha}}(\xi)\}$ in the graded lexicographic order.

6.2.2 *Numerical implementation issues*

In the basis function generation, we have to compute the expectations in a correlated parameter space, which involves evaluating the moments $\mathbb{E}[\xi^{\boldsymbol{\alpha}}]$ up to order $2p$. Some techniques can be used to speed up the computation.

For general cases, we may estimate the moments by changing the integrand and the density function:

$$\mathbb{E}[\xi^{\boldsymbol{\alpha}}] = \int_{\mathbb{R}^d} g_\alpha(\xi)\widehat{\rho}(\xi)d\xi, \text{ with } g_\alpha(\xi) = \frac{\xi^\alpha \rho(\xi)}{\widehat{\rho}}(\xi). \tag{6.9}$$

Here $\widehat{\rho}(\xi) = \prod_{k=1}^{d} \widehat{\rho}_k(\xi_k)$ is a new joint density function of independent random parameters. Then, standard quadrature methods such as sparse grid [32] or tensor product Gauss quadrature can be used to evaluate the integration. The tensor-train-based method in [33] can be used to reduce the integration cost when d is large. The potential drawback is that it may be non-trivial to obtain highly accurate results if $g_{\boldsymbol{\alpha}}(\xi)$ is highly nonlinear or even non-smooth. Note that we only need to use a high-order quadrature rule in an independent parameter space and repeatedly evaluate some cheap closed-form functions here, and we do not need to perform expensive device or circuit simulations when we compute the basis functions.

In practice, one often fits a Gaussian mixture model from measurement data to describe a non-Gaussian correlated density function:

$$\rho(\xi) = \sum_{i=1}^{n} r_i \mathcal{N}(\xi|\mu_i, \Sigma_i), \text{ with } r_i > 0, \sum_{i=1}^{n} r_i = 1. \tag{6.10}$$

Here $\mathcal{N}(\xi|\mu_i, \Sigma_i)$ denotes a multivariate Gaussian distribution with mean $\mu_i \in \mathbb{R}^d$ and a covariance matrix $\Sigma_i \in \mathbb{R}^{d\times d}$. With a Gaussian mixture, the moments can be written as the weighted sum of the moments of some Gaussian disributions:

$$\mathbb{E}[\xi^\alpha] = \sum_{i=1}^{n} r_i q_{\boldsymbol{\alpha},i}, \text{ with } q_{\boldsymbol{\alpha},i} = \int_{-\infty}^{\infty} \xi^\alpha \mathcal{N}(\xi|\mu_i, \Sigma_i) d\,\xi. \tag{6.11}$$

Each term $q_{\boldsymbol{\alpha},i}$ can be computed as follows [19,34]. Firstly, we can first use a linear transform to change the variables to some independent and standard Gaussian variables η, and the integrand function becomes a polynomial of η. Secondly, starting from $|\alpha| = 1$, we can calculate all terms $q_{\boldsymbol{\alpha},i}$ exactly with tensor-train operations. The details can be found in [34].

6.3 Toward a stochastic collocation framework

After constructing the basis functions, we still need to calculate $c_{\boldsymbol{\alpha}}$. Since all basis functions are orthogonal and normalized, we compute the unknown model parameters via stochastic collocation, which uses a projection step:

$$c_{\boldsymbol{\alpha}} = \mathbb{E}[y(\xi)\Psi_{\boldsymbol{\alpha}}(\xi)] = \int_{\mathbb{R}^d} y(\xi)\Psi_{\boldsymbol{\alpha}}(\xi)\rho(\xi)d\xi \approx \widetilde{c}_{\boldsymbol{\alpha}} = \sum_{k=1}^{M} y(\xi_k)\Psi_{\boldsymbol{\alpha}}(\xi_k)w_k \tag{6.12}$$

where $\{\xi_k\}_{k=1}^M$ are the quadrature nodes, and $\{w_k\}_{k=1}^M$ are the corresponding quadrature weights. The key of stochastic collocation is to choose proper basis functions and an excellent quadrature rule, such that M is as small as possible.

The choice of quadrature samples and weights depend on the density function of ξ. Various numerical quadrature rules such as tensor product methods and sparse grids have been widely used for independent random parameters. However, they do not work when the elements of ξ are non-Gaussian correlated. In [18,31], we have developed an optimization method to determine the quadrature samples and weights accurately in a non-Guassian correlated parameter space.

6.3.1 Optimization-based quadrature rule

The quadrature samples and weights should be selected such that they can compute the integration (6.12) accurately for a wide range of output function $y(\xi)$. While the specific output function $y(\xi)$ can significantly vary from one case to another, we assume that $y(\xi)$ is smooth enough such that it can be well approximated by a polynomial function bounded by order p. In this case, we can design a general quadrature rule that works well for a broad range of functions $y(\xi)$.

Our key idea is as follows: $y(\xi)\Psi_{\boldsymbol{\alpha}}(\xi)$ can be well approximated by a linear combination of some orthogonal polynomials with degree $\leq 2p$ if $y(\xi)$ can be well approximated by a polynomial function bounded by order p. If the numerical integration is highly accurate for all basis involved functions up to order $2p$, then the numerical integration in (6.12) should have a small error.

Assume that $\{\Psi_j(\xi)\}_{j=1}^{N_{2p}}$ denotes all orthonormal polynomial basis functions constructed by our method. Note that here we consider more basis functions than those used in approximating $y(\xi)$. Ideally, we seek for the quadrature nodes and weights $\{\xi_k, w_k\}_{k=1}^M$ by matching the integration of basis functions up to order $2p$:

$$\mathbb{E}[\Psi_j(\xi)] = \int_{\mathbb{R}^d} \Psi_j(\xi)\rho(\xi)d\xi = \sum_{k=1}^{M} \Psi_j(\xi_k)w_k, \ \forall j = 1, \ldots, N_{2p}. \tag{6.13}$$

Here, $N_{2p} = \binom{2p+d}{d}$ denotes the total number of basis functions with their total order bounded by $2p$. In practice, we can compute the quadrature samples and weights by solving the following nonlinear least-square problem

$$\min_{\bar{\xi},\mathbf{w}\geq 0} \ || \boldsymbol{\Phi}(\bar{\xi})\mathbf{w} - \mathbf{e}_1||_2^2, \ \text{s.}t. \ \sum_{i=1}^{M} w_i = 1, \tag{6.14}$$

Where $\bar{\xi} = [\xi_1^T, \ldots, \xi_M^T]^T \in \mathbb{R}^{Md}$, $\mathbf{w} = [w_1, \ldots, w_M]^T \in \mathbb{R}^M$, $\mathbf{e}_1 = [1, 0, \ldots, 0]^T \in \mathbb{R}^{N_{2p}}$, $\Phi(\bar{\xi})$ is a matrix of size $N_{2p} \times M$ with the (j,k)-th element being $(\Phi(\bar{\xi}))_{jk} = \Psi_j(\xi_k)$, $||\cdot||_2$ denotes the Euclidean norm. Here, we require the quadrature weights to be non-negative, which is an natural extension of the one-dimensional Gauss quadrature rule [26]. Note that the jth element of $\mathbf{e}_1$ is

$\mathbb{E}(\Psi_j(\xi))$, which equals 1 for $j = 1$ and 0 otherwise since all basis functions are orthogonal and normalized.

The above optimization problem can be solved via a proper numerical method (e.g., block coordinate descent), and we even does not need to get a global or minimal optimal solution. Instead, we can stop the numerical iteration process of the iteration when the objective function is small enough. Here we use the stopping criteria

$$||\Phi(\bar{\xi})\mathbf{w} - \mathbf{e}_1||_1 \leq \varepsilon \tag{6.15}$$

since it helps establishing our theoertical error bound of this method.

6.3.2 *Error and complexity bounds*

Rigorous theoretical bounds for the computational complexity and prediction accuracy of the stochastic collocation with optimization-based quadrature have been provided in [18]. The key results are briefly summarized below.

Simulation complexity. The number of simulation samples required in our stochastic collocation is decided by two factors: the required integration accuracy in (6.12), and the number of basis functions we used to approximate $y(\xi)$. Assume we want an order- $2p$ integration accuracy, then the number of simulation samples, M, has a lower and upper bounds:

$$N_p \leq M \leq N_{2p}, \tag{6.16}$$

where N_p and N_{2p} are the number of basis functions with degree upper bounded by p and $2p$, respectively.

We can determine M automatically by weighted clustering and repeatedly solving (6.14) [18]. The weighted clustering is used to gradually reduce the number of quadrature samples, and (6.14) is solved repeatedly to refine the quadrature samples and weights for every reduced M until the number of quadrature samples cannot be further reduced. To illustrate this process, we consider the following synthetic smooth random function

$$y(\xi) = exp(\xi_1) + 0.1cos(\xi_1)sin(\xi_2). \tag{6.17}$$

we assume that the random parameters follow a Gaussian mixture distribution

$$\xi = \xi_0 + \frac{1}{10}\Delta\xi, \text{ where } \Delta\xi \sim \frac{1}{2}\mathcal{N}(\mu_1, \Sigma_1) + \frac{1}{2}\mathcal{N}(\mu_2, \Sigma_2). \tag{6.18}$$

Here, the mean values $\mu_1 = \mathbf{1}$, $\mu_2 = -\mathbf{1}$; the positive definite covariance matrices $\mathbf{\Sigma}_1$ and $\mathbf{\Sigma}_2$ are randomly generated. We use $\mathbf{1}$ to denote a vector of a compatible size with all elements being one. Assume that we want to approximate $y(\xi)$ by a forth-order expansion. Firstly, 90 random samples are generated via Monte Carlo. Secondly, these points are grouped into 30 clusters via a weighted linkage clustering approach, and they are used as the initial samples and weights to solve (6.14). The obtained solution samples and weights were further reduced to

$M - 1 = 29$ via weighted clustering, and used as an initial guess to solve (6.14) again. This process is repeated until the number of quadrature nodes is reduced to 17 automatically and cannot be further reduced. The process of generating quadrature samples and weights is shown in Figure 6.1. Note that this process only depends on the probability density function and the basis functions, and is independent of $y(\xi)$.

While (6.16) only provides a lower and upper bound of the simulation samples, the practical result is more promising: the actual number M is very close to the lower bound N_p in many cases. Table 6.1 shows the number of simulation samples used in the uncertainty quantification of some practical circuit benchmarks. Here

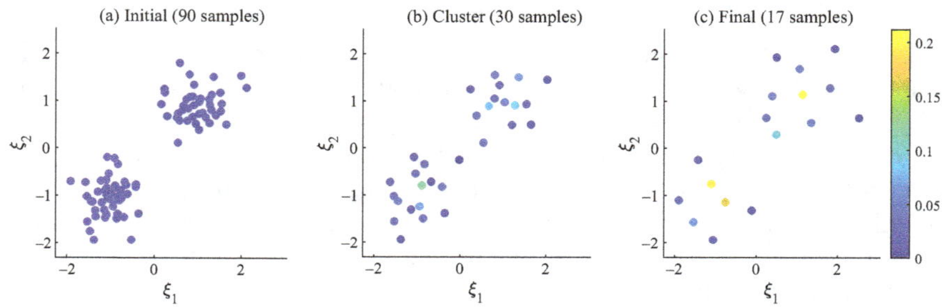

Figure 6.1 The process of generating quadrature samples and weights for the synthetic example. The quadrature weights are shown by the color bar. (a) Initial candidate points generated via Monte Carlo; (b) clustered samples via a weighted clustering method; and (c) the final optimized quadrature nodes. This figure is re-produced from [18] with permission

Table 6.1 The number of quadrature samples used in all experiments. Here, p denotes the maximal order of basis functions, d is the number of random parameters. This table is reproduced from [18] with permission

			Proposed		
Benchmarks	p	d	**# Samples**	**Lower bound**	**Upper bound**
Synthetic	1	2	3	3	6
	2	2	6	6	15
	3	2	10	10	28
	4	2	17	15	45
	5	2	66	21	66
CMOS ring	2	6	33	28	210
Optical filter	22	411	16,139	1,578	701,365
AWG	2	11	127	78	1365

"synthetic" means the synthetic random function specified in (6.17); "CMOS Ring" is a three-stage CMOS ring oscillator consisting of one p-MOS and one n-MOS transistors per stage; "Optical filter" is a three-stage parallel-coupled ring resonator optical filter; "AWG" is an arrayed waveguide grating device.

Error bound. The error of the stochastic collocation algorithm comes from two parts. The first part is the order-p approximation of $y(\xi)$ using orthonormal polynomial basis functions. The actual $y(\xi)$ may not be a polynomial or may be a polynomial with a degree higher than p, so an order-p expansion may not exactly describe $y(\xi)$. The second part is the unavoidable numerical error of computing the quadrature samples and weights $\{\xi_j, w_j\}_{j=1}^{M}$ by solving the optimization problem (6.14). Let $\widehat{c}_{\boldsymbol{\alpha}}$ be the numerical results of (6.12) computed with the optimization-based quadrature rule, then the model error is

$$\|y(\xi) - \sum_{|\boldsymbol{\alpha}|\le p} \widehat{c}_{\boldsymbol{\alpha}}\Psi_{\boldsymbol{\alpha}}(\xi)\|_2 = \sqrt{\mathbb{E}\left(y(\xi) - \sum_{|\boldsymbol{\alpha}|\le p} \widehat{c}_{\boldsymbol{\alpha}}\Psi_{\boldsymbol{\alpha}}(\xi)\right)^2} \le C_1\sigma + C_2\varepsilon \tag{6.19}$$

where ε is the threshold in (6.15), σ is the gap between $y(\xi)$ and its best possible order- p approximation, C_1 and C_2 are two constants. Note that σ indicates the smoothness of $y(\xi)$. Therefore, when p is small, the first term dominate the error, and the approximation error almost decays exponentially as p increases. When p is large, the 2nd term plays an important role, and the stochastic collocation will have a lower convergence rate. This is clearly shown in Figure 6.2, which

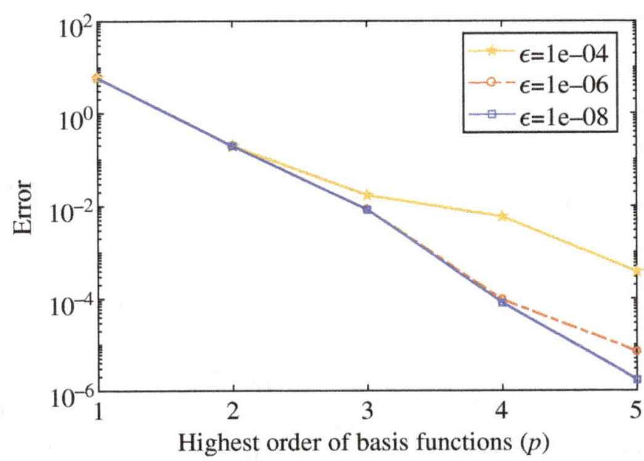

Figure 6.2 Convergence rate for the synthetic example specified in (6.17). Here ε is the numerical error of optimization defined in (6.15). This figure demonstrates the error estimated in (6.19): the stochastic collocation algorithm shows a nearly exponential convergence rate as p increases and before ε dominates the error. This figure is reproduced from [18] with permission

investigates the convergence rate of the proposed method on the the synthetic function in (6.17). Clearly, the ℓ_2 error of the obtained surrogate model nearly exponentially decay as the order p increases and before optimization threshold ε dominates the error.

6.3.3 Practical example: CMOS ring oscillator

Several practical electronic and photonic IC examples were used to verify the proposed algorithm in [18]. For the three-stage CMOS ring oscillator in Figure 6.3, the relative threshold voltage variations of six transistors are described as Gaussian mixture model. Then the proposed optimization-based stochastic collocation method is used to build a second-order surrogate model to approximate its oscillation frequency. The proposed method only needs 33 pairs of quadrature samples and weights to accurately quantify the performance uncertainty. Figure 6.4 shows

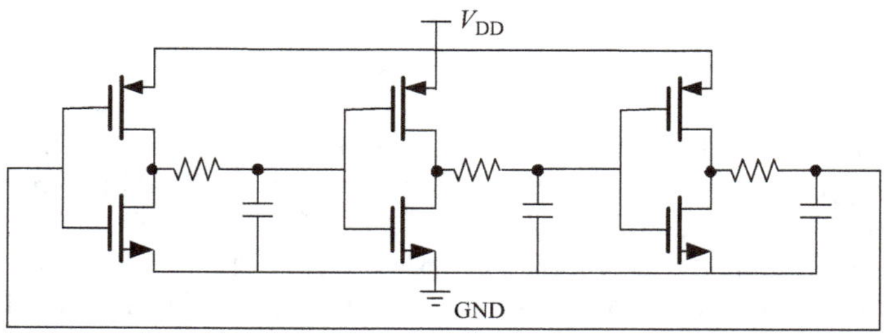

Figure 6.3 Schematic of a three-stage CMOS ring oscillator. Reproduced from [18] with permission

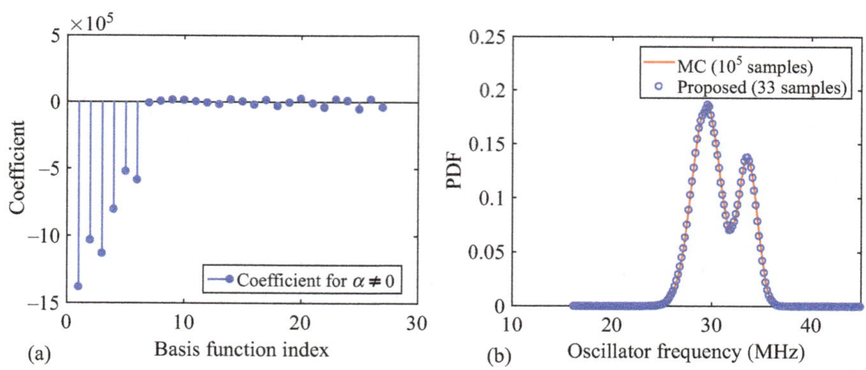

Figure 6.4 Numerical results of the CMOS ring oscillator. (a) Obtained coefficients/weights of our basis functions; and (b) probability density functions of the oscillator frequency obtained by our proposed method and Monte Carlo. Reproduced from [18] with permission

the obtained weights of all basis functions and the probability density function. The proposed method can achieve the similar level of accuracy (i.e., with a precision of 0.01 MHz when estimating the mean frequency) compared with Monte Carlo using 10^5 random simulations.

6.3.4 Extension of the framework

The developed basis functions allow easy extraction of statistical moments, and they serve as a foundation to solve many more complicated cases. Two representative extensions are summarized below:

- *High-dimensional cases* [19,34]. In large-scale electronic and photonic design, the circuit often has massive components and many uncertain parameters. In this case, the optimization-based quadrature rule requires excessive sample parameters. Since the developed basis functions are orthogonal and normalized in the stochastic parameter space, they can be easily used in a compressed sensing framework [35] to produce sparse surrogate models. A key requirement of the compressed sensing is the restricted isometry property, which can be easily satisfied with the proposed basis functions when the simulation samples are randomly generated. In [19,34], adaptive sampling methods are developed to reduce the sample complexity. It is worth investigating if the proposed methods of non-Gaussian correlated cases can be extended to other high-dimensional solvers such as tensor methods [36,37] and analysis of variance [16].
- *Design problems with mixed-type uncertainties* [38]. While most existing uncertainty quantification frameworks only consider continuous random variables, practical electronic and photonic designs often involve discrete random variables. This can become more common in mixed-signal design and in architecture-level design. To handle this issue, the optimization-based quadrature rule is modified in [38] which use a mixed-integer programming to compute the quadrature samples in a coupled parameter space with both continuous and discrete variables.

6.4 Yield-aware chance-constraint design optimization

The built surrogate model can be used to efficiently optimize electronic and photonic circuit design under various process variations [20,21]. Instead of directly optimizing the yield of a circuit, our recent works [20,21] have proposed some alternative methods that can better trade off the performance and yield of an electronic or photonic IC design.

6.4.1 Yield optimization and yield-aware optimization

We denote the design variables by $\mathbf{x} = [x_1, x_2, \ldots, x_{d_1}]^T \in X$, and the process variations by random parameters $\xi = [\xi_1, \xi_2, \ldots, \xi_{d_2}]^T \in \Xi$. Let $\{y_i(\mathbf{x}, \xi)\}_{i=1}^n$ denote n performance metrics that are considered in yield estimation, $\{u_i\}_{i=1}^n$ denote their corresponding upper bounds specifying the design requirements. An indicator function is defined as

$$I(\mathbf{x},\xi)=\begin{cases}1, & y_i(\mathbf{x},\xi)\le u_i, \forall i=[n];\\ 0, & \text{otherwise.}\end{cases} \tag{6.20}$$

Here $[n]=\{1,2,\cdots,n\}$. The yield at a certain design choice $\mathbf{x}$ is defined as

$$Y(\mathbf{x})=\mathbb{P}_\xi(\mathbf{y}(\mathbf{x},\xi)\le\mathbf{u})=\mathbb{E}_\xi[I(\mathbf{x},\xi)]. \tag{6.21}$$

In conventional yield optimization, one often tries to achieve the best possible yield. This often requires losing remarkable design performance $f(\mathbf{x},\xi)$ in order to achieve a high yield.

Simply maximizing the yield may lead to an *over-conservative* design. As an example shown in Figure 6.5, one may lose lots of performance while just getting a very marginal yield improvement. In order to avoid an over-conservative design, we will not simply maximize the yield. Instead, we propose a chance-constrained optimization in [20]:

$$\max_{\mathbf{x}\in X} \quad \mathbb{E}_\xi[f(\mathbf{x},\xi)] \tag{6.22a}$$

$$\text{s.t.} \quad \mathbb{P}_\xi(y_i(\mathbf{x},\xi)\le u_i)\ge 1-\varepsilon_i, \forall i=[n]. \tag{6.22b}$$

where $f(\mathbf{x},\xi)$ is the performance metric that we aim to optimize, and $\varepsilon_i\in[0,1]$ is a risk level to control the probability of meeting each design constraint. Instead of simply maximizing the yield, the chance-constrained optimization tries to achieve a good balance between yield and performance: one can optimize the performance $f(\mathbf{x},\xi)$ while ensuring a high yield. The circuit yield can be controlled by ε_i: reducing ε_i ensures a lower failure rate and thus a higher yield.

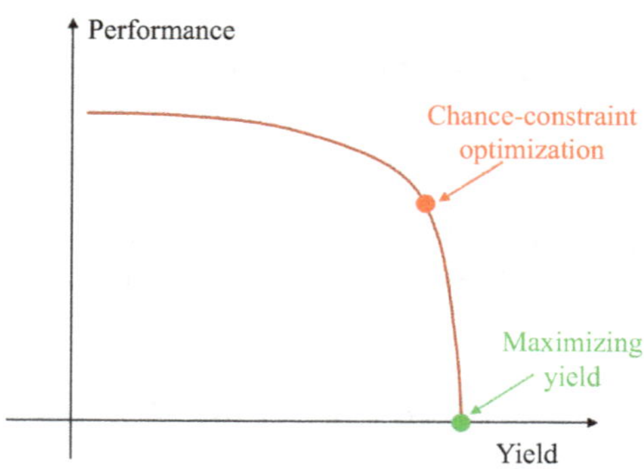

Figure 6.5 The trade-off between yield and an objective design performance. Simply maximizing the yield can lead to over-conservative performance

6.4.2 Moment-bounding chance-constraint optimization

The chance-constrained optimization (6.22) is generally hard to solve. The feasible set produced by the chance constraints is often non-convex and hard to estimate. It is also expensive to estimate the design objective function $f(\mathbf{x}, \xi)$ and design constraint function $y_i(\mathbf{x}, \xi)$ due to the lack of analytical expressions. Fortunately, a moment-bounding method was combined with uncertainty quantification techniques in [20] to make the problem tractable.

Employing the Chebyshev-Cantelli inequality, one can ensure the chance constraint via a moment-bounding technique [20,39]. Specifically, with the first- and second-order statistical moments of the constraint function, one can convert the probabilistic constraint in (6.22) to a stronger deterministic constraint, leading to the following optimization:

$$\begin{array}{ll} \underset{\mathbf{x}\in\mathcal{X}}{max} & \mathbb{E}_{\xi}[f(\mathbf{x}, \xi)] \\ \text{s.t.} & \mathbb{E}_{\xi}[y_i(\mathbf{x}, \xi)] + \gamma_{\varepsilon_i}\sqrt{\text{Var}_{\xi}[y_i(\mathbf{x}, \xi)]} \leq u_i, \forall i \in [n], \end{array} \tag{6.23}$$

where constant $\gamma_{\varepsilon_i} = \sqrt{\frac{1-\varepsilon_i}{\varepsilon_i}}$. When the objective and constraint functions are described by certain surrogate models such as gPC [18,40], one can easily extract their mean and variances. This can greatly simplify the problem and reduce the computational cost: we do not need to evaluate the objective function and yield via additional sampling-based simulations any more.

Specifically, using the stochastic collocation developed in [18], we can construct surrogate models for both the constraint and objective functions:

$$y_i(\mathbf{x}, \xi) \approx \sum_{|\alpha|+|\beta|=0}^{p} c^i_{\alpha,\beta}\Phi_\alpha(\mathbf{x})\Psi_\beta(\xi), \quad f(\mathbf{x}, \xi) \approx \sum_{|\alpha|+|\beta|=0}^{p} h_{\alpha,\beta}\Phi_\alpha(\mathbf{x})\Psi_\beta(\xi), \tag{6.24}$$

where $\Phi_{\boldsymbol{\alpha}}(\mathbf{x})$ and $\Psi_{\boldsymbol{\beta}}(\xi)$ are the basis functions for design variables $\mathbf{x}$ and process variation parameters ξ, respectively. The mean value and variance of $y_i(\mathbf{x}, \xi)$ can be approximated based on the surrogate models:

$$\mathbb{E}_{\xi}[y_i(\mathbf{x}, \xi)] \approx \sum_{|\boldsymbol{\alpha}|=0}^{p} c^i_{\boldsymbol{\alpha},0}\Phi_{\boldsymbol{\alpha}}(\mathbf{x}), \quad \text{var}_{\xi}[y_i(\mathbf{x}, \xi)] \approx \sum_{|\boldsymbol{\beta}|=1}^{p}\left(\sum_{|\boldsymbol{\alpha}|=0}^{p-|\boldsymbol{\beta}|} c^i_{\boldsymbol{\alpha},\boldsymbol{\beta}}\Phi_{\boldsymbol{\alpha}}(\mathbf{x})\right)^2. \tag{6.25}$$

The mean value of the objective function $f(\mathbf{x}, \xi)$ can be evaluated in the same way.

With the above mean and variance information, the chance-constraint yield-aware optimization (6.23) can be converted to a deterministic formulation:

$$\begin{array}{ll} \underset{\mathbf{x}\in X}{\min} & \sum_{|\boldsymbol{\alpha}|=0}^{p} h_{\boldsymbol{\alpha},0}\Phi_{\boldsymbol{\alpha}}(\mathbf{x}) \\ \text{s.t.} & \gamma_{\varepsilon_i}\sqrt{\sum_{|\boldsymbol{\beta}|=1}^{p}\left(\sum_{|\boldsymbol{\alpha}|=0}^{p-|\boldsymbol{\beta}|} c^i_{\boldsymbol{\alpha},\boldsymbol{\beta}}\Phi_{\boldsymbol{\alpha}}(\mathbf{x})\right)^2} + \sum_{|\boldsymbol{\alpha}|=0}^{p} c^i_{\boldsymbol{\alpha},0}\Phi_{\boldsymbol{\alpha}}(\mathbf{x}) \leq u_i, \quad \forall i \in [n]. \end{array}$$

This problem can be solved via global polynomial optimization solver based on semi-definite programming [41–43].

An illustrative synthetic example. We consider a synthetic example with two design variables $\mathbf{x} = [x_1, x_2]$ that admit a uniform distribution $\mathcal{U}[-1, 1]^2$ and two non-Gaussian correlated random parameters $\xi = [\xi_1, x_2]$ that follow a Gaussian mixture distribution. Assume that the yield criterion is $(x_1 + \xi_1)^2 \pm (x_2 + \xi_2) \leq 1$. Our goal is to maximize $\mathbb{E}_\xi[3(x_1 + \xi_1) + (x_2 + \xi_2)]$. We formulate the yield into chance constraints and have the following problem:

$$\begin{aligned} \max_{\mathbf{x}} \quad & \mathbb{E}_\xi[3(x_1 + \xi_1) - (x_2 + \xi_2)] \\ \text{s.t.} \quad & \text{Prob}_\xi\left((x_1 + \xi_1)^2 - (x_2 + \xi_2) \leq 1\right) \geq 1 - \varepsilon, \\ & \text{Prob}_\xi\left((x_1 + \xi_1)^2 + (x_2 + \xi_2) \leq 1\right) \geq 1 - \varepsilon. \end{aligned} \tag{6.26}$$

We consider the distribution with a non-Gaussian positive correlation $\frac{1}{2}N(\mathbf{0}.01, 10^{-4}\Sigma) + \frac{1}{2}N(-\mathbf{0}.01, 10^{-4}\Sigma)$ with $\boldsymbol{\Sigma} = \begin{pmatrix} 1 & 0.75 \\ 0.75 & 1 \end{pmatrix}$. We first build the surrogate models for both the objective and constraints by the second-order polynomial basis functions. Our optimization-based quadrature and weighted clustering produces $M = 19$ optimized samples and weights. The results for different risk tolerance levels ε are shown in Table 6.2. A smaller ε results in a smaller feasible domain, and generates a higher yield but a smaller objective value.

Mach–Zehnder interferometer (MZI). This chance-constraint yield optimization has been applied to several photonic circuits. Let us consider third-order MZI consisting of three port coupling and two arms, as shown in Figure 6.6(a). We hope to optimize the design variables $\mathbf{x} = [g_1, g_2, g_3]$ in the interval of [100 nm, 300 nm]^3. The random variables ξ follow a Gaussian mixture distribution and they describe the uncertainty of the coupling coefficients. We consider three performance metrics of the MZI: the 3dB bandwidth (BW, in GHz), the crosstalk (XT, in dB), and the attenuation (α, in dB) of the peak transmission. The yield is defined through the crosstalk and the attenuation. The yield-aware optimization

Table 6.2 The optimal solution for the synthetic experiment under difference risk threshold ε. Here $\mathbf{x}^*$ is the obtained numerical optimal solution. This table is reproduced from [20] with permission

Algorithm	$\mathbf{x}^*$		**Objective**	**Yield (%)**
Height 1.0 pt Proposed ($\varepsilon = 0.01$)	0.8630	−0.1172	2.4717	100
Proposed ($\varepsilon = 0.05$)	0.9379	−0.0522	2.7616	100
Proposed ($\varepsilon = 0.10$)	0.9587	−0.0402	2.8360	99.42
Proposed ($\varepsilon = 0.15$)	0.9689	−0.0351	2.8717	93.84
Proposed (ε =0.20)	0.9751	−0.0293	2.8959	87.49

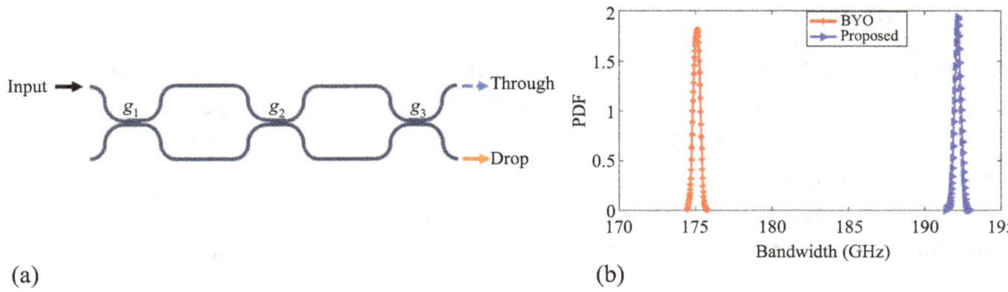

Figure 6.6 (a) The schematic of a photonic MZI circuit. (b) The optimized bandwidth of the MZI by the BYO and our proposed method, respectively. The expectation bandwidth of the BYO is 175.4 GHz while our proposed method with $\varepsilon = 0.05$ *can get 186.4 GHz. This figure is generated based on the results in [20]*

Table 6.3 Optimization result for the MZI. This table is reproduced from [20] with permission

Algorithm	**Simulations**	$\mathbb{E}_{\xi}[\mathrm{BW}]$ **(GHz)**	**Yield (%)**
Proposed ($\varepsilon = 0.03$)	36	188.8	100
Proposed ($\varepsilon = 0.05$)	36	192.2	100
Proposed ($\varepsilon = 0.07$)	36	194.5	100
Proposed ($\varepsilon = 0.10$)	36	195.0	87.7
BYO [44]	2,020	175.0	100

is formulated as:

$$\begin{aligned} &\max_{\mathbf{x}} \quad \mathbb{E}_{\xi}[\mathrm{BW}(\mathbf{x},\xi)] \\ &\text{s.t.} \quad \mathrm{Prob}_{\xi}(\mathrm{XT}(\mathbf{K},\xi) \leq \mathrm{XT}_0) \geq 1-\varepsilon, \\ &\mathrm{Prob}_{\xi}(\alpha(\mathbf{x},\xi) \leq \alpha_0) \geq 1-\varepsilon, \end{aligned} \tag{6.27}$$

where the yield risk level is ε. The threshold crosstalk (XT_0) and attenuation (α_0) are −4 dB and 2 dB, respectively. We used only 36 simulation samples to build the second-order polynomial surrogate models for BW, XT and α by our stochastic collocation framework. We compare this method with a recent Bayesian yield optimization (BYO) method [44] in Table 6.3. Our method requires 56× fewer simulation points than BYO, which is a great advantage for design cases with the time-consuming simulations. Clearly, smaller ε produces higher circuit yield. Figure 6.6(b) shows the probability density of the optimized bandwidth by our chance-constrained optimization and the BYO, respectively. It clearly shows that our proposed method produces higher bandwidth.

6.4.3 The PoBO method [21]

The moment constraint in (6.23) is a sufficient but unnecessary condition of the original chance constraint in (6.22). Therefore, any feasible solution of (6.23) should satisfy the probability constraint of (6.22). The feasible set produced by a moment bounding is much smaller than exact one [45]. This usually leads to an over-conservative solution in some design cases. When the risk level is very small, the moment-bounding method may even produce an empty feasible set, leading to an unsolvable problem. This has motivated us to develop a less conservative polynomial-bounding chance-constraint optimization (PoBO) [21]. This method is based on the assumption that the random parameters ξ are bounded.

The PoBO method is based on the following theoretical upper bound for the probability constraint function in (6.22b):

$$\mathbb{P}_{\xi}(y_i(\mathbf{x},\xi) > u_i) \leq \int_{\Xi} \kappa(y_i(\mathbf{x},\xi) - u_i)\rho(\xi)d\xi. \tag{6.28}$$

Here $\kappa(z)$ is a **Kinship function** defined for any $z \in [-1,+\infty)$, and it satisfies the following constraints:

- $\kappa(z) = 1$ when $z = 0$;
- $\kappa(z) \geq 0$ for any $z \in [-1,+\infty)$;
- $\kappa(z)$ is non-decreasing for $z \in [-1,+\infty)$;

If we set $z = y_i(\mathbf{x},\xi)$, the optimal Kinship function can be found via

$$\kappa^*(\cdot) := \underset{\kappa(\cdot)\in\mathcal{K}}{\operatorname{argmin}} \int_{-1}^{0} \kappa(z)dz. \tag{6.29}$$

The optimal Kinship function tries to minimize the gap caused by the integration in the range of $[-1,0]$. In order to make the chance-constraint optimization efficient to solve, we can choose the Kinship function from a proper family of polynomial functions. Specifically, we may choose a kinship function from a family of order- ρ polynomials $\mathcal{K}_\rho := \{\kappa(z)|\kappa(z) = \sum_{i=0}^{\rho} \zeta_i z^i\}$. The optimal polynomial kinship function $\kappa_\rho(\cdot)$ can be obtained by solving the following optimization problem:

$$\begin{aligned}
&\min_{\zeta_0,\ldots,\zeta_\rho} \quad \int_{-1}^{0} \kappa_\rho(z)dz \\
&\text{s.t.} \quad \kappa_\rho(z) = \zeta_0 + \zeta_1 z + \cdots + \zeta_\rho z^\rho \in \mathcal{K}_\rho, \\
&\qquad \kappa_\rho(0) = 1, \ \ \kappa_\rho(-1) = 0, \\
&\qquad \kappa_\rho'(z) \geq 0, \ \ \forall z \geq -1.
\end{aligned} \tag{6.30}$$

After obtaining the Kinship function, we can replace the original chance constraint with

$$\int_{\Xi} \kappa_p(y_i(\mathbf{x}, \xi) - u_i)\rho(\xi)d\xi \leq \varepsilon_i. \tag{6.31}$$

This provides a more accurate feasible set than the moment-bounding method. If we have built gPC-type surrogate models $\widehat{f}(\mathbf{x}, \xi) \approx f(\mathbf{x}, \xi)$ and $\widehat{y}_i(\mathbf{x}, \xi) \approx y_i(\mathbf{x}, \xi)$ for the objective and constraint function, then the original chance-constraint function will be approximate as:

$$\begin{aligned} \max_{\mathbf{x} \in \mathbf{X}} \quad & \mathbb{E}_\xi[\widehat{f}(\mathbf{x}, \xi)] \\ \text{s.t.} \quad & \sum_{l=1\widehat{M}} w_l \kappa_p(y_i(\mathbf{x}, \xi_l) - u_i) \leq \varepsilon_i, \quad \forall i \in [n]. \end{aligned} \tag{6.32}$$

Here $\{\xi_l, w_l\}_{l=1}^{\widehat{M}}$ are a set of numerical quadrature samples and weights that are computed via our optimization-based method to compute the integration in (6.31).

Result of the MZI photonic circuit. Consider again the same photonic MZI circuit shown in Figure 6(a) with the threshold crosstalk (XT_0) and attenuation (α_0) as −4 dB and 1.6 dB, respectively, in the optimization setting (6.27). When $\varepsilon = \varepsilon_1 = \varepsilon_2 = 0.1$, the moment-based chance-constraint optimization produces an expected bandwidth of 189.35 dB. On the other hand, the PoBO framework can produce a better design with an expected bandwidth 193.28 dB due to the less conservative treatment of the chance constraints.

6.5 Conclusions

Handling non-Gaussian correlated uncertainty was a long-standing challenge in electronic design automation. This chapter has briefly summarized our efforts to address the challenge, including:

- Numerical methods (e.g., Gram–Schmidt and Choselsky factorization methods) to construct orthogonal and normalized polynomial basis functions.
- Optimization-based numerical quadrature methods for designing stochastic collocation algorithms, as well as the theoretical bound of error and simulation complexity.
- Some new data efficient techniques to use the build surrogate models in uncertainty-aware design optimization. Specifically, we have proposed to solve chance-constraint optimization in order to avoid over-conservative design performance. This has been implemented via two methods: moment bounding and polynomial Kinship bounding.

Acknowledgements

The author would like to thank Dr. Chunfeng Cui and Mr. Zichang He, who have contributed to most the technical work described in this chapter.

Mr. Kaikai Liu also contributed to the moment-bounding chance-constraint yield-aware optimization. This work was partly supported by NSF grants #1763699 and #1846476.

References

[1] Xiu D, Hesthaven JS., High-order collocation methods for differential equations with random inputs. *SIAM Journal on Scientific Computing*. 2005; 27(3): 1118–1139.

[2] Ghanem R, Spanos P., *Stochastic finite elements: a spectral approach*. Berlin: Springer; 1991.

[3] Zhang Z, El-Moselhy TA, Elfadel IAM, *et al.*, Stochastic testing method for transistor-level uncertainty quantification based on generalized polynomial chaos. *IEEE Trans actions on Computer-Aided Design of Integrated Circuits and Systems*. 2013; 32(10): 1533–1545.

[4] Ahadi M, Roy S., Sparse linear regression (SPLINER) approach for efficient multidimensional uncertainty quantification of high-speed circuits. *IEEE Transactions on Computer-Aided Design of Integrated Circuits and Systems*. 2016; 35(10): 1640–1652.

[5] Manfredi P, Grivet-Talocia S., Rational polynomial chaos expansions for the stochastic macromodeling of network responses. *IEEE Transactions on Circuits and Systems I, Regular Papers*. 2019; 67(1): 225–234.

[6] Manfredi P, Ginste DV, De Zutter D, *et al.*, Decoupled polynomial chaos for nonlinear circuits with many random parameters. *IEEE Microwave and Wireless Components Letters*. 2015; 25(8): 505–507.

[7] Yucel AC, Bagci H, Michielssen E., An ME-PC enhanced HDMR method for efficient statistical analysis of multiconductor transmission line networks. *IEEE Transactions on Components, Packaging and Manufacturing Technology*. 2015; 5(5): 685–696.

[8] Wang F, Cachecho P, Zhang W, *et al.*, Bayesian model fusion: large-scale performance modeling of analog and mixed-signal circuits by reusing early-stage data. *IEEE Transactions on Computer-Aided of Design Integrated and Circuits Systems*. 2015; 35(8): 1255–1268.

[9] Zhang S, Lyu W, Yang F, *et al.*, An efficient multi-fidelity Bayesian optimization approach for analog circuit synthesis. In: Proceedings of the Design Automation Conference; 2019. pp. 1–6.

[10] Zhang Z, Weng TW, Daniel L., Big-data tensor recovery for high-dimensional uncertainty quantification of process variations.*IEEE Transactions on Components, Packaging and Manufacturing Technology*. 2017; 7(5): 687–697.

[11] Li X., Finding deterministic solution from underdetermined equation: large-scale performance variability modeling of analog/RF circuits. *IEEE Transactions on Computer-Aided Design of Integrated Circuits and Systems*. 2010; 29(11): 1661–1668.

[12] Ahadi M, Prasad AK, Roy S., Hyperbolic polynomial chaos expansion (HPCE) and its application to statistical analysis of nonlinear circuits. In: Proceedings of the IEEE Workshop of Signal and Power Integrity; 2016. pp. 1–4.

[13] Ma X, Zabaras N., An adaptive high-dimensional stochastic model representation technique for the solution of stochastic partial differential equations. *Journal of Computational Physics*. 2010; 229(10): 3884–3915.

[14] Yang X, Choi M, Lin G, *et al.*, Adaptive ANOVA decomposition of stochastic incompressible and compressible flows. *Journal of Computational Physics*. 2012; 231(4): 1587–1614.

[15] El-Moselhy T Daniel L., Variation-aware interconnect extraction using statistical moment preserving model order reduction. In: Proceedings of the Design, Automation and Test in Europe; 2010. pp. 453–458.

[16] Zhang Z, Yang X, Marucci G, *et al.*, Stochastic testing simulator for integrated circuits and MEMS: hierarchical and sparse techniques. In: Proceedings of the IEEE Custom Integrated Circuits Conference; 2014. pp. 1–8.

[17] Zhang Z, Yang X, Oseledets IV, *et al.*, Enabling high-dimensional hierarchical uncertainty quantification by ANOVA and tensor-train decomposition. *IEEE Transactions on Computer-Aided Design of Integrated Circuits and Systems*. 2014; 34(1): 63–76.

[18] Cui C, Zhang Z., Stochastic collocation with non-Gaussian correlated process variations: Theory, algorithms, and applications. *IEEE Transactions on Components, Packaging and Manufacturing Technology*. 2018; 9(7): 1362–1375.

[19] Cui C, Zhang Z., High-dimensional uncertainty quantification of electronic and photonic IC with non-Gaussian correlated process variations. *IEEE Transactions on Computer-Aided Design of Integrated Circuits and Systems*. 2019; 39(8): 1649–1661.

[20] Cui C, Liu K, Zhang Z., Chance-constrained and yield-aware optimization of photonic ICs with non-Gaussian correlated process variations. *IEEE Transactions on Computer-Aided Design of Integrated Circuits and Systems*. 2020; 39(12): 4958–4970.

[21] He Z, Zhang Z., PoBO: A polynomial bounding method for chance-constrained yield-aware optimization of photonic ICs. arXiv preprint, arXiv:210712593; 2021.

[22] Xiu D, Karniadakis GE., The Wiener-Askey polynomial chaos for stochastic differential equations. *SIAM Journal on Scientific Computing*. 2002; 24(2): 619–644.

[23] Soize C, Ghanem R., Physical systems with random uncertainties: chaos representations with arbitrary probability measure. *SIAM Journal on Scientific Computing*. 2004; 26(2): 395–410.

[24] Gautschi W., On generating orthogonal polynomials.*SIAM Journal on Scientific Computing*. 1982; 3(3): 289–317.

[25] Ghanem RG, Spanos PD., Stochastic finite element method: response statistics. In: *Stochastic finite elements: a spectral approach*. Berlin: Springer; 1991. pp. 101–119.

[26] Golub GH, Welsch JH., Calculation of Gauss quadrature rules. *Mathematics of Computation*. 1969; 23: 221–230.
[27] Barthelmann V, Novak E, Ritter K., High dimensional polynomial interpolation on sparse grids. *Advances in Computational Mathematics*. 2000; 12 (4): 273–288.
[28] Wold S, Esbensen K, Geladi P., Principal component analysis. *Chemometrics and Intelligent Laboratory Systems*. 1987; 2(1–3): 37–52.
[29] Singh J, Sapatnekar S., Statistical timing analysis with correlated non-Gaussian parameters using independent component analysis. In: Proceedings of Design Automation; 2006. pp. 155–160.
[30] Weng TW, Zhang Z, Su Z, *et al.*, Uncertainty quantification of silicon photonic devices with correlated and non-Gaussian random parameters. *Optics Express*. 2015; 23(4): 4242–4254.
[31] Cui C, Gershman M, Zhang Z., Stochastic collocation with non-Gaussian correlated parameters via a new quadrature rule. In: Proceedings of the IEEE Conference on Electrical Performance of Electronics Packaging Society. San Jose, CA; 2018. pp. 57–59.
[32] Nobile F, Tempone R, Webster CG., A sparse grid stochastic collocation method for partial differential equations with random input data. *SIAM Journal of Numerical Analysis*. 2008; 46(5): 2309–2345.
[33] Zhang Z, Osledets I, Yang X, *et al.*, Enabling high-dimensional hierarchical uncertainty quantification by ANOVA and tensor-train decomposition. *IEEE Transactions on Computer-Aided Design of Integrated Circuits and Systems*. 2015; 34(1): 63–76.
[34] Cui C, Zhang Z., Uncertainty quantification of electronic and photonic ICs with non-Gaussian correlated process variations. In: Proceedings of the International Conference on Computer-Aided Design. San Diego, CA; 2018. pp. 1–8.
[35] Donoho DL., Compressed sensing. *IEEE Transactions on Information Theory*. 2006; 52(4): 578–594.
[36] He Z, Zhang Z., High-dimensional uncertainty quantification via active and rank-adaptive tensor regression. In: Proceedings of the IEEE Conference on Electrical Performance of Electronics Packaging Society; 2020. pp. 1–3.
[37] He Z, Zhang Z., High-dimensional uncertainty quantification via rank-and sample-adaptive tensor regression. arXiv preprint, arXiv:210317236; 2021.
[38] He Z, Cui W, Cui C, *et al.*, Efficient uncertainty modeling for system design via mixed integer programming. In: IEEE/ACM International Conference on Computer-Aided Design; 2019. pp. 1–8.
[39] Calafiore GC, El Ghaoui L., On distributionally robust chance-constrained linear programs. *Journal of Optimization Theory and Applications*. 2006; 130(1): 1–22.
[40] Zhang Z, El-Moselhy TA, Elfadel IM, *et al.*, Stochastic testing method for transistor-level uncertainty quantification based on generalized polynomial chaos. *IEEE Transactions on Computer-Aided Design of Integrated Circuits and Systems*. 2013; 32(10): 1533–1545.

[41] Waki H, Kim S, Kojima M, *et al.*, Sums of squares and semidefinite program relaxations for polynomial optimization problems with structured sparsity. *SIAM Journal on Optimization*. 2006; 17(1): 218–242.
[42] Nie J., An exact Jacobian SDP relaxation for polynomial optimization. *Mathematical Programming*. 2013; 137(1–2): 225–255.
[43] Henrion D, Lasserre JB, Löfberg J., GloptiPoly 3: moments, optimization and semidefinite programming. *Optimization Methods and Software*. 2009; 24(4–5): 761–779.
[44] Wang M, Yang F, Yan C, *et al.*, Efficient Bayesian yield optimization approach for analog and SRAM circuits. In: Proceedings of the Design Automation Conference; 2017. pp. 1–6.
[45] Van Parys BP, Goulart PJ, Kuhn D., Generalized Gauss inequalities via semidefinite programming. *Mathematical Programming*. 2016; 156(1–2): 271–302.

Chapter 7

Machine learning approaches towards uncertainty quantification

Riccardo Trinchero[1] *and Flavio Canavero*[1]

This chapter deals with the application of advanced Machine Learning (ML) techniques to the uncertainty quantification (UQ) of stochastic nonlinear systems. In particular, ML regressions will be adopted as *black-box* techniques with the aim to construct a surrogate model able to mimic the actual stochastic behavior of the system under modeling. Several regression techniques will be presented along the Chapter, starting from the classical ones (e.g., the least squares, Ridge and least absolute shrinkage and selection operator (LASSO) regressions) to the more recent ML-based regressions, such as: support vector machine, least squares-support vector machine and Gaussian process regression. All the above techniques will be presented along with their advantages and limitations with the help of several illustrative examples. Moreover, their applicability to the UQ will be investigated by considering realistic application examples belonging to the electronic field.

7.1 What is a surrogate model?

This section aims to quickly present the key-idea behind the surrogate models and to define the main keywords that will be used hereafter in this Chapter.

First of all, let us introduce the concept of *(full-)computational model.* A computational model, hereafter denoted as $M(\cdot)$, is the most accurate deterministic description of the actual behavior of the system under modeling, able to provide, for any configuration of the uncertain parameters, a prediction of the output of interest. The above model can have different levels of fidelity and complexity going from a simple closed-form analytical approximation to a more sophisticated physical-based description of the system under consideration [1]. The computational model is used to generate the so-called *training samples* or *training dataset.* The training dataset consist of a deterministic set of pairs collecting a set of configurations of the input parameters and the corresponding output values. Such

[1]EMC Group, Department of Electronics and Telecommunications, Politecnico di Torino, Torino, Italy

samples are used as input for a regression or an interpolation technique during the training of a *surrogate model* [1–3].

A surrogate model $\widetilde{M}(\cdot)$, also known as a metamodel, is a closed-form and fast-to-evaluate approximation of the input–output behavior of the computational model, constructed by using a limited amount of training data. The model accuracy is usually investigate by comparing the model predictions with the ones provided by the *test samples* or *test dataset*. Similar to the training set, the test dataset collects a set of deterministic configurations of the input parameters and the corresponding outputs calculated via the computational model, which has never been used during the model training.

An accurate surrogate model can be adopted for either UQ [3] or optimization purposes [4]. In this chapter, the surrogate models will be used within the simulation flow of a Monte Carlo (MC) simulation to inexpensively predict the system output for a large set of configurations of the uncertain parameters, thus providing an efficient alternative to the plain MC simulation based on the computation model [1].

7.2 Learning paradigm

This section addresses in a rigorous way, the problem of constructing a metamodel via a generic regression or interpolation technique based on the so-called learning paradigm [5]. For the sake of simplicity, hereafter we will consider the case of computational model $M(\cdot)$, describing a single-output noisy nonlinear and static system with possible uncertain parameters*:

$$y = M(\mathbf{x}) + \varepsilon = \bar{y}(\mathbf{x}) + \varepsilon, \tag{7.1}$$

where $\mathbf{x} = [x_1, \ldots, x_d]^T \subset \mathcal{X}$, with $\mathcal{X} \in \mathbb{R}^d$, is a set of (uncertain) input design parameters, $\bar{y}(\mathbf{x}) = M(\mathbf{x}) \in \mathbb{R}$ is the system output, ε is a random error or noise with zero mean and unknown variance [1] and $y \in \mathbb{R}$ is the noisy response. We do assume that all d uncertain parameters in $\mathbf{x}$ are independent and equally significant. Techniques are available to possibly reduce the set of input parameters when they are correlated [6].

Given the noisy responses y provided by the computational model $M(\mathbf{x})$, we are looking for the best configuration of the parameters $\boldsymbol{\beta}$ of a metamodel $\widetilde{M}(\mathbf{x}; \boldsymbol{\beta})$ (e.g., the coefficients of a linear model $\widetilde{M}(\boldsymbol{\beta}) = \beta_0 + \beta_1 x$), which minimizes the following *generalized risk functional* (GRF) [5]:

$$R(\boldsymbol{\beta}) = \int_{(\mathbf{x},y)} \mathcal{L}\left(\widetilde{M}(\mathbf{x}; \boldsymbol{\beta}), y, \mathbf{x}\right) P(\mathbf{x}, y) d\mathbf{x} dy \tag{7.2}$$

where $P(\mathbf{x}, y) = P(\mathbf{x})P(y|\mathbf{x})$ is the joint probability of getting a given pair $(\mathbf{x}, y)$ and $\mathcal{L}$ is the cost (loss) function. The latter quantifies the "error" (discrepancy) of the

*The following mathematical formulation can be also generalised to the case of a multioutput system.

model predictions. A probabilistic relationship between $\mathbf{x}$ and y is assumed in order to account for the noise (e.g., two or more different values of y could be obtained for the same $\mathbf{x}$) [2].

With the GRF, we are looking for the best set of the model parameters $\boldsymbol{\beta}$, able to minimize the average error defined as the integral of the loss function $\mathcal{L}$ computed for all possible values of the parameters $\mathbf{x} \in \mathcal{X}$ and model output $y = M(\mathbf{x}) + \varepsilon$ weighted by their joint probability $P(\mathbf{x}, y)$. This means that the solution of the above minimization problem would require a full knowledge of all the possible responses of the computational model M computed for all the possible parameter configurations and error values. In other words, the knowledge of $P(\mathbf{x}, y)$ would require an infinite number of simulations with the computational model M by considering all the possible configuration of the input $\mathbf{x}$ and the realizations of the error ε. Obviously, in the real world it is impossible to run an *infinite* number of simulations. In addition, if such huge amount of data were available, there would be no reason to construct a metamodel!

Usually, a small set of training pairs $\mathcal{D} = \{(\mathbf{x}_i, y_i)\}_{i=1}^{L}$ calculated with the expensive computational model can be used for the training of a surrogate model $\widetilde{M}(\mathbf{x}; \boldsymbol{\beta})$ (i.e., to estimate the unknown parameters $\boldsymbol{\beta}$ of the surrogate model). In the above case, the parameters $\boldsymbol{\beta}$ can be estimated by minimizing the following *empirical risk funcional* (ERF) [5],

$$R(\boldsymbol{\beta}) \approx \underbrace{\frac{1}{L}\sum_{i=1}^{L} \mathcal{L}\left(\widetilde{M}(\mathbf{x}_i; \boldsymbol{\beta}), y_i, \mathbf{x}_i\right)}_{R_{\mathrm{emp}}(\boldsymbol{\beta})}. \tag{7.3}$$

This means that we are trying to minimize the average discrepancy, in terms of a given loss function $\mathcal{L}$, between the available output samples y_i belonging to the training set $\mathcal{D}$, with respect to the corresponding values predicted by the model $\widetilde{M}(\mathbf{x}_i)$. Therefore, we are looking for the optimum set of model parameters $\boldsymbol{\beta}^*$ that minimizes the average cost (7.3) on the available set of training samples $\mathcal{D}$, that is,

$$\boldsymbol{\beta}^* = \arg\min_{\boldsymbol{\beta}}\{R_{\mathrm{emp}}(\boldsymbol{\beta})\} = \arg\min_{\boldsymbol{\beta}}\left\{\frac{1}{L}\sum_{i=1}^{L} \mathcal{L}\left(\widetilde{M}(\mathbf{x}_i; \boldsymbol{\beta}), y_i, \mathbf{x}_i\right)\right\} \tag{7.4}$$

Several kinds of loss functions $\mathcal{L}$ are available [1,7], of which the most common ones are the following:

- *Squared Loss*:

$$\mathcal{L}_S\left(\widetilde{M}(\mathbf{x}_i; \boldsymbol{\beta}), y_i, \mathbf{x}_i\right) = (y_i - \widetilde{M}(\mathbf{x}_i; \boldsymbol{\beta}))^2. \tag{7.5}$$

  The squares loss is the most popular regression loss function (e.g., is the one used by the least squares regression). The main advantage is that such function is differentiable everywhere, but it is also too sensitive to outliers and noise;

- *Absolute Loss*

$$\mathcal{L}_A\left(\widetilde{M}(\mathbf{x}_i;\boldsymbol{\beta}), y_i, \mathbf{x}_i\right) = |y_i - \widetilde{M}(\mathbf{x}_i;\boldsymbol{\beta})|. \tag{7.6}$$

The absolute loss is a quite common regression loss function. Its main advantage is that it is less sensitive to the outliers compared to the squared loss, but it is not differentiable in 0;

- *Huber Loss*

$$\mathcal{L}_H\left(\widetilde{M}(\mathbf{x}_i;\boldsymbol{\beta}), y_i, \mathbf{x}_i\right) = \begin{cases} \left(y_i - \widetilde{M}(\mathbf{x}_i;\boldsymbol{\beta})\right)^2, \text{ if } |y_i - \widetilde{M}(\mathbf{x}_i;\boldsymbol{\beta})| < \delta \\ \delta \cdot \left(|y_i - \widetilde{M}(\mathbf{x}_i;\boldsymbol{\beta})| - \dfrac{\delta}{2}\right), \text{ otherwise.} \end{cases} \tag{7.7}$$

The Huber loss function takes the best of the squared and absolute loss functions (i.e., it behaves as squared loss, when the model error is small and as the absolute loss, when the error is large). On the other hand, the parameter δ must be tuned by the user.

The behavior of the above loss functions as a function of the model error is shown in Figure 7.1 in which $e_i = y_i - \widetilde{M}(\mathbf{x}_i)$.

In the framework of UQ, the training set $\mathscr{D}$ is obtained by drawing samples $\mathbf{x}_i$ of the input variables randomly or pseudo-randomly, and computing the corresponding system responses y_i. As an example, the latin hypercube sampling (LHS) represents a widely used and effective strategy to randomly sample the design space with good exploration properties [8].

Once the model $\widetilde{M}(\mathbf{x}_i)$ is built via the training set $\mathscr{D}$, its accuracy is usually investigated on the test set $\mathscr{T} = \{(\mathbf{x}_1, y_1), \ldots, (\mathbf{x}_T, y_T)\}$, which collects a set of sample pairs which were not been used during the training phase, that is, $\mathscr{D} \cap \mathscr{T} = \varnothing$ [9].

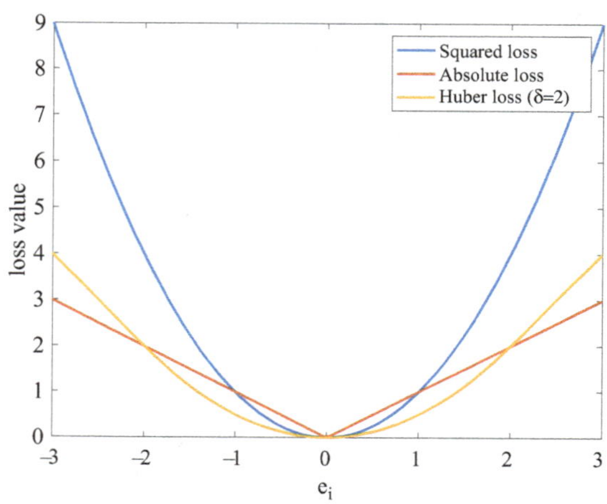

Figure 7.1 Plot of the common loss functions

7.3 Ordinary least squares

Let us consider the specific case in which the ERF minimization is applied to train a metamodel via the ordinary least squares (OLS) regression. In this specific case, we are looking for a metamodel $\widetilde{M}(\mathbf{x};\boldsymbol{\beta})$ with n basis functions, which writes [10]:

$$\widetilde{M}(\mathbf{x};\boldsymbol{\beta}) = \sum_{n=1}^{N} \phi_n(\mathbf{x})\beta_n, \tag{7.8}$$

where $\boldsymbol{\phi}_n(\mathbf{x})$ are the basis functions and β_n are the corresponding regression coefficients (i.e., $\boldsymbol{\beta} = [\beta_1, \ldots, \beta_N]^T$).

The OLS regression uses the squared loss function in (7.5). Substituting (7.8) and (7.5) into the empirical risk functional (7.3), leads to:

$$R_{\text{emp}}(\boldsymbol{\beta}) = \frac{1}{L}\sum_{i=1}^{L}\left(\widetilde{M}(\mathbf{x}_i;\boldsymbol{\beta}) - y_i\right)^2 = \frac{1}{L}\sum_{i=1}^{L}\left(\sum_{n=1}^{N}\beta_n\phi_n(\mathbf{x}_i) - y_i\right)^2. \tag{7.9}$$

According to (7.4), the optimal set of regression coefficients $\boldsymbol{\beta}^*$ is obtained by minimizing the above ERF:

$$\begin{aligned}\boldsymbol{\beta}^* &= \arg\min_{\boldsymbol{\beta}} \frac{1}{L}\sum_{i=1}^{L}\left(\sum_{n=1}^{N}\beta_n\phi_n(\mathbf{x}_i) - y_i\right)^2 \\ &= \arg\min_{\boldsymbol{\beta}} \frac{1}{L}\sum_{i=1}^{L}\left(\boldsymbol{\phi}(\mathbf{x}_i)^T\boldsymbol{\beta} - y_i\right)^2,\end{aligned} \tag{7.10}$$

Where $\boldsymbol{\phi}(\mathbf{x}) = \left[\phi_1(\mathbf{x}), \ldots, \phi_N(\mathbf{x})\right]^T \in \mathbb{R}^{N\times 1}$ is a vector collecting the N basis functions.

The above equation can be suitably recasted in terms of the following matrix relation:

$$\boldsymbol{\beta}^* = \arg\min_{\boldsymbol{\beta}} \frac{1}{L}(\boldsymbol{\Phi}\boldsymbol{\beta} - \mathrm{y})^T(\boldsymbol{\Phi}\boldsymbol{\beta} - y), \tag{7.11}$$

where $\boldsymbol{\Phi} \in \mathbb{R}^{(L\times N)}$ is a matrix collecting the basis functions evaluated on the training inputs, such that $\boldsymbol{\Phi}_{ij} = \boldsymbol{\phi}_j(\mathbf{x}_i)$ and $\mathbf{y} = [y_1, \ldots, y_L]^T \in \mathbb{R}^{(L\times 1)}$ is column vector collecting the training output.

This means that we are looking for the optimal set of coefficients $\boldsymbol{\beta}$, which minimizes the average of the squared error computed on a set of training samples $\mathscr{D}$. The optimization problem in (7.11) can be easily solved in a closed-form by computing the derivative of the ERF with respect to the model parameters $\boldsymbol{\beta}$ [10]:

$$\frac{dR_{\text{emp}}(\boldsymbol{\beta})}{d\boldsymbol{\beta}} = \frac{d}{d\boldsymbol{\beta}}\left[(\boldsymbol{\Phi}\boldsymbol{\beta} - \mathbf{y})^T(\boldsymbol{\Phi}\boldsymbol{\beta} - \mathbf{y})\right] = 2\boldsymbol{\Phi}^T\boldsymbol{\Phi}\boldsymbol{\beta} - 2\boldsymbol{\Phi}^T\mathbf{y} = 0, \tag{7.12}$$

providing the following solution:

$$\beta^* = (\mathbf{\Phi}^T\mathbf{\Phi})^{-1}\mathbf{\Phi}^T\mathbf{y}. \tag{7.13}$$

The OLS regression is extremely flexible, since the only restriction for the selection of the basis functions is that the squares matrix $(\mathbf{\Phi}^T\mathbf{\Phi})$ must be invertible. Moreover, the model parameters for the OLS regression can be estimated in a closed-form via the simple matrix relation in (7.13). The above features motivate the popularity of the OLS techniques. On the other hand, such simple approach has two main limitations, restricting its applicability and effectiveness in complex regression problems. First of all, the OLS regression provides a parametric model in which the number of regression unknowns (i.e., the regression coefficients β_n) is equal to the number of basis functions [11]. This leads to the infamous *curse of dimensionality* [12], making the application of such method impractical for systems with many parameters or requiring a large set of basis functions. In addition to that, the OLS metamodels suffer from high variance and are usually affected by overfitting (additional details on the above issues will be provided in Section 7.4).

7.3.1 Illustrative example

A synthetic example is presented here to illustrate the pros and cons of a standard OLS regression. Let us consider the following univariate (i.e., $d = 1$) noiseless computational model:

$$M(x) = 0.5|x - 3| + \sin(x) \tag{7.14}$$

where $x \in [0, 10]$ is modeled as a uniform distributed random variable $x \sim \mathcal{U}([0, 10])$.

The above computational model M is used to generate the training dataset collecting L training pairs, that is, $\mathcal{D} = \{(\mathbf{x}_i, y_i)\}_{i=1}^{L}$. The training set $\mathcal{D}$ is used, along with the OLS in (7.13), to construct a metamodel based on a polynomial expansion with maximum degree N_{degree}, which reads:

$$\widetilde{M}(\mathbf{x}; \boldsymbol{\beta}) = \sum_{n=0}^{N_{\text{degree}}} \beta_n x^n, \tag{7.15}$$

where, for the case of a polynomial expansion, the basis functions specified in (7.8) are defined as $\phi_n(x) = x^n$.

The accuracy of the OLS regression is evaluated on $N_{\text{test}} = 1{,}000$ test samples $\mathcal{T} = \{(\mathbf{x}_i, y_i)\}_{i=1}^{N_{\text{test}}}$ by considering the following root mean squares error (RMSE):

$$\text{RMSE} = \sqrt{\frac{1}{N_{\text{test}}} \sum_{i=1}^{N_{\text{test}}} (y_i - \widetilde{M}(x_i))^2}. \tag{7.16}$$

Specifically, the above error will be used in the following analysis to evaluate the performance of the OLS regression in (7.15) with respect to the polynomial degree, that is, $N_{\text{degree}} = 1, \ldots, 20$ and the number of training samples, that is, $L = 1, \ldots, 100$.

Figure 7.2 shows the RMSE computed on the training samples by considering the predictions of several OLS metamodels with an increasing model degree and constructed via an increasing number of training samples. According to the plots, for a given size of the training set, the RMSE computed on the training samples decreases by increasing the model complexity (i.e., the order of the polynomial regression). Also, there is a straight line in the 3D plot defined as $|\boldsymbol{\beta}| = L$, along which the error is almost zero. That line defines a curve along which the number of coefficients $\boldsymbol{\beta}$ (i.e., the cardinality $|\boldsymbol{\beta}|$) is exactly equal to the number of training samples L. Along such line, the matrix $\boldsymbol{\Phi}$ is a square matrix and the solution of (7.13) is equivalent of a standard matrix inversion, that is, $\boldsymbol{\beta} = (\boldsymbol{\Phi}^T)^{-1}\mathbf{y}$.

Figure 7.3 presents a similar analysis on the test samples. In this case the behavior of the RMSE is less regular. In fact, different from what happens with the training error, increasing the model order does not improve the model error.

Without loss of generality, Figure 7.4 investigates the accuracy of the OLS-based surrogate, built with a fixed set of training samples $L = 30$, on both the training and the test samples. Specifically, the RMSE, calculated for an increasing regression order N_{degree}, is the average of the error computed for five realizations of the training set. The plot clearly highlights that the model error on the training samples decreases when the number of model coefficients is larger than the number of samples, whilst we see an opposite behavior on the test samples. This means that an increasing of the model complexity does not imply an improvement of the model accuracy, but on the contrary, it can decrease the model accuracy on the test samples. By increasing the model complexity, the resulting model is specialized for a given set of training samples, and does not generalize well on the test set. Questions: "Where does this error come from?" and "Are we able of explaining such error?"

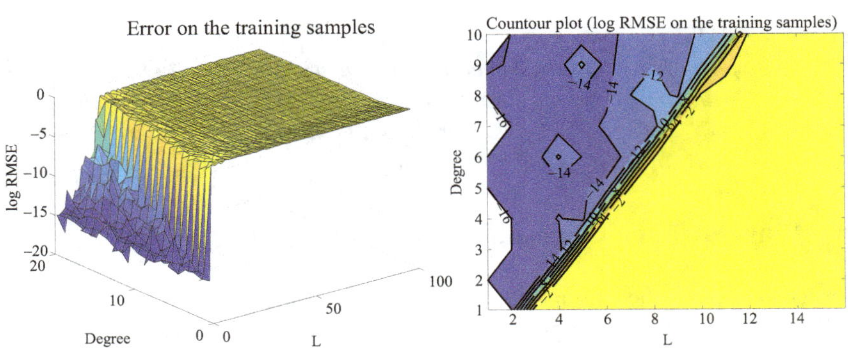

Figure 7.2 RMSE computed on the training samples considering the proposed polynomial metamodel constructed via the OLS regression for an increasing size of the training set L and polynomial degree N_{degree}

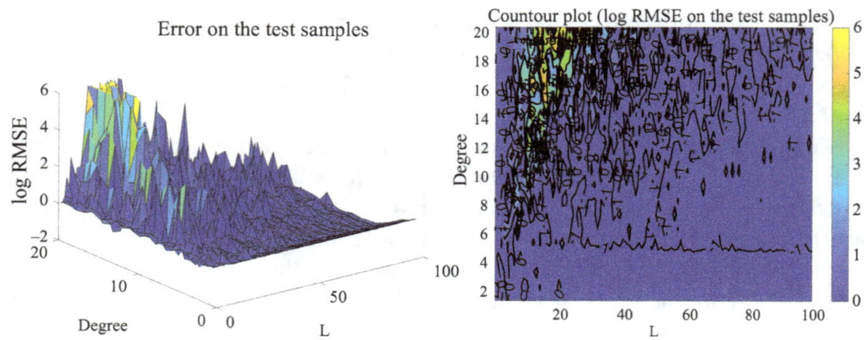

Figure 7.3 *RMSE computed on* $N_{\text{test}} = 1,000$ *test samples considering the proposed polynomial metamodel constructed via the OLS regression for an increasing size of the training set L and polynomial degree* N_{degree}

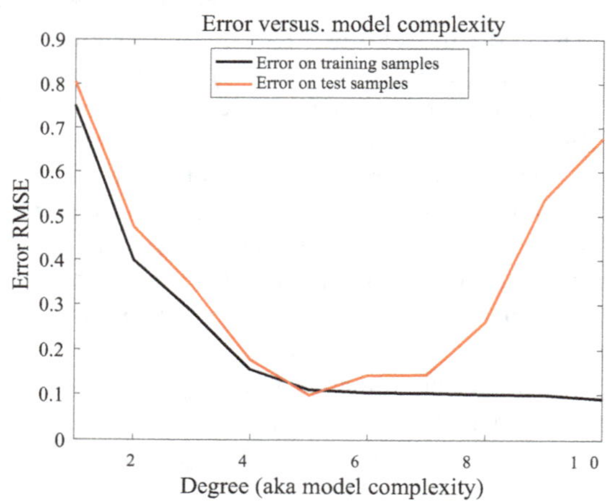

Figure 7.4 *RMSE computed on both the training and test samples considering the proposed polynomial metamodel constructed via the OLS regression with* $L = 30$ *training samples and for an increasing polynomial degree* N_{degree}. *The RMSE has been computed as the average error over* 20 *realizations of the training samples*

7.4 Bias-variance trade-off

In order to explain the origin of the model error on the test samples, we need to recall the learning paradigm presented in Section 7.2. Let us consider the problem of building a metamodel $\widetilde{M}(\cdot)$ based on a generic algorithm $\mathscr{A}$ and a computational model $M(\cdot)$, which reads [1,7,10]:

$$y(\mathbf{x}) = \bar{y}(\mathbf{x}) + \varepsilon(\mathbf{x}) \tag{7.17}$$

where the latent function $\bar{y}(\mathbf{x}) = M(\mathbf{x})$ and $\varepsilon(\mathbf{x})$ is a random process representing the additive noise.

According to Section 7.2, the metamodel $\widetilde{M}(\cdot)$ is trained by means of the algorithm $\mathscr{A}$ via the following procedure:

- Drawn a set of L training samples $\mathscr{D} = \{(\mathbf{x}_1, y_1), \ldots, (\mathbf{x}_L, y_L)\}$ from the distribution $P(\mathbf{x}, y)$ (the output samples $\{y_i\}_{i=1}^{L}$ are noisy samples computed via the computational model).
- Use the training samples to train the metamodel $\widetilde{M}_{\mathscr{D}}$ with the algorithm $\mathscr{A}$, that is, $\widetilde{M}_{\mathscr{D}} = \mathscr{A}(\mathscr{D})$.
- Select a test sample $\mathbf{x}_0 \in \mathcal{X}$ and use the computational model to get the output $y_0 = \bar{y}(\mathbf{x}_0) + \varepsilon(\mathbf{x}_0)$.
- Measure the mean square error (MSE) of the metamodel prediction $\widetilde{M}_{\mathscr{D}}(\mathbf{x}_0)$ computed on the test samples $\mathbf{x}_0$ with respect to the actual value of the latent function $\bar{y}(\mathbf{x}_0)$.

Given the above procedure, we would like to understand and to define via a set of equations, which terms mainly contribute to the model error. Let us define the *expected test error*, that will be used hereafter to evaluate the performance of the proposed algorithm $\mathscr{A}$:

$$E_{(\mathbf{x},y),\mathscr{D}}\left[(\widetilde{M}_{\mathscr{D}}(\mathbf{x}) - y)^2)\right] = \int_{\mathscr{D}}\int_{\mathbf{x}}\int_{y} (\widetilde{M}_{\mathscr{D}}(\mathbf{x}) - y)^2 P(\mathbf{x}, y) P(\mathscr{D}) d\mathbf{x} dy d\mathscr{D}. \tag{7.18}$$

This means that we are computing the average (i.e., the expected value) of the MSE† computed between the metamodel predictions and the corresponding noisy ones computed by the computational model over all possibile test points $(\mathbf{x}, y)$ for all the possible realizations of the training set $\mathscr{D}$. In order to simplify the expected test error in (18), we need to define some useful quantities.

Expected Value $\bar{y}(\mathbf{x})$

Given a test sample $\mathbf{x}_0$, the expected value $\bar{y}(\mathbf{x}_0)$ is the average of all the possible realizations $y = \bar{y}(\mathbf{x}_0) + \varepsilon(\mathbf{x}_0)$ with respect to the additive noise $\varepsilon(\mathbf{x}_0)$ (i.e., we are looking for the latent function $\bar{y}(\mathbf{x}_0)$ by canceling out the noise):

$$\bar{y}(\mathbf{x}_0) = E_{y|\mathbf{x}_0}[y] = \int_{y} y P(y|\mathbf{x}_0) dy. \tag{7.19}$$

Expected Test Error

Given a metamodel $\widetilde{M}_{\mathscr{D}} = \mathscr{A}(\mathscr{D})$, the expected test error provide the average of the MSE of the metamodel $\widetilde{M}_{\mathscr{D}}$ on the test samples y:

$$E_{(\mathbf{x},y)}\left[(\widetilde{M}_{\mathscr{D}}(\mathbf{x}) - y)^2\right] = \int_{\mathbf{x}}\int_{y} \left(\widetilde{M}_{\mathscr{D}}(\mathbf{x}) - y\right)^2 P(\mathbf{x}, y) d\mathbf{x} dy. \tag{7.20}$$

†The MSE is considered just for simplicity, any cost functions or error can be used as well.

Expected Regression $\bar{M}$

The expected regression $\bar{M}$ is the expected model obtained as the average of the metamodel $\widetilde{M}_{\mathcal{D}} = A(\mathcal{D})$ overall the possible realizations of the training set $\mathcal{D}$ with a fixed number of samples L:

$$\bar{M}(\mathbf{x}) = E_{\mathcal{D}}[A(\mathcal{D})] = \int_{\mathcal{D}} \widetilde{M}(\mathbf{x})P(\mathcal{D})d\mathcal{D}. \tag{7.21}$$

The expected test error in (1.18) for the algorithm $\mathcal{A}$ can be rewritten as follows:

$$\begin{aligned} E_{(\mathbf{x},y),\mathcal{D}}\left[(\widetilde{M}_{\mathcal{D}}(\mathbf{x}) - y)^2\right] &= E_{(\mathbf{x},y),\mathcal{D}}\left[\left((\widetilde{M}_{\mathcal{D}}(\mathbf{x}) - \bar{M}(\mathbf{x})) + (\bar{M}(\mathbf{x}) - y)\right)^2\right] \\ &= E_{\mathbf{x},\mathcal{D}}\left[\left(\widetilde{M}_{\mathcal{D}}(\mathbf{x}) - \bar{M}(\mathbf{x})\right)^2\right] + E_{\mathbf{x},y}\left[(\bar{M}(\mathbf{x}) - y)^2\right] \\ &\quad + 2E_{(\mathbf{x},y),\mathcal{D}}\left[\left(\widetilde{M}_{\mathcal{D}}(\mathbf{x}) - \bar{M}(\mathbf{x})\right) \cdot (\bar{M}(\mathbf{x}) - y)\right]. \end{aligned} \tag{7.22}$$

Let us focus on the last term:

$$\begin{aligned} &E_{(\mathbf{x},y),\mathcal{D}}\left[\left(\widetilde{M}_{\mathcal{D}}(\mathbf{x}) - \bar{M}(\mathbf{x})\right) \cdot (\bar{M}(\mathbf{x}) - y)\right] \\ &= E_{x,y}\left[E_{\mathcal{D}}\left[\widetilde{M}_{\mathcal{D}}(\mathbf{x}) - \bar{M}(\mathbf{x})\right] \cdot (\bar{M}(\mathbf{x}) - y)\right] = 0, \end{aligned} \tag{7.23}$$

since, from the definition of the expected regression in (7.21):

$$E_{\mathcal{D}}\left[\widetilde{M}_{\mathcal{D}}(\mathbf{x}) - \bar{M}(\mathbf{x})\right] = \bar{M}(\mathbf{x}) - \bar{M}(\mathbf{x}) = 0. \tag{7.24}$$

Also, the term $E_{\mathbf{x},y}\left[(\bar{M}(\mathbf{x}) - y)^2\right]$ can be simplified as follows:

$$\begin{aligned} E_{\mathbf{x},y}\left[(\bar{M}(\mathbf{x}) - y)^2\right] &= E_{\mathbf{x},y}\left[((\bar{M}(\mathbf{x}) - \bar{y}(\mathbf{x})) + (\bar{y}(\mathbf{x}) - y))^2\right] \\ &= E_{\mathbf{x},y}\left[(\bar{M}(\mathbf{x}) - \bar{y}(\mathbf{x}))^2\right] + E_{\mathbf{x},y}\left[(\bar{y}(\mathbf{x}) - y)^2\right] \\ &\quad + 2E_{\mathbf{x},y}[(\bar{y}(\mathbf{x}) - y)], \end{aligned} \tag{7.25}$$

where the last term is qual to 0, since:

$$\begin{aligned} E_{\mathbf{x},y}[(\bar{y}(\mathbf{x}) - y)] &= E_{\mathbf{x}}\left[(\bar{M}(\mathbf{x}) - \bar{y}(\mathbf{x}))E_{y|\mathbf{x}}(\bar{y}(\mathbf{x}) - y)\right] \\ &= E_{\mathbf{x}}[(\bar{M}(\mathbf{x}) - \bar{y}(\mathbf{x}))(\bar{y}(\mathbf{x}) - \bar{y}(\mathbf{x}))] = 0. \end{aligned} \tag{7.26}$$

Substituting (7.26), (7.25) and (7.26) into (7.24),

$$E_{(\mathbf{x},y),\mathscr{D}}\left[(\widetilde{M}_{\mathscr{D}}(\mathbf{x})-y)^2\right]=\underbrace{E_{\mathbf{x},y}\left[(\bar{M}(\mathbf{x})-\bar{y}(\mathbf{x}))^2\right]}_{\text{BIAS}^2}+\underbrace{E_{\mathbf{x},y}\left[(\bar{y}(\mathbf{x})-y)^2\right]}_{\text{NOISE}}+ \\ +\underbrace{E_{\mathbf{x},\mathscr{D}}\left[\left(\widetilde{M}_{\mathscr{D}}(\mathbf{x})-\bar{M}(\mathbf{x})\right)^2\right]}_{\text{VARIANCE}}, \tag{7.27}$$

where, the "Bias^2" term provides the average error between the best-fitting model $\bar{M}$ constructed over all the possible data set $\mathscr{D}$ and the actual values of the computational model M. The "Variance" represents the sensitivity of the model on the particular choice of the training samples. While, the "Noise" term measures the impact of the additive noise on the model accuracy.

An illustrative interpretation of the above results is shown in Figure 7.5. The picture highlights two different regions for the model error [9]:

- *Underfitting*: the model presents high bias and low variance. This means that the model is too simple to capture the complex behavior of the computational model presenting high discrepancies on both the training and the test samples (large bias). On the other hand, the model accuracy is independent from training dataset used during the training of the model (low variance).
- *Overfitting*: the model presents low bias and high variance. The model error is low on the training samples, but it is high on the test samples. This means that the model is able to capture the complex behavior of the computational model (low bias), but due to its complexity is over-specialized to a particular training set and does not generalize well on the test data.

7.4.1 Illustrative example

For the sake illustration let us consider the following computational model:

$$y=\bar{y}(x)+\varepsilon(x) \tag{7.28}$$

where $x\in[0,4\pi]$, $\bar{y}(x)=M(x)=\sin(x)$ and $e(x)\sim\mathscr{U}([-0.4,0.4])$ is an uniform distributed additive error.

Again, we build a polynomial metamodel $\widetilde{M}(x)=\sum_{i=0}^{N_{\text{degree}}}\beta_i x^i$ via the OLS algorithm (i.e., $\mathbb{R}$) by using 300 realizations of 44 training samples $\mathscr{D}=\{\mathscr{D}_1,\ldots,\mathscr{D}_{300}\}$, where $\mathscr{D}_j=\left\{(x_{j,i},y_{j,i})\right\}_{i=1}^{44}$ for $j=1,\ldots,300$. The training samples are used to train a set of surrogates models via the OLS polynomial regression. The expected error is computed on 63 test samples $\mathscr{T}=\left\{(x_{t,n},y_{t,n})\right\}_{n=1}^{63}$ by considering a set of metamodels with an increasing degree (i.e., $N_{\text{degree}}=1,\ldots,11$).

Figure 7.5 shows the expected RMSE computed on the test samples by the proposed metamodels. As we aspected from (27), there are two regions. The underfitting region corresponding to models with low complexity (e.g., low polynomial order), in which the expected error is dominated by the bias term. On the

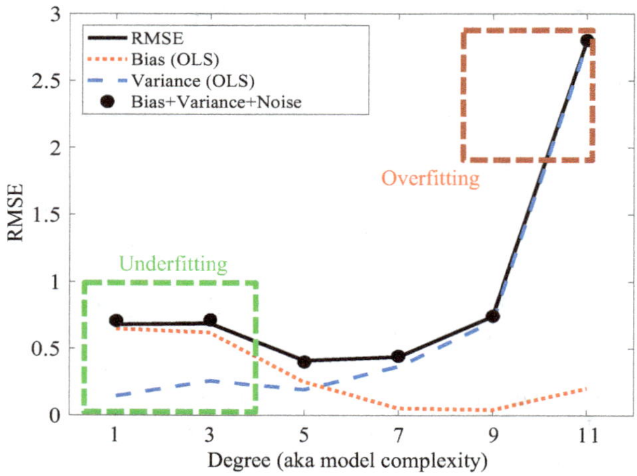

Figure 7.5 RMSE, bias and variance computed on test samples considering the proposed polynomial metamodel constructed via the OLS regression with $L = 44$ *training samples and for an increasing polynomial degree* N_{degree}. *The above terms have been computed by considering* 300 *realizations of the training samples*

other hand, the overfitting region corresponds to models with high-complexity (e.g., high polynomial order), in which the model error is dominated by the variance term.

The results show that when the complexity of the model built via the OLS regression increased, the obtained model presents low bias, but large variance (i.e., it specifies too much on the training samples, but it does not generalize well on the test samples). We can try to overcome the above limitation by shrinking the value of the regression coefficients via a *regularizer*. In this way, we will increase the model bias, but at the same time we will reduce the model variance [10].

7.5 Regularizer

The *regularization process* or *regularizer* is a penalty, which is added to the loss function to limit the value of the regression coefficients, and thus preventing overfitting [10]. In order to account for the regularizer term, the optimization problem used for the empirical risk functional in (7.3) needs to be modified as follows:

$$\min_{\boldsymbol{\beta}} R(\boldsymbol{\beta}) = \underbrace{\min_{\boldsymbol{\beta}} \frac{1}{L} \sum_{i=1}^{L} \mathcal{L}\left(\widetilde{M}(\mathbf{x}_i; \boldsymbol{\beta}), y_i, \mathbf{x}_i\right)}_{\text{Empirical Risk}} + \underbrace{\lambda r(\boldsymbol{\beta})}_{\text{Regularizer}}, \tag{7.29}$$

where again $\mathcal{L}(\cdot)$ is the loss functions and $r(\cdot)$ is the so-called regularizer.

A common idea is to use as regularizer the p-norm of the vector $\boldsymbol{\beta}$ collecting the regression coefficients, such that:

$$\min_{\boldsymbol{\beta}} R(\boldsymbol{\beta}) = \min_{\boldsymbol{\beta}} \underbrace{\frac{1}{L}\sum_{i=1}^{L} \mathcal{L}\left(\widetilde{M}(\mathbf{x}_i;\boldsymbol{\beta}), y_i, \mathbf{x}_i\right)}_{\text{Empirical Risk}} + \underbrace{\lambda\|\boldsymbol{\beta}\|_q^q}_{\text{Regularizer}}, \tag{7.30}$$

where:

$$\|\boldsymbol{\beta}\|_q = \left(\sum_{i\in N} \beta_i^q\right)^{1/q}. \tag{7.31}$$

According to (7.30), the regularizer penalizes the accuracy of the model on the training samples (i.e., we are increasing the model bias) by limiting the norm of the regression coefficints $\boldsymbol{\beta}$, in order to keep the model as flat as possible. Usually, two different cases are considered [10]:

- $q = 2$ Ridge regression, also known as Tikhonov regularizations; and
- $q = 1$ LASSO.

7.5.1 *Ridge regression* (q = 2)

Given a set of training samples $\mathcal{D} = \{(\mathbf{x}_i, y_i)\}_{i=1}^{L}$, the Ridge regression looks for the optimal set of coefficients $\boldsymbol{\beta}^*_{\text{RIDGE}}(\lambda)$ of a linear surrogate model $\widetilde{M}^{(\lambda)}(\mathbf{x}) = \sum_{n=1}^{N} \beta_n \phi_n(\mathbf{x})$, by minimizing the optimization problem [10]:

$$\boldsymbol{\beta}^*_{\text{RIDGE}}(\lambda) = \arg\min_{\boldsymbol{\beta}} \frac{1}{L}\sum_{i=1}^{L}\left(\sum_{n=1}^{N} \beta_n \phi_n(\mathbf{x}_i) - y_i\right)^2 + \lambda\sum_{n=1}^{N}\beta_n^2, \tag{7.32}$$

where a squared loss function and a L2 regularized (i.e., $q = 2$) are used, whereas λ is the regularizer parameter, which must carefully tuned by the user.

Similar to the OLS, for any given value of λ, the optimal coefficients $\boldsymbol{\beta}^*(\lambda)$ of a generic linear Ridge regression can be computed in a closed-form as[‡]:

$$\boldsymbol{\beta}^*_{\text{RIDGE}}(\lambda) = (\boldsymbol{\Phi}^T\boldsymbol{\Phi} + \lambda\mathbf{I})^{-1}\boldsymbol{\Phi}^T\mathbf{y} \tag{7.33}$$

Where $I \in \mathbb{R}^{(N+1)\times(N+1)}$ is the identity matrix and $\lambda \in \mathbb{R}$ is a scalar free parameter used by the regularizer.

From (7.32), we see that by increasing the value of the regression parameter λ, we are limiting the maximum norm allowed by the regression coefficients, penalizing the error on the training samples. In such a way, by changing the value of λ, we are able to tune the model bias and variance.

[‡]Similar to the OLS regression, the ERF is minimized by finding the value of $\boldsymbol{\beta}^*_{\text{RIDGE}}(\lambda)$, for which the partial derivative of the ERF (1.35) Lagrangian are zero.

7.5.2 LASSO regression (q = 1)

Again, given a set of training samples $\mathscr{D}$, we are looking for the optimal set of coefficients $\boldsymbol{\beta}^*_{\text{LASSO}}(\lambda)$ of a generic linear metamodel $\widetilde{M}(\mathbf{x}) = \sum_{n=1}^{N} \beta_n \phi_n(\mathbf{x})$ as a function of the regularizer parameter λ, by minimizing the following ERF [10]:

$$\boldsymbol{\beta}^*_{\text{LASSO}}(\lambda) = \arg\min_{\boldsymbol{\beta}} \frac{1}{L} \sum_{i=1}^{L} \left(\sum_{n=1}^{N} \beta_n \phi_n(\mathbf{x}) - y_i \right)^2 + \lambda \sum_{n=1}^{N} |\beta_n|. \tag{7.34}$$

The above optimization problem must be solved numerically, since the configuration $\boldsymbol{\beta}^*_{\text{LASSO}}(\lambda)$ cannot be calculated in a closed-form.

Similar to the Ridge regression, the parameter λ allows finding the right trade-off between the bias and the variance of the metamodel. Moreover, a sufficiently large value of the parameter λ will cause some of the coefficients collected in $\boldsymbol{\beta}$ to be exactly zero, leading to a sparse solution.

7.6 Tuning of the Hyperparameter

The tuning of the parameter λ, usually referred to as *hyperparameter*, is a key aspect during the metamodel construction. In fact, the hyperparameter of the regularizer allows providing a trade-off between the model flatness and the capability of the metamodel of following the training samples. Different from the regression coefficients $\boldsymbol{\beta}$, the hyperparamater λ cannot be tuned directly by solving the optimization problem of the Ridge (7.32) or LASSO (7.34) regression, otherwise the optimum value of λ would always be $\lambda^* = 0$.

For the above reasons, two different strategies are usually adopted to tune the value of the hyperparamter λ: the validation set and the cross validation (CV) [10,13].

7.6.1 Validation set

Since the hyperparameter λ cannot be tuned directly on the training samples, we can introduced the so-called validation set. The underlying idea is to partition the available dataset into three sets: the training set, the test set and the validation set. The regression coefficients are estimated via the training set, whilst the model hyperparameter λ is tuned on the validation set. Then, the test samples are used for the final assessment of the model accuracy.

Summarizing,

- *Step 1*: We split the available sample into the training set $\mathscr{D} = \{(\mathbf{x}_i, y_i)\}_{i=1}^{L}$, the validation set $\mathscr{V} = \{(\mathbf{x}_i, y_i)\}_{i=1}^{V}$ and the test set $\mathscr{T} = \{(\mathbf{x}_i, y_i)\}_{i=1}^{T}$, such that $\mathscr{D} \cap \mathscr{V} \cap \mathscr{T} = \emptyset$.
- *Step 2*: We start defining a possible set of values for the hyperparameter λ (e.g., log-spaced values $\lambda = 10^{-2}, 10^{-1}, 1, \ldots, 10^5$).
- *Step 3*: We use the training set to build a metamodel $\widetilde{M}^{(\lambda)}(\mathbf{x})$ for each value of λ defined in the previous step.

- *Step 4*: For each metamodel $\widetilde{M}^{(\lambda)}(\mathbf{x})$, we compute the error on the validation set V, for example, the MSE:

$$MSE^{(\lambda)}_{\text{validation}} = \frac{1}{V} \sum_{(\mathbf{x},y)\in\mathscr{V}} (y - \widetilde{M}^{(\lambda)}(\mathbf{x}))^2 \tag{7.35}$$

- *Step 5*: Select the optimal λ^* among the λ values defined at *Step 2*, such that the error on the validation samples is minimized

$$\lambda^* = \arg\min_{\lambda} MSE^{(\lambda)}_{\text{validation}}; \tag{7.36}$$

- *Step 6*: Evaluate the accuracy (e.g., the MSE) of the resulting metamodel $\widetilde{M}^{(\lambda^*)}(\mathbf{x})$ on the test set $\mathscr{T}$.

The main limitation of the above approach is that by partitioning the available data into three sets, we are drastically reducing the number of samples which can be used for learning the model, and thus the model performance can depend on a particular choice of the random samples in the train and validation set [10,13].

7.6.2 Cross validation

CV can be seen as an effective alternative to the above strategy, for which the validation set is no longer needed. The basic approaches are the k-fold and leave-one-out (LOO) CV. Such strategies can be computationally expensive, but they do not waste too many data, which is a major advantage in problems where the number of samples is "small."

In the k-fold CV, the training set is split into k smaller sets, called "folds." Then for each of the k folds, a metamodel is trained using $k-1$ folds as training data and the remaining folder is used as validation set. The above scheme is iterated for all the k-fold. For each value of the hyperparameter λ, the overall model performance calculated by the k-fold CV is the average of the values computed during the k iterations. The common choice for k is usually 5 or 10, as these values have been shown empirically to yield test error rate estimates that suffer neither from excessively high bias nor from very high variance. When $k = 1$ the k-fold validation is called LOO CV. An illustration of the k-fold CV for the case in which we have $L = 100$ training samples and five folds (i.e., $k = 5$) is shown in Figure 7.6.

The hyperparameter selection via the k-fold CV can be summarized as follows:

- *Step 1*: We partition the training data $\mathscr{D} = \{(\mathbf{x}_i, y_i)\}_{i=1}^{L}$ into K separate datasets of equal size $\mathscr{D}_k = \left\{(\mathbf{x}_{k,i}, y_{k,i})\right\}_{i=1}^{\widetilde{L}}$ with $\widetilde{L} = L/K$, such that $\mathscr{D} = \cup_{k=1}^{K}\mathscr{D}_k$.
- *Step 2*: For each $k = 1\ldots, K$, we fit the model $M^{(\lambda)}_{-k}$ by considering the training set $\mathscr{D}$ excluding the kth fold.
- *Step 3*: We use the model $\widetilde{M}^{(\lambda)}_{-k}$ to compute the fitted values on the observations $\mathscr{D}_k$ (the model $\widetilde{M}^{(\lambda)}_{-k}$ has been trained excluding this fold).

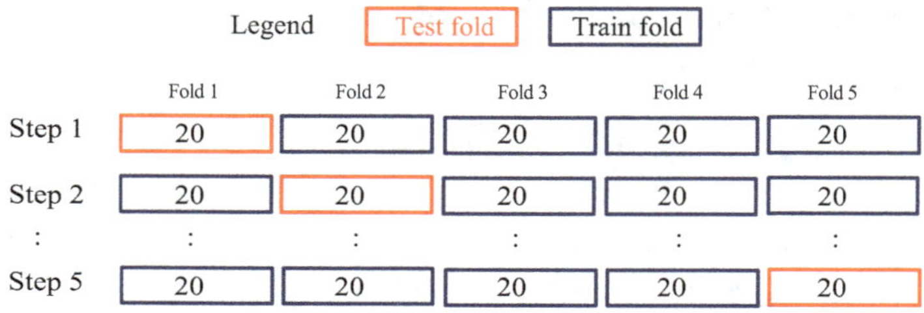

Figure 7.6 Example: a k-fold CV with K=5 and L=100 training samples

- *Step 4*: We compute the CV error for the k-fold:

$$CV_{\text{error},k}^{(\lambda)} = \frac{1}{\widetilde{L}} \sum_{(\mathbf{x},y)\in\mathscr{D}_k} \left(y - \widetilde{M}_{-k}^{(\lambda)}(\mathbf{x})\right)^2 \tag{7.37}$$

- *Step 5*: The overall model CV error for a given value of the hyperparameter λ is defined as the average of the CV error computed on the available folds, and writes:

$$\text{CV}_{\text{error}}^{(\lambda)} = \frac{1}{K}\sum_{k=1}^{K} \text{CV}_{\text{error},k}^{(\lambda)} \tag{7.38}$$

- *Step 6*: The optimum value of the hyperparameter λ^* is selected as the one that minimizes the corresponding overall CV error $\text{CV}_{\text{error}}^{(\lambda)}$, that is,

$$\lambda^* = \arg\min_{\lambda} \text{CV}_{\text{error}}^{(\lambda)} \tag{7.39}$$

- *Step 7*: After selecting the optimal value λ^*, the performances of the meta-model $\widetilde{M}^{(\lambda^*)}$ are evaluated by computing the error on the test samples.

7.6.3 Illustrative example

For the sake of illustration, let us consider the following synthetic computational model:

$$y = \bar{y}(x) + \varepsilon(x) \tag{7.40}$$

where the latent function $\bar{y}(x) = 0.5|x-3| + \sin(x)$, $x \in [0, 10]$ and the additive noise $\varepsilon(x) \sim \mathscr{U}([-0.2, 0.2])$ is modeled as a uniform distributed random variable.

The computational model M is used to generate a set of $L = 15$ random samples $\mathscr{D} = \{(x_i, y_i)\}_{i=1}^{L}$, in which the input samples $x_i \sim \mathscr{U}([0, 10])$. The OLS, the Ridge and the LASSO regression have been used to estimate the coefficients of a polynomial model for an increasing degree. For the Ridge and LASSO regressions,

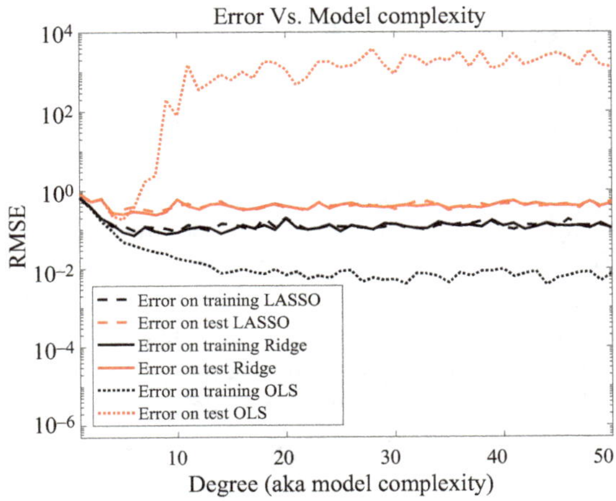

Figure 7.7 RMSE computed on both the training and test samples by considering three polynomial metemodels built with the OLS, ridge and LASSO regression, respectively. Each metamodel has been trained by using $L = 15$ *samples. The RMSE has been computed as the average error over* 20 *realizations of the training set*

the optimal value of the regularizer parameter λ in the log-spaced range $\left[0, 10^5\right]$ is estimated via a LOO CV. The performance of each regression have been assessed by considering the average RMSE computed for an increasing polynomial degree (i.e., $N_{\text{degree}} = 1 : 2 : 50$) over 20 realizations of the training samples.

Figure 7.7 shows the resulting RMSE computed by comparing the computational model results with the predictions of the Ridge, the LASSO and the OLS regressions on the training sample and on 1,000 test samples. As we aspected, the OLS regression turns out to be the most accurate approach when the RMSE is computed on the training samples. This is due to the fact that the OLS does not use the regularizer, therefore, the model bias on the training samples is usually very low. On the other hand, when the RMSE on the test samples is considered, the results clearly highlight the improved accuracy of the Ridge and LASSO regression with respect to the plain OLS regression. In fact, thanks to the regularizer, the metamodels constructed via Ridge and LASSO regression provide the lowest RMSE on test dataset, which is at least three order of magnitude smaller than the corresponding one provided by the OLS regression. Moreover, thanks to the L1-regularizer, the LASSO regression leads to a sparse model with an improved interpretability (e.g., in the considered example, the average sparsity of a polynomial LASSO regression with degree 50 computed over 20 realizations is 38%).

7.7 Summary

OLS regression can be considered as the most straightforward way to construct a surrogate model. The OLS regression problem can be solved in a closed-form without requiring the tuning of regression hyperparameters. The metamodel estimated via the OLS has a low bias, but high variance, since the model accuracy strongly depends on the random samples used to train it. The regularizer can be used to mitigate the above issue. In fact, the regularizer is adopted by the Ridge and the LASSO regressions to increase the model bias and to decrease the model variance. This allows us to construct a metamodel that generalizes well on the test samples (better than the corresponding ones built via the OLS regression). It is important to remark that, due to the regularizer, the training of the LASSO and the Ridge regression requires the tuning of a hyperparameter via either the validation set or CV.

The main limitation of such kind of regression algorithms is that they provide a "parametric model." A parametric model is a regression model in which the number of regression unknowns is equal to the number of basis functions considered by the model [11]. This means that the such algorithms suffer the so-called "curse of dimensionality" (i.e., an exponential increasing of the model complexity when the dimensionality d of the input space or the number of basis functions increases) [12,14]. As an example a surrogate model built via the Ridge regression of order $p = 2$ with $d = 500$ input parameters would require the estimation of $(p + d)!/(p!d!) > 100\text{k}$ coefficients. This means that we need to construct $> 100\text{k}$ basis functions, thus making the model construction extremely expensive and cumbersome (the same comment holds also for the case of the OLS and LASSO regression). Machine learning techniques can be seen as a promising and effective alternative to the state-of-the-art regression techniques for the construction of accurate, compact and fast-to-evaluate surrogate models.

7.8 Machine learning based metamodels

This section introduces machine learning regression tools for the constructions of metamodels for the UQ. Specifically, the support vector machine (SVM) regression, the least squares-support vector machine (LS-SVM) regression and the Gaussian process regression (GPR) are presented, along with their mathematical formulation, with the aim of stressing their advantages and key properties.

7.8.1 SVM regressions

Let us consider the problem of approximating a set of L training data $\{(\mathbf{x}_i, y_i)\}_{i=1}^{L}$ provided by a generic nonlinear system $y = M(\mathbf{x})$, in which the vector $\mathbf{x} = [x_1, \ldots, x_d]^T \in \mathbb{R}^d$ represents the input parameters, via the following SVM regression in its primal space formulation [5,15]:

$$\widetilde{M}_{\mathrm{SVM}}(\mathbf{x}) = \langle \mathbf{w}, \mathbf{\Phi}(\mathbf{x}) \rangle + b = \sum_{i=1}^{D} w_i \mathbf{\Phi}_i(\mathbf{x}) + b, \tag{7.41}$$

where $\mathbf{\Phi}(\mathbf{x}) = [\phi_1(\mathbf{x}), \ldots, \phi_D(\mathbf{x})]^T$ is a nonlinear map $\mathbf{\Phi}(\cdot) : \mathbb{R}^d \to \mathbb{R}^D$ which maps the parameter space of dimension d into the corresponding *feature space* of dimension D, $\mathbf{w} \in \mathbb{R}^D$ is a vector collecting the regression coefficients, $b \in \mathbb{R}$ is the bias term and $\langle \mathbf{w}, \mathbf{\Phi}(\mathbf{x}) \rangle = \sum_{i=1}^{D} w_i \mathbf{\Phi}_i(\mathbf{x})$ defines the inner product in $\mathbb{R}^D$.

It is important to remark that the definition of the primal space formulation of the SVM regression in (7.41) turns out to be equivalent to a classical linear regression, like the ones presented in the previous sections (e.g., OLS, Ridge and LASSO regression). Indeed, also in this case the number of regression unknown (i.e., the cardinality of $\mathbf{w}$) is equal to the dimensionality of the feature space (i.e., D) defined by the nonlinear map $\mathbf{\Phi}(\mathbf{x})$.

To understand the main differences between the SVM and the classical regression techniques presented in the previous sections, we need to have a look to the optimization problem used to estimate the coefficients in $\mathbf{w}$ and the bias term b in (7.41). For the primal space formulation of the SVM regression the best combination of the parameters $(\mathbf{w}, b)$ in (1.44) is obtained by minimizing the following functional:

$$\min_{\mathbf{w},b} \underbrace{\frac{1}{2}\|\mathbf{w}\|_2^2}_{\text{Regularizer}} + \underbrace{\frac{C}{L}\sum_{i=1}^{L} |y_i - \widetilde{M}_{\mathrm{SVM}}(\mathbf{x}_i; \mathbf{w}, b)|_\varepsilon}_{\text{EmpiricalRisk}} \tag{7.42}$$

$$= \min_{\mathbf{w},b} \frac{1}{2}\|\mathbf{w}\|_2^2 + \frac{C}{L}\sum_{i=1}^{L} |y_i - (\langle \mathbf{w}, \mathbf{\Phi}(\mathbf{x}_i) \rangle + b)|_\varepsilon, \tag{7.43}$$

where, different from the Ridge regression, the above optimization uses as loss function the linear ε-insensitive loss function denoted as $|\cdot|_\varepsilon$, defined as [5]:

$$|y_i - \widetilde{M}_{\mathrm{SVM}}(\mathbf{x}_i)|_\varepsilon = \begin{cases} 0, & \text{if } |y_i - \widetilde{M}_{\mathrm{SVM}}(\mathbf{x}_i)| \le \varepsilon \\ |y_i - \widetilde{M}_{\mathrm{SVM}}(\mathbf{x}_i)| - \varepsilon, & \text{otherwise.} \end{cases} \tag{7.44}$$

The above loss function is shown in Figure 7.9 (right panel). It is a good compromise between the squared and the absolute loss functions. If the absolute value of the model error is less than ε, the loss function does not add any penalization to the empirical risk functional. On the other hand, when the model error is larger than ε, the ε-insesitive loss function behaves as the absolute loss function. The region $[-\varepsilon, +\varepsilon]$ is called ε-insensitive zone. The parameter ε can be seen as the regression tolerance, thus making such kind of regression extremely useful when we need to deal with noisy samples.

Minimizing the risk function (7.43) is equivalent to find the best combination of the parameters $(\mathbf{w}, b)$ that minimizes the deviation of the model predictions from the training samples outside the ε-insentitive zone. This can be done via the following optimization problem in the primal space [5]:

$$\text{minimize}\ \frac{1}{2}||\mathbf{w}||^2 + C\sum_{i=1}^{L}(\xi_i + \xi_i^*)\ \text{subject to} \begin{cases} y_i - \langle \mathbf{w}, \mathbf{\Phi}(\mathbf{x}_i)\rangle - b \leq \varepsilon + \xi_i \\ \langle \mathbf{w}, \mathbf{\Phi}(\mathbf{x}_i)\rangle + b - y_i \leq \varepsilon + \xi_i^* \\ \xi_i, \xi_i^* \geq 0,\ \text{for} i = 1, \ldots, L \end{cases} \tag{7.45}$$

where ξ_i, ξ_i^* are the so-called slack variables and C is a parameter, chosen by the user, which provides a trade-off between the model accuracy and its flatness (i.e., it is the reciprocal of the regularizer parameter λ used by the Ridge and LASSO

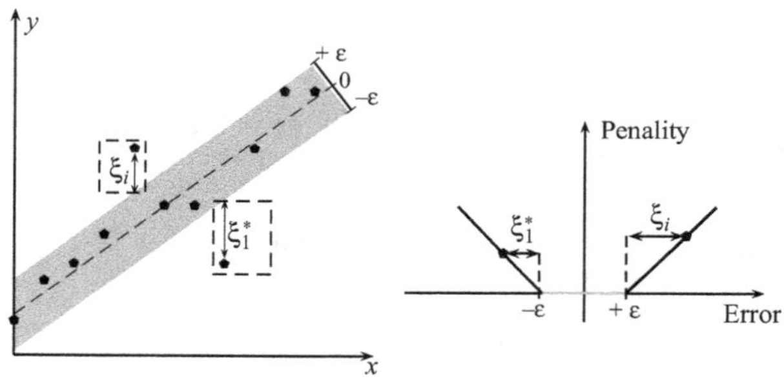

Figure 7.8 Graphical interpretation of the minimization problem (7.43) and of the role of ε-intensive loss function in (7.44) (inspired by [16,42])

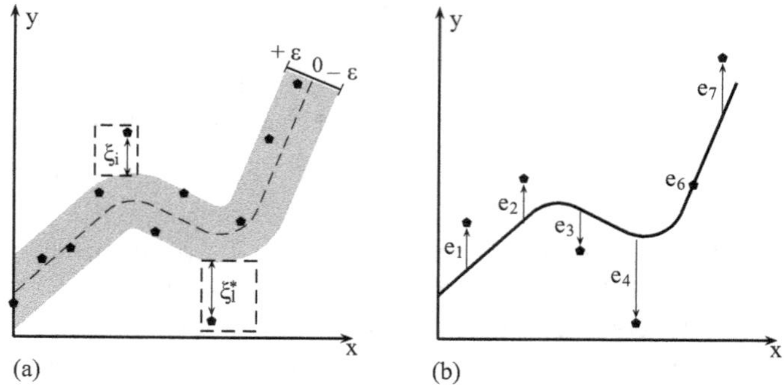

Figure 7.9 Panel (a): Graphical interpretation of the SVM regression optimization problem in (7.45) (inspired by [16,42]). Panel (b): Illustration of the corresponding least square formulation in (7.52) for the LS-SVM regression (inspired by [43])

regression) [16]. Figure 7.8 (left panel) provides a graphical interpretation of the above optimization problem. The underlying idea is to minimize the upper ξ_i and lower ξ_i^* deviations of the training samples which lay outside the ε-insensitive zone (gray area), but at the same time with the help of a Ridge regularizer, we are also maximizing the model flatness to avoid overfitting.

The nonlinear optimization problem in (7.45) can be solved by minimizing the following Lagrangian function:

$$\begin{aligned}\mathcal{L}(\mathbf{w}, b, \xi, \xi; \alpha, \alpha^*, \eta, \eta^*) = \frac{1}{2}||\mathbf{w}||^2 + C\sum_{i=1}^{L}(\xi_i + \xi_i^*)\\ - \sum_{i=1}^{L}\alpha_i(\varepsilon + \xi_i - y_i + \langle\mathbf{w}, \mathbf{\Phi}(\mathbf{x}_i)\rangle + b)\\ - \sum_{i=1}^{L}\alpha_i^*(\varepsilon + \xi_i^* + y_i - \langle\mathbf{w}, \mathbf{\Phi}(\mathbf{x}_i)\rangle - b)\\ - \sum_{i=1}^{L}(\eta_i\xi_i + \eta_i^*\xi_i^*),\end{aligned} \tag{7.46}$$

where the parameters $\alpha, \alpha^*, \eta, \eta^* \geq 0$ are the Lagrangian multipliers (dual parameters) related to the constrained optimization problem. The saddle point of the Lagrangian (7.46) is obtained by computing the partial derivatives w.r.t. the primal variables:

$$\frac{\partial\mathcal{L}}{\partial\mathbf{w}} = 0 \rightarrow \mathbf{w} = \sum_{i=1}^{L}(\alpha_i - \alpha_i^*)\mathbf{\Phi}(\mathbf{x}_i) \tag{7.47a}$$

$$\frac{\partial\mathcal{L}}{\partial b} = 0 \rightarrow \sum_{i=1}^{L}(\alpha_i - \alpha_i^*) = 0 \tag{7.47b}$$

$$\frac{\partial\mathcal{L}}{\partial\xi_i} = 0 \rightarrow C - \alpha_i - \eta_i = 0 \tag{7.47c}$$

$$\frac{\partial\mathcal{L}}{\partial\xi_i^*} = 0 \rightarrow C - \alpha_i^* - \eta_i^* = 0. \tag{7.47d}$$

Substituting (7.47a), (7.47b), (7.47c) and (7.47d) into (7.45) leads to the following dual optimization problem:

$$\text{maximize}\begin{cases} -\frac{1}{2}\sum_{i,j=1}^{L}(\alpha_i - \alpha_i^*)(\alpha_j - \alpha_j^*)\mathcal{K}(\mathbf{x}_i, \mathbf{x}_j) + \\ -\varepsilon\sum_{i=1}^{L}(\alpha_i + \alpha_i^*) + \sum_{i=1}^{L}y_i(\alpha_i + \alpha_i^*)\end{cases} \text{subject to}\begin{cases}\sum_{i=1}^{L}(\alpha_i - \alpha_i^*) = 0\\ \alpha_i, \alpha_i^* \in [0, C],\end{cases} \tag{7.48}$$

where $\mathcal{K}(\mathbf{x}_i, \mathbf{x}_j) = \langle \mathbf{\Phi}(\mathbf{x}_i), \mathbf{\Phi}(\mathbf{x}_j) \rangle$ is the kernel function defined as the inner product in the feature space between the function $\mathbf{\Phi}(\mathbf{x})$ evaluated at the training samples $\mathbf{x}_i$ and $\mathbf{x}_j$. Therefore, the primal space formulation of the SVM regression in (7.41) can be rewritten in the dual space as follows:

$$\widetilde{M}_{SVM}(\mathbf{x}) = \sum_{i=1}^{L} (\alpha_i - \alpha_i^*) \langle \mathbf{\Phi}(\mathbf{x}_i), \mathbf{\Phi}(\mathbf{x}) \rangle + b = \sum_{i=1}^{L} (\alpha_i - \alpha_i^*) \mathcal{K}(\mathbf{x}_i, \mathbf{x}) + b. \tag{7.49}$$

The solution of the optimization problem in (7.48) allows estimating the parameters α_i and α_i^*, while the bias term b is computed by exploiting the Karush-Kuhn-Tucker conditions [17]. The direct computation of the inner product $\langle \mathbf{\Phi}(\mathbf{x}_i), \mathbf{\Phi}(\mathbf{x}) \rangle$ can be extremely inefficient for a high-dimensional feature space. However, since both the dual optimization problem in (7.48) and the SVM expansion in (7.49) do not require an explicit calculation of the inner product in the D-space, the latter can be efficiently estimated by operating directly in the parameter space, which has a finite dimensionality (i.e., $\mathbb{R}^d$), by considering the kernel as a functions $\mathcal{K}(\cdot, \cdot) : \mathbb{R}^d \times \mathbb{R}^d \to \mathbb{R}$. This means that the dual space SVM expansion is fully defined by the kernel function $\mathcal{K}$ without requiring an explicit definition of the nonlinear transformation $\mathbf{\Phi}$. This is the so-called *kernel trick*.

Thanks to the kernel trick, the number of regression unknowns in (7.49) (i.e., the coefficients α_i and α_i^*) to be estimated by solving the SVM optimization in (7.48) is always equal to the number of training pairs L. Therefore, the dual space formulation of the SVM regression of (7.49) is a *non-parametric* model in which the model unknowns are independent from the number of input parameters d and the dimensionality of the feature space D. Such property can be extremely useful, since it alleviates the effects of the curse of dimensionality. Also, it is important to note, the resulting SVM expansion turns out be sparse. Indeed, when $|y_i - M_{SVM}(\mathbf{x}_i)| \leq \varepsilon$, the corresponding Lagrange multipliers α_i, α_i^* turn out to be equal to 0. The training configurations $\mathbf{x}_i$ for which either α_i or α_i^* do not vanish are called *support vectors* [17].

Several kernel can be adopted within the dual space formulation of the SVM regression in (7.49). Among them, the most common kernel functions are listed below:

- linear: $\mathcal{K}(\mathbf{x}_i, \mathbf{x}) = \mathbf{x}_i^T \mathbf{x}$;
- polynomial of order q: $\mathcal{K}(\mathbf{x}_i, \mathbf{x}) = (1 + \mathbf{x}_i^T \mathbf{x})^q$; and
- Gaussian radial basis function (RBF): $\mathcal{K}(\mathbf{x}_i, \mathbf{x}) = \exp\left(-||\mathbf{x}_i - \mathbf{x}||^2 / 2\sigma^2\right)$.

The SVM regression algorithm is already available in MATLAB. A generic SVM regression can be suitably trained via the command fitrsvm and the resulting surrogate model can be evaluated for an arbitrary value of the input parameter $\mathbf{x}$ via the command predict [18].

7.9 LS-SVM regression

The LS-SVM regression provides an alternative solution to the standard SVM regression, which allows recasting the convex nonlinear optimization problem used by the SVM regression in (7.41), in terms of a more standard least squares formulation [19].

Given a set of training samples $\mathcal{D} = \{(\mathbf{x}_i, y_i)\}_{i=1}^{L}$ generated by a computational model (i.e., $y_i = M(\mathbf{x}_i)$), the primal space formulation of the LS-SVM regression writes:

$$\widetilde{M}_{\text{LS-SVM}}(\mathbf{x}) = \langle \mathbf{w}, \mathbf{\Phi}(\mathbf{x}) \rangle + b. \tag{7.50}$$

Similar to the Ridge regression, the minimization problem for the LS-SVM regression uses a squared loss function and a Tihkonov regularization:

$$\min_{\mathbf{w},b} \underbrace{\frac{1}{2}||\mathbf{w}||_2^2}_{\text{Regularizer}} + \underbrace{\frac{\gamma}{L}\sum_{i=1}^{L}\left(y_i - \widetilde{M}_{\text{LS-SVM}}(\mathbf{x}_i; \mathbf{w}, b)\right)^2}_{\text{Empirical Risk Function}}, \tag{7.51}$$

Which can be rewritten as follows:

$$\min_{\mathbf{w},b,\mathbf{e}} \frac{1}{2}||\mathbf{w}||_2^2 + \gamma\frac{1}{2}\sum_{i=1}^{\mathcal{L}} e_i^2 \tag{7.52}$$

subject to $y_i = \langle \mathbf{w}, \mathbf{\Phi}(\mathbf{x}_i) \rangle + b + e_i$, for $i = 1, \ldots, L$, where the terms $e_i = y_i - \widetilde{M}_{\mathcal{L}S\text{-}SVM}(\mathbf{x}_i; \mathbf{w}, b)$ with $i = 1, \ldots, L$ are the error variables and γ is an empirical parameter which provides a trade-off between the accuracy of the model and its flatness, playing the same role of the parameter C in the SVM primal optimization of (7.45).

The underlying idea is to minimize the squares of the regression errors e_i, which are defined as the deviation of the model prediction from the corresponding training samples, but at the same time by keeping the model as flat as possible, thus reducing the overfitting. For the sake of illustration, Figure 7.9 compares the error function used by the SVM (panel (a)) and the LS-SVM regression (panel (b)).

The optimization problem of the LS-SVM regression in (7.52) can be solved by minimizing the following Lagrangian:

$$\mathcal{L}(\mathbf{w}, b, \mathbf{e}; \boldsymbol{\alpha}) = \frac{1}{2}||\mathbf{w}||^2 + \gamma\frac{1}{2}\sum_{i=1}^{L} e_i^2 - \sum_{i=1}^{L} \alpha_i\{\langle \mathbf{w}, \mathbf{\Phi}(\mathbf{x}_i) \rangle + b + e_i - y_i\} \tag{7.53}$$

where $\alpha_k \geq 0$ are the Lagrange multipliers.

By computing the partial derivatives of the above Lagrangian leads to:

$$\frac{\partial \mathcal{L}}{\partial \mathbf{w}} = 0 \rightarrow \mathbf{w} = \sum_{i=1}^{L} \alpha_i \mathbf{\Phi}(\mathbf{x}_i) \tag{7.54a}$$

$$\frac{\partial \mathcal{L}}{\partial b} = 0 \rightarrow \sum_{i=1}^{L} \alpha_i = 0 \tag{7.54b}$$

$$\frac{\partial \mathcal{L}}{\partial e_i} = 0 \rightarrow \alpha_i = \gamma e_i \tag{7.54c}$$

$$\frac{\partial \mathcal{L}}{\partial \alpha_k} = 0 \rightarrow \langle \mathbf{w}, \mathbf{\Phi}(\mathbf{x}_i) \rangle + b + e_i - y_i = 0. \tag{7.54d}$$

By substituting (7.54a), (7.54b) and (7.54c) into (7.54d), we can write the following system of equations in the dual space:

$$\begin{cases} \sum_{i=1}^{L} \alpha_i = 0 \\ \sum_{i=1}^{L} \alpha_i \mathcal{K}(\mathbf{x}_i, \mathbf{x}_1) + b + \frac{\alpha_1}{\gamma} - y_1 = 0 \\ \vdots \\ \sum_{i=1}^{L} \alpha_i \mathcal{K}(\mathbf{x}_i, \mathbf{x}_L) + b + \frac{\alpha_L}{\gamma} - y_L = 0 \end{cases} \tag{7.55}$$

where again $\mathcal{K}(\mathbf{x}_i, \mathbf{x}_j) = \langle \mathbf{\Phi}(\mathbf{x}_i), \mathbf{\Phi}(\mathbf{x}_j) \rangle$ is the kernel function previously defined for the case of the SVM regression.

The linear system of equations in (7.55) can be recasted via the following matrix formulation:

$$\begin{bmatrix} 0 & \mathbf{1}^T \\ \mathbf{1} & \Omega + \mathbf{I}/\gamma \end{bmatrix} \begin{bmatrix} b \\ \boldsymbol{\alpha} \end{bmatrix} = \begin{bmatrix} 0 \\ \mathbf{y} \end{bmatrix} \tag{7.56}$$

where $\boldsymbol{\alpha} = [\alpha_1, \ldots, \alpha_L]^T$, $\mathbf{y} = [y_1, \ldots, y_L]^T$, $\mathbf{1}^T = [1, \ldots, 1] \in \mathbb{R}^{1 \times L}$, $I \in \mathbb{R}^{L \times L}$ is the identity matrix and $\Omega \in \mathbb{R}^{L \times L}$ is the kernel matrix for which the element $\Omega_{ij} = \mathcal{K}(\mathbf{x}_i, \mathbf{x}_j)$ for any $i, j = 1, \ldots, L$. Similar to the SVM regression, the most common kernel functions adopted in the dual space formulation of the LS-SVM regression are the linear, polynomial and RBF kernel functions.

The solution of the square linear system of equations in (7.56) for $\boldsymbol{\alpha}$ and b leads to the following nonlinear regression in the dual space:

$$\widetilde{M}_{\text{LS-SVM}}(\mathbf{x}) = \sum_{i=1}^{L} \alpha_i \mathcal{K}(\mathbf{x}_i, \mathbf{x}) + b. \tag{7.57}$$

Similar to the SVM regression, the above LS-SVM formulation in the dual space is a non-parametric model for which the number of regression unknowns α_i is independent from both the dimensionality of the dual space D and from the number of parameters d, but it is equal to the number of training samples L.

The LS-SVM regression is already implemented within LS-SVMLab Toolbox version 1.8 [20], which is fully compatible with the MATLAB environment.

7.10 Gaussian process regression

All the parametric and non-parametric regression techniques presented so far provide as a result a *deterministic* metamodel, that is, a deterministic function able to predict the output variable of interest for any configuration of the input parameters. The main limitation of such kind of models is that they do not provide any information about the reliability and accuracy of their predictions. Indeed, the error for the model predictions is known only for the set of configurations of the input parameter for which an equivalent simulation with the computational model has been run (i.e., the training and test samples), but it is completely unknown when the model is evaluated for a generic point in the parameter space.

GPR, also known as Kriging model, represents a possible solution to the above challenging problem [21]. Starting from a limited set of training samples, the GPR allows building a *probabilistic* metamodel, which not only accurately estimates the model output, but it also provides useful statistical information on the reliability of its prediction for any configuration of the input parameters, such as the confidence interval (CI).

Under the assumption that a generic nonlinear computational model $y = M(\mathbf{x})$ follows a Gaussian Process (GP) prior, a GPR reads [21]:

$$y = M(\mathbf{x}) \sim \underbrace{GP(m(\mathbf{x}), k(\mathbf{x}, \mathbf{x}'))}_{\widetilde{M}_{\mathrm{GPR}}(\mathbf{x})} \tag{7.58}$$

where $\widetilde{M}_{\mathrm{GPR}}(\mathbf{x}) \sim GP(m(\mathbf{x}), k(\mathbf{x}, \mathbf{x}'))$ is a GP defined by the trend function $m(\mathbf{x})$ and covariance function $k(\mathbf{x}, \mathbf{x}')$.

A GP extends the concept of a Gaussian distribution from numbers to functions. The trend $m(\mathbf{x})$ provides the average function among the ones drawn from the GP prior, whilst the covariance provides the correlation between the values of such functions at different points (i.e., $\mathbf{x}$ and $\mathbf{x}'$) in the parameters space.

The above GP is called prior distribution, since it fixes the properties of the unknown nonlinear model, before looking at the training data [21]. In fact, different from the determinist regressions (e.g., the SVM regression, least squares regression, Ridge regression, etc.), in which the candidate functions of the model are restricted to a specific class of functions (e.g., polynomial, linear, etc.), the GPR model considers as candidate functions for our model $\widetilde{M}_{\mathrm{GPR}}$ all the possible nonlinear functions drawn from the GP prior by letting data "speak" and assigns a probability to each of them [22]. This means that we are implicitly assuming, as candidate for our model, all the possible functions obtained by drawing a sample from an infinite dimensional multivariate Gaussian distribution.

Now, we can try to incorporate into the prior distribution the information on training samples $\mathcal{D} = \{(\mathbf{x}_i, y_i)\}_{i=1}^{L}$. For the sake of compactness, the input training vectors $\{\mathbf{x}_i\}_{i=1}^{L}$ can be recasted into the design matrix $\mathbf{X} = [\mathbf{x}_1, \ldots, \mathbf{x}_L]^T$ of dimension $L \times d$ and the corresponding training output $\{y_i\}_{i=1}^{L}$ can be aggregated into the column vector $\mathbf{y} = [y_1, \ldots, y_L]^T$, so we can write $\mathcal{D} = (\mathbf{X}, \mathbf{y})$. According to the theory of the GP, the joint distribution of the training outputs $\mathbf{y}$ combined with the value of a generic test output y_* for $\mathbf{x}_* \in \mathcal{X}$ with $\mathbf{x}_* \neq \mathbf{x}_i$, follows a $L+1$ Gaussian distribution:

$$\begin{bmatrix} \mathbf{y} \\ y_* \end{bmatrix} \sim P(\mathbf{y}, y_* | \mathbf{X}, \mathbf{x}_*) = \mathcal{N}\left(\begin{bmatrix} \mathbf{0} \\ 0 \end{bmatrix}, \begin{bmatrix} \mathbf{K} + \sigma_n^2 \mathbf{I} & \mathbf{k}_*^T \\ \mathbf{k}_* & k_{**} \end{bmatrix} \right), \tag{7.59}$$

where $\mathbf{I} \in \mathbb{R}^{L \times L}$ is the identity matrix, $\mathbf{K} \in \mathbb{R}^{L \times L}$ is the correlation matrix evaluated on all the training configurations with $K_{ij} = k(\mathbf{x}_i, \mathbf{x}_j)$, $\mathbf{k}_* = [k(\mathbf{x}_*, \mathbf{x}_1), \ldots, k(\mathbf{x}_*, \mathbf{x}_L)] \in \mathbb{R}^{1 \times L}$ is a vector in which the covariance is evaluated on the training and test set, $k_{**} = k(\mathbf{x}_*, \mathbf{x}_*)$ is the covariance evaluated on the test sample $\mathbf{x}_*$ and σ_n^2 is an hyperparameters representing the variance of a possible additive Gaussian noise corrupting the training set [21].

It is important to remark that for the case of a GPR model trained with noisy observations (i.e., $\sigma_n \neq 0$), the formulation of the GPR turns out to be equivalent to the ones of the LS-SVM regression [2]. Also, the formulation of the prior joint distribution of (7.59) can be generalized to the case of n_* test points. In this case, the scalar y_* becomes a $n_* \times 1$ vector, the vector $\mathbf{x}_*$ becomes a $n_* \times d$ matrix $\mathbf{X}_*$, the vectors k_{**} becomes a $n_* \times n_*$ matrix $\mathbf{K}_{**}$ and similarly for the other entries $\mathbf{k}_*$ and $\mathbf{k}_*^T$.

The posterior distribution $P(y_*)$ is computed over all the possible functions drawn from the prior by restricting the joint distribution of (7.59) to contain only those functions which agree with the observed data points. Ideally, we should generate an infinite number of functions from the prior and throw out the ones that disagree with the observations (i.e., training data). In probabilistic terms this operation is extremely simple, and corresponds to conditioning the joint Gaussian prior distribution on the observations. To this aim, we need to condition the probability $P(\mathbf{y}, y_* | \mathbf{X}, \mathbf{x}_*)$ in (7.59) in order to get the following posterior probability:

$$y_* \sim p(y_* | \mathbf{y}, \mathbf{X}, \mathbf{x}_*) = \mathcal{N}(\mu_*, \sigma_*^2), \tag{7.60}$$

where the posterior mean $\mu_{\mathbf{x}_*}$ and variance $\sigma_{\mathbf{x}_*}^2$ writes:

$$\mu_{\mathbf{x}_*} = m(\mathbf{x}_*) + \mathbf{k}_* (\mathbf{K} + \sigma_n^2 \mathbf{I})^{-1} \mathbf{y}, \tag{7.61a}$$

$$\sigma_{\mathbf{x}_*}^2 = k_{**} - \mathbf{k}_* (\mathbf{K} + \sigma_n^2 \mathbf{I})^{-1} \mathbf{k}_*^T. \tag{7.61b}$$

With the posterior distribution we are computing the probability $P(y_*)$ by considering, among all the functions drawn from the prior, only the ones that pass

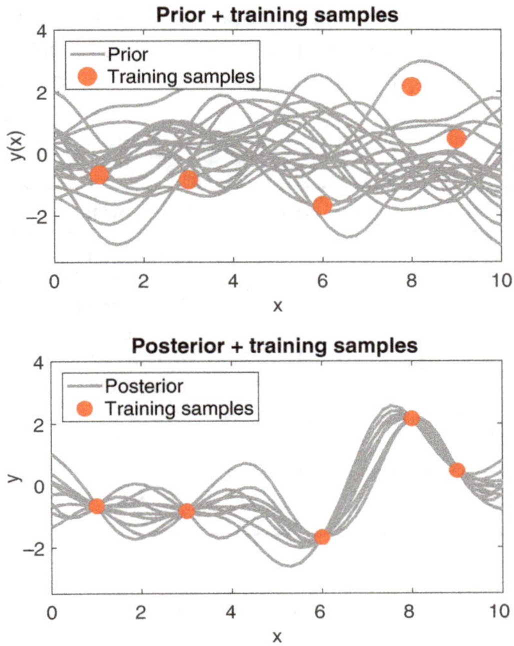

Figure 7.10 Illustration of 10 functions drawn from the prior (top panel) and of 10 functions drawn from the posterior (bottom panel)

thought the training samples and by assigning a probability to each of them. Figure 7.10 provides a graphical interpretation of the above process.

The GPR requires to specify both the covariance k and the trend functions m used by the prior $GP(m(x), k(\mathbf{x}, \mathbf{x}'))$. Two different formulations of the GP trend $m(\mathbf{x})$ are usually adopted:

- *Fixed mean function*, that is, $f(\mathbf{x}) \sim GP(\widetilde{M}_{det}(\mathbf{x}), k(\mathbf{x}, \mathbf{x}'))$, where $\widetilde{M}_{\rm det}(\mathbf{x})$ can be any generic surrogates constructed via a regression techniques (such that the error on the training sample is usually $\neq 0$) [21,23–25]. The resulting model is obtained via a two-step strategy: (*i*) use the training sample to build the deterministic model $\widetilde{M}_{\rm det}(\mathbf{x})$; and (*ii*) use the error $e_i = y_i - \widetilde{M}_{det}(\mathbf{x}_i)$ for $i = 1, \ldots, L$ computed on the training set to build the final model based on the GPR.
- *Fixed basis function*, that is, $f(\mathbf{x}) \sim GP(\boldsymbol{\beta}^T\boldsymbol{\phi}(\mathbf{x}), k(\mathbf{x}, \mathbf{x}'))$, where the GP trend $\boldsymbol{\beta}^T\boldsymbol{\phi}(\mathbf{x})$ is the linear combination of a set of basis functions $\boldsymbol{\phi}(\mathbf{x}) = [\phi_1(\mathbf{x}), \ldots, \phi_P(\mathbf{x})]^T$ with the corresponding weights $\boldsymbol{\beta} = [\beta_1, \ldots, \beta_P]^T$ to be estimated during the training of the GPR. As an example the MATLAB tool for the GPR works directly with constant, linear and polynomial bases functions [26].

Moreover, similar to the kernel functions adopted by the SVM and LS-SVM regressions, several covariance functions can be used by the GPR[§]. Table 7.11 provides a coincise overview of the most common ones. Additional details can be found in [21].

Name	Covariance functions	Description
Squared exponential	$\sigma_f^2 \exp\left(-\frac{r^2}{2\sigma_l^2}\right)$	Generates infinitely differentiable (i.e., smooth) functions
Exponential	$\sigma_f^2 \exp\left(-\frac{r}{\sigma_l}\right)$	Generates continuous, but non-differentiable functions
Matern 3/2	$\sigma_f^2\left(1+\frac{\sqrt{3}r}{\sigma_l}\right)\exp\left(\frac{-\sqrt{3}r}{\sigma_l}\right)$	Generates one-time differentiable functions
Matern 5/2	$\sigma_f^2\left(1+\frac{\sqrt{5}r}{\sigma_l}+\frac{5r^2}{3\sigma_l^2}\right)\exp(-\sqrt{5}r)$	Generates two-time differentiable functions
Periodic	$\sigma_f^2 \exp\left(-\frac{2}{\sigma_l^2}\sin\left(\frac{\pi r}{p}\right)\right)$	Generates periodic functions with period p
$(r = \|\|\mathbf{x}-\mathbf{x}'\|\|)$		

The probabilistic interpretation in (7.60) provided by the GPR allows estimating the CI of its predictions. Specifically, under the assumption that the actual value of the function under modeling is approximated by the posterior mean predicted by the GPR (i.e., $y(\mathbf{x}_*) \approx \mu_*$), the CI provides useful statistical information on the reliability of the model prediction. At any test point $\mathbf{x}_*$, the CI proposes a range for which there is a $100 \times (1-\alpha)\%$ chance of containing all the possible realizations of the output $y(\mathbf{x}_*)$ and writes:

$$P\left(\left(\mu_{\mathbf{x}_*} - z_{1-\frac{\alpha}{2}}\sigma_{\mathbf{x}_*}\right) \leq y(\mathbf{x}_*) \leq \left(\mu_{\mathbf{x}_*} + z_{1-\frac{\alpha}{2}}\sigma_{\mathbf{x}_*}\right)\right) = 1-\alpha \tag{7.62}$$

or

$$y(\mathbf{x}_*) \in \left[\mu_{\mathbf{x}_*} - z_{1-\frac{\alpha}{2}}\sigma_{\mathbf{x}_*}, \mu_{\mathbf{x}_*} + z_{1-\frac{\alpha}{2}}\sigma_{\mathbf{x}_*}\right] \tag{7.63}$$

with a probability of $100(1-\alpha)\%$, where z denotes the inverse of the Gaussian cumulative distribution function evaluated at $1-\frac{\alpha}{2}$.

7.11 Application examples

This section presents two applications of the ML regression techniques introduced in the previous section (i.e., the SVM regression, the LS-SVM regression and the GPR) to the construction of metamodels for the UQ of complex systems.

[§] Either isotropic or anisotropic covariance functions can be adopted [21].

7.11.1 High-speed communication link

As a first application test case, the SVM regression is applied to the UQ of the realistic high-speed communication link as a function of 4 stochastic parameters [18]. The network, depicted in Figure 7.11, consists of the interconnection of three PCB traces and lumped elements. The traces are modeled as three uncoupled sections of lossy transmission lines [27]. The value of some lumped components is parametric, namely $C_1(x_1) = (5 + 2.5x_1)\,\mathrm{pF}$, $C_2(x_2) = (5 + 2.5x_2)\,\mathrm{pF}$, $C_3(x_3) = (5 + 2.5x_3)\,\mathrm{pF}$ and $R(x_4) = (50 + 25x_4)\,\Omega$. The parameters $\mathbf{x} = [x_1, x_2, x_3, x_4]^{\mathrm{T}}$ are uniformly distributed in $\mathcal{X} = [-1, 1]^4$ with a variation of $\pm 50\%$ around their mean value.

The transfer function between the output voltage V_{out} and the excitation $E(f)$ is considered, that is,

$$y(f;\mathbf{x}) = M(f;\mathbf{x}) = \left|\frac{V_{\mathrm{out}}(f;\mathbf{x})}{E(f)}\right|, \tag{7.64}$$

Figure 7.12 shows the large spread of the transfer function (7.64) resulting from the high variability of the four stochastic parameters, and obtained by a MC simulation in the frequency range $f \in [10\,\mathrm{MHz}, 1\,\mathrm{GHz}]$.

A small subset of the MC realizations has been used to train the SVM surrogate model denoted as $\widetilde{M}_{\mathrm{SVM}}(f;\mathbf{x})$. In order to assess the convergence rate of the metamodel based on the SVM regression, the SVM model $\widetilde{M}_{\mathrm{SVM}}(f;\mathbf{x})$ is trained with an increasing number of samples, that is, $L = 30, 50, 100, 150$. The SVM metamodel uses a polynomial kernel of degree $q = 5$, with $C = 1$ and $\varepsilon = \mathrm{eps}$ to maximize the model accuracy. Moreover, in order to evaluate the performance of the SVM regression, an alternative PC surrogate model $\widetilde{M}_{\mathrm{PC}}(f;\mathbf{x})$ maximum degree $p = 5$ based on Legendre polynomials has been constructed via a OLS regression by using the same training samples [28]. Since the accuracy of both the

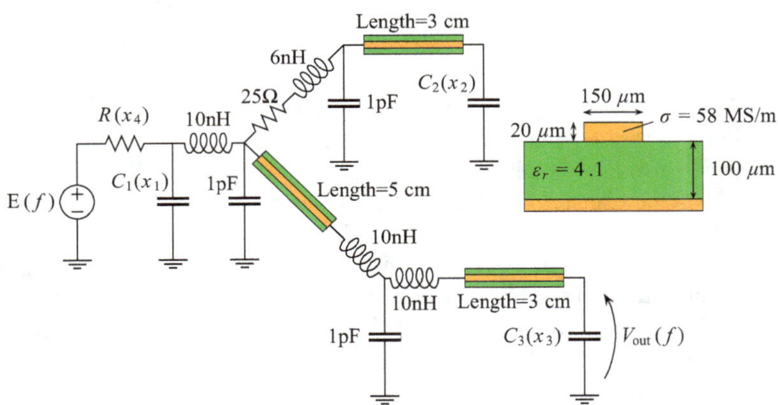

Figure 7.11 Schematic of the high-speed link with lossy transmission lines

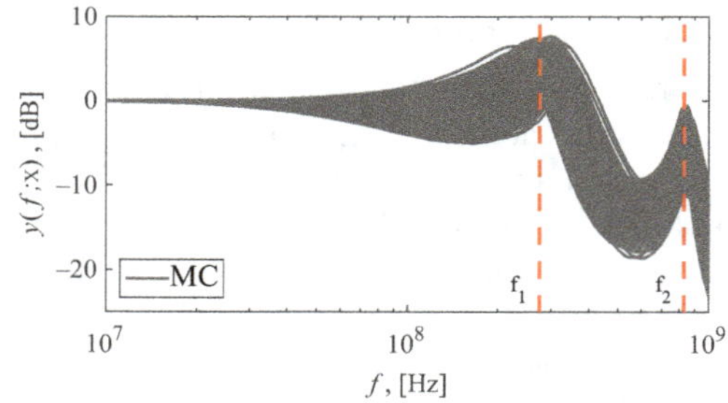

Figure 7.12 Variability of the frequency-domain response of the circuit of Figure 7.11

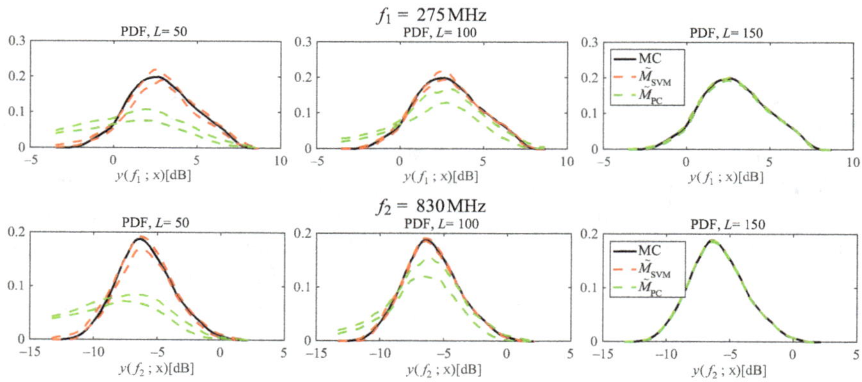

Figure 7.13 PDF of the transfer function (1.67) evaluated at $f_1 = 275$ MHz and $f_2 = 830$ MHz. The result obtained from 10,000 MC simulations (solid black curves) is compared with the predictions from the SVM regression (dashed red curves) and the PC expansion (dashed green curves) for an increasing number of training pairs

PC and SVM regression depends on the training samples 10 different realizations of the training set are examined for a given size L.

Next, the surrogate models are used to compute the probability density functions (PDF) of the transfer function (7.64) at the frequencies $f_1 = 275\,\text{MHz}$ and $f_2 = 830\,\text{MHz}$, which are indicated by the dashed vertical lines in Figure 7.12. The results are collected in Figure 7.13 and highlight again the better accuracy of the SVM regression (dashed red lines) with respect to the PC expansion (dashed green line) in predicting the reference MC result (solid black line) from a very small training set size (e.g., $L = 30$ or 50). Indeed, the PC model starts to converge to the actual PDF only when the number of training samples is greater than the number of unknowns in the regression problem, that is, for $L > 126$ in this case. For a large number of training samples, both models converge to the actual PDF.

Table 7.1 Comparison of the computational cost and accuracy of the PC and SVM macromodels at $f_2 = 830\,MHz$

Method	# of Training samples L	Model construction (s)	Model evaluation (s)	Max rel. error (%)
	30	0.08	1.60	100
SVM	50	0.01	1.63	99
(Order 5)	100	0.09	1.64	33
	150	4.30	1.66	13
	30	0.07	0.04	>100
PC	50	0.11	0.04	>100
(Order 5)	100	0.22	0.03	>100
	150	0.33	0.04	17

The convergence rate and the computational cost of the two metamodels are investigated in Table 7.1. The table considers the computational time required to build and evaluate the PC and the SVM models at a given point $\mathbf{x}$ in the parameter space $\mathcal{K}$, as well as the maximum error of the parametric models over 10 different realizations of the training set at the frequency $f_2 = 830\,\text{MHz}$. The figures highlight the better efficiency of the PC approach, but also confirm the fast convergence of the SVM regression with respect to the training set size.

Finally, to investigate the robustness of the presented techniques to noisy data, the training samples are corrupted by a Gaussian noise as follows:

$$\widetilde{y}_i(f;\mathbf{x}_i) = y_i(f;\mathbf{x}_i) \times (1 + \xi_n), \tag{7.65}$$

where $\xi_n \sim \mathcal{N}(0, \sigma_n^2)$ is a Gaussian variable with $\sigma_n = 0.025$.

As in the previous case, the surrogate models $\widetilde{M}_{PC}(f;\mathbf{x})$ and $\widetilde{M}_{SVM}(f;\mathbf{x})$ are estimated with an increasing number of training samples, that is, $\mathcal{L} = 50, 100, 150, 200$, by considering 10 different realizations for each case. In the presence of noise, it is useful to set a non-negligible value of ε in the SVM regression. Indeed, according to the ε-insensitive loss function (7.47), a deviation of the training samples within the bound $[-\varepsilon, +\varepsilon]$ does not affect the SVM model. The empirical relation $\varepsilon = 3\sigma_n\sqrt{log(L)/L}$ is suggested in [16] to minimize the perturbation of the model accuracy due to noise. In such a way, the width of the ε-insensitive zone decreases by increasing the number of training samples.**

Figure 7.14 compares the PDFs previously obtained with MC (without noise) against the ones predicted by the PC and SVM surrogates constructed from noisy training samples $\{(\mathbf{x}_i, \widetilde{y}_i(f_k))\}_{i=1,k=1}^{L,2}$ with $f_1 = 275\,MHz$ and $f_2 = 830\,MHz$. These results confirm the assessment of the noiseless case. Indeed, the SVM regression provides again a better convergence than the PC expansion when a small set of training samples is available (e.g., $L = 50$ and 100). However, as expected,

**If the noise amplitude is unknown, the optimal width of the ε-insensitive zone can be estimated via an optimization scheme, for example, via CV.

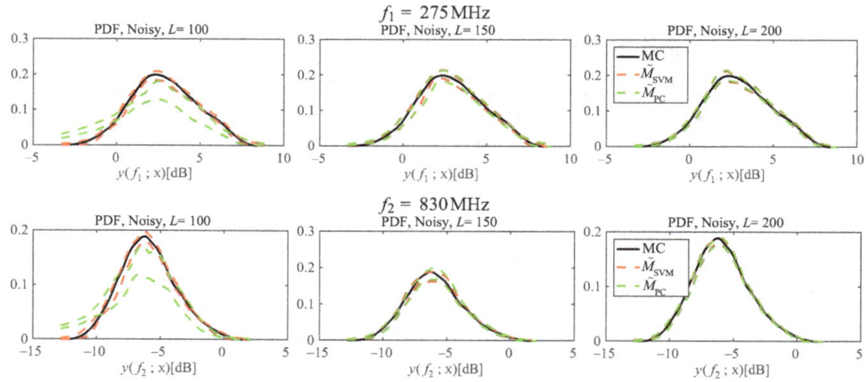

Figure 7.14 PDF of the transfer function (7.64). The result obtained from noise-free MC samples (solid black curves) is compared with the predictions from the SVM regression (dashed red curves) and the PC expansion (dashed green curves) constructed from noisy training samples of increasing size

when the number of samples is increased, the two models converge to the same solution.

7.11.2 Switching converter

As a second example, the ML techniques presented in this chapater have been applied to build a set of surrogate models for the UQ of the output voltage spectral envelope of the switching converter in Figure 7.15, as a function of 17 stochastic parameters [23]. The converter is a 12V:5V switching buck converter with its feedback network (see [29] for additional details) operating at a switching frequency of 100 kHz through a sawtooth signal defined between 0 V and 5 V.

The values of all the components specified in the schematic of Figure 7.15 have been considered as Gaussian stochastic variables centered at their nominal value and with a standard deviation of 10% around their mean value, leading to 17 uncorrelated Gaussian parameters (i.e., $\mathbf{x} \in \mathbb{R}^{17}$). The full-computational model used to evaluate the global effect of the uncertainty parameters on the stochastic behavior of the spectral envelope of the voltage output of the converter is based on a parametric transient simulation in LTspice. Specifically, the spectral envelope of the output voltage v_{out}, namely $V_{\text{out},E}(f;\mathbf{x}_*)$, for a generic configuration of the circuit parameters $\mathbf{x}_*$ is calculated through the full-computational model via the following procedure: (i) use the input parameter configuration to run the corresponding transient simulation in LTspice; (ii) compute the spectrum $\widehat{V}_{\text{out}}(f;\mathbf{x}_*)$ by applying the Fast Fourier transform (FFT) on the steady-state portion

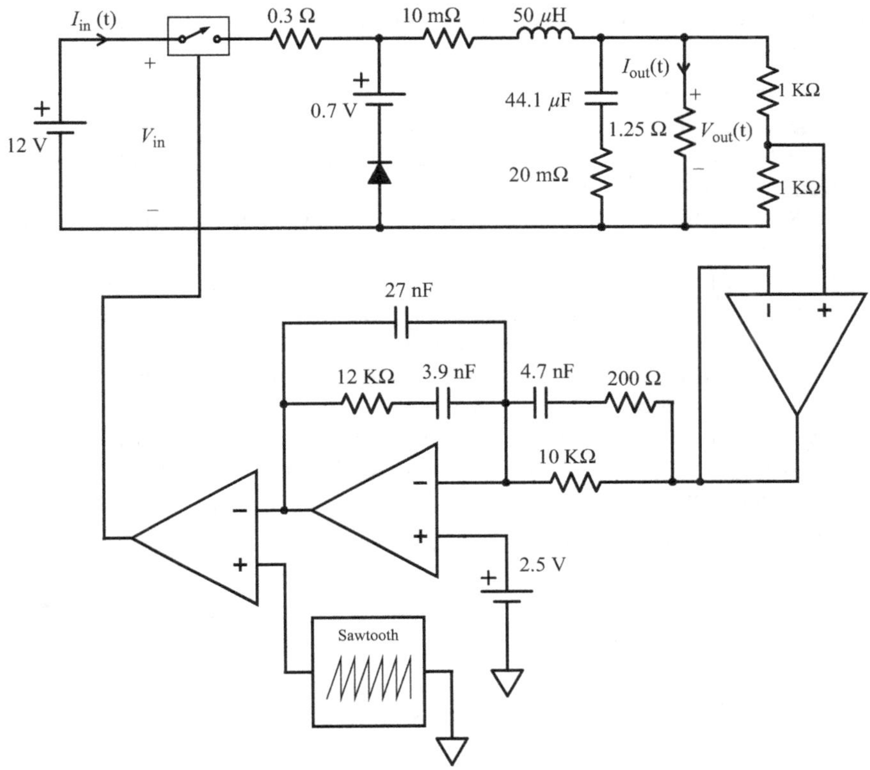

Figure 7.15 Schematic of the considered buck converter [29]. For each component, the nominal values are indicated

of the voltage waveform $v_{\text{out}}(t; \mathbf{x}_*)$; and (iii) compute the magnitude of the peak spectral envelope $V_{\text{out},E}(f; \mathbf{x}_*)$ via the MATLAB function envelop and convert the resulting spectrum in dB.

For any configuration of the input parameters, the transient simulation has been run in the time window $[0, 3]$ ms with a time step of 10 ns. In order to ensure that all the waveforms have reached the steady-state, the FFT has been applied only to the last three switching periods of the voltage waveform $v_{\text{out}}(t; \mathbf{x}_*)$. The resulting spectral envelope $V_{\text{out},E}(f_k; \mathbf{x}_*)$ with $k = 1, \ldots, N_f$ covers a frequency bandwidth from DC to 30MHz via a set of $N_f = 91$ linearly spaced frequency samples.

For each of the frequency component f_k, the above simulation scheme (i.e., the full-computational model) has been used to generate a set of L training samples $\{(\mathbf{x}_i, y_i(f_k))\}_{i=1}^{L}$, where the set of input parameter configurations $\{\mathbf{x}_1, \ldots, \mathbf{x}_L\}$ have been drawn based on the LHS scheme [8] and $y_i(f_k) = V_{\text{out},E}(f_k; \mathbf{x}_i))$. The training samples have been used to train seven different surrogate models, in which the GPR surrogates are built with the automatic relevance determination (ARD) covariance [21,26,30]:

- LS-SVM (linear): LS-SVM surrogate with linear kernel.
- LS-SVM (RBF): LS-SVM surrogate with RBF kernel.
- GPR (constant trend): noiseless (i.e., $\sigma_n = 0$) GPR-based surrogate with constant trend and ADR Matern 5/2 covariance function.
- GPR (linear trend): noiseless GPR-based surrogate with linear trend and ADR Matern 5/2 covariance function.
- GPR (poly trend order 2): noiseless GPR-based surrogate with an second-order polynomial trend and ADR Matern 5/2 covariance function.
- LS-SVM (linear)+GPR: GPR-based surrogate with a fixed mean function consisting of a LS-SVM surrogate with linear kernel and ADR Matern 5/2 covariance function.
- LS-SVM (RBF)+GPR: GPR-based surrogate with a fixed mean function consisting of a LS-SVM surrogate with RBF kernel and ADR Matern 5/2 covariance function.

The ARD Matern 5/2 covariance writes:

$$k(\mathbf{x}, \mathbf{x}'|\theta) = \sigma_f^2\left(1 + \sqrt{5}r + \frac{5}{3}r^2\right)\exp(-\sqrt{5}r), \tag{7.66}$$

with,

$$r = \sqrt{\sum_{m=1}^{d} \frac{(x_m - x'_m)^2}{\sigma_m^2}}. \tag{7.67}$$

in which σ_f and σ_m for $m = 1, \ldots, d$ are the so-called hyperparameters collected in the vector θ estimated through optimization from the available information on the training samples.

The ARD covariance function has the main advantage of increasing the model flexibility, since different from the standard Matern 5/2 kernel, it adds to the model a new set of hyperparameters σ_m. On the other hand, such kernel increases the complexity of the GPR model, thus compromising the model accuracy, when a small set of training samples is available.

The surrogate models should be able to predict the effect of the uncertain parameters on the spectral envelope of the converter output voltage. To this aim, each of the above regression techniques is used to build a different metamodel $\widetilde{M}_k$ for each considered frequency f_k, such as:

$$V_{\text{out},E}(f_k; \mathbf{x}) \approx \widetilde{M}(f_k; \mathbf{x}) = \widetilde{M}_k(\mathbf{x}), \tag{7.68}$$

For any $\mathbf{x}$ and $k = 1, \ldots, N_f$.

The accuracy of each model is illustrated in Figure 7.16. The plots provide a comparison among the accuracy provided by each of the considered surrogate

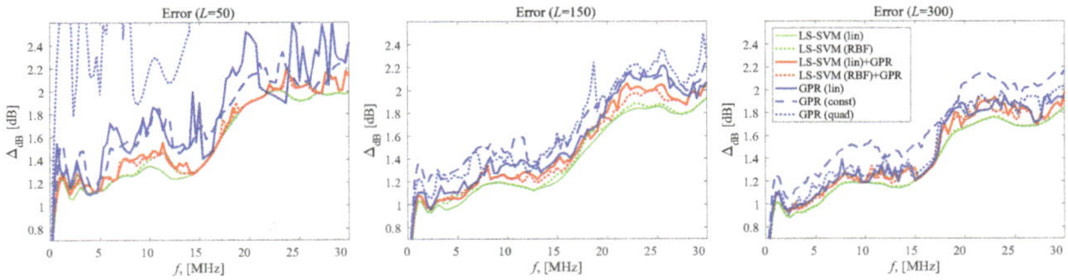

Figure 7.16 Average absolute error $\Delta_{dB}(f)$ calculated by comparing the predictions of the surrogate models based on the LS-SVM (solid and dashed green curves), the LS-SVM+GPR (solid and dashed red curves) and the plain GPR (blue solid, dashed and dotted curves) with the corresponding ones obtained via a MC simulation with 10,000 samples for an increasing number of training samples $L = 50, 150$ *and* 300

models in terms of the average absolute error spectrum $\Delta_{\text{dB}}(f_k)$ defined for $k = 1, \ldots, N_f$, as follows

$$\Delta_{\text{dB}}(f_k) = \sum_{i=1}^{N_{\text{MC}}} \frac{\left| V_{\text{out},E}(f_k; \mathbf{x}_i) - \widetilde{V}_{\text{out},E}(f_k; \mathbf{x}_i) \right|}{N_{\text{MC}}}, \tag{7.69}$$

where $V_{\text{out},E}(f_k; \mathbf{x}_i)$ corresponds to the envelope amplitude in dB obtained via the LTspice simulations for each configuration $\mathbf{x}_i$ of the uncertain parameters considered in a MC simulation with $N_{\text{MC}} = 10{,}000$ samples, while $\widetilde{V}_{\text{out},E}(f_k; \mathbf{x}_i)$ is the corresponding value estimate by a given surrogate. The above error is computed by using each of the considered surrogate built with an increasing number of training samples $L = 50, 150$ and 300.

From the curves of Figure 7.16, the most accurate models are the ones based on the deterministic LS-SVM regression with linear and RBF kernel (solid and dashed green lines), however, the proposed modeling scheme based on the LS-SVM with linear and RBF kernel+GPR (see the solid and dashed red curves) provides the most accurate probabilistic surrogate models for all the considered set of training samples (i.e., $L = 50$, 150 and 300). Indeed, the curves related to the LS-SVM+GPR are usually below the ones related to standard GPR (solid, dashed and dotted blue lines). The results also show the improved convergence of the LS-SVM-based deterministic and probability models when a small set of training samples are available, that is, $L = 50$, [18,31,32].

As a further validation, Figure 7.17 provides a graphical comparison between the scattering plots obtained by comparing the results of a MC simulation with 10,000 samples for all the $N_f = 91$ frequency points, with the corresponding ones provided by the deterministic surrogate models based on LS-SVM regression with linear kernel (green dots) and the mean values predicted by the proposed LS-SVM (RBF)+GPR statistical model (red dots). The plots highlight the capability of the

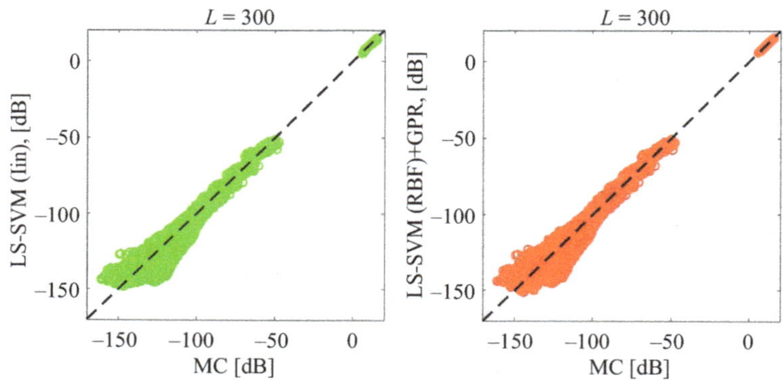

Figure 7.17 Scatter plot (10,000 samples) comparing, for all the considered frequency points, the voltage spectral envelope $V_{\text{out},E}$ predicted by the deterministic model based on the LS-SVM with linear kernel (green dots; left panel) and by the statistical surrogate model based on LS-SVM (RBF)+GPR (red dots; right panel) against the MC samples generated by the full-computational model

two models to accurately predict the actual value of the MC simulation, since the samples are very close to the dashed line, which represents the perfect agreement between the model and the reference samples.

In addition, Figure 7.18 compares the PDFs of the spectral envelope magnitude at $f_0 = 100\,\text{kHz}$ provided by the deterministic LS-SVM regression with linear kernel (solid green line) and the mean values of the LS-SVM (RBF)+GPR (solid red line) in (7.61a) with the PDF resulting from 10,000 MC samples (gray bins). The results highlight once again the excellent capability of the two models to capture the main feature of the reference PDF resulting from the MC simulations.

As a final comparison between the two models, Figure 7.19 shows two realizations of the spectral envelope randomly selected among the results of the MC simulation (black curve) along with the corresponding predictions provided by both the deterministic surrogate model based on the LS-SVM with linear kernel (dashed green curve) and results of the proposed probabilistic model (i.e., the mean values and the 99% CI) based on the LS-SVM (RBF)+GPR model (red vertical bars) for $L = 300$. The results clearly highlight the advantages of the proposed modeling techniques with respect to the classical deterministic regression. In fact, different from the deterministic model based on the LS-SVM regression, for any given configuration of the input parameters, the proposed probabilistic one built via the LS-SVM+GPR does not only provide an approximation of the envelope spectra (red dots), but it also provides the users with a statistical information on the model error and reliability by means of the 99% CI (red error bars). The accuracy of such CI can be easily appreciated by noticing that the actual spectral envelope provided by the full-computational model (black curve) lays between the CI estimated by the proposed models. It is important to remark that the CI shown in Figure 7.19 cannot

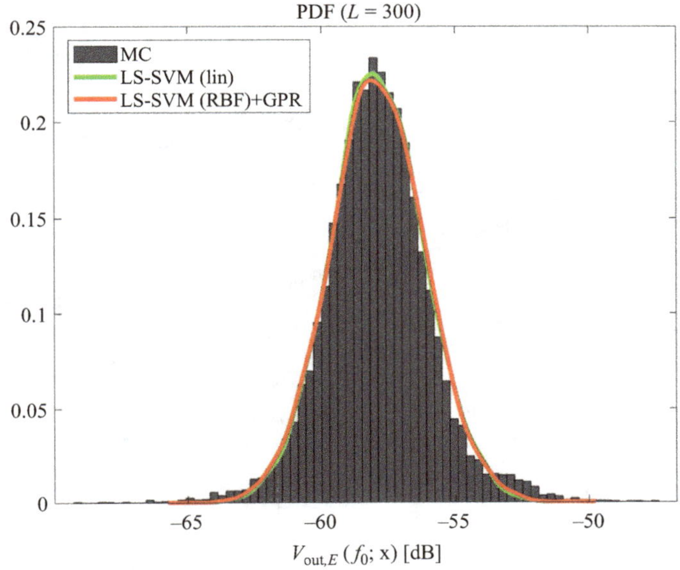

Figure 7.18 Comparison among the PDFs computed for the realizations of the spectral envelop $V_{\mathrm{out},E}(f_0;\mathbf{x})$ at $f_0 = 100\,kHz$ obtained from the mean values of the probabilistic model based on the LS-SVM (RBF)+ GPR (solid red line) and the LS-SVM regression with linear kernel (solid green line) with the histogram of 10,000 MC samples (black bins)

be computed from the results of the MC simulation, since they are not related to any statistical quantity (e.g., statistical moments, quartiles, confidence limits, etc.) associated to the uncertain responses of the system under modeling. Indeed, the CI provide the user with a statistical information on the model error, only.

Table 7.2 provides a detailed summary of the computational cost required to build each of the considered surrogate models t_{model} and the computational cost t_{pred} required by each model to predict 10,000 realizations of the envelope spectra. All the simulations have been performed on a MacBook Pro with an Intel Core i5 CPU running at 3.1GHz and 16GB of RAM. The individual simulation with the full-computational model takes 1.7s, therefore the computational time required to generate the $L = 50, 150$ and 300 training samples is about 85, 255 and 510 s. However, even for the maximum number of training samples (i.e., $L = 300$), the most complex surrogate model can be generated in less than 465 s (i.e., less than 8 min). The results highlight the advantage of the proposed technique based on the surrogate models with respect to the standard MC simulation, since each of the considered models allow predicting the envelope spectra for 10,000 realizations of the uncertain parameters in less than 30 s, whilst the corresponding MC simulation requires about 4 h 43 min.

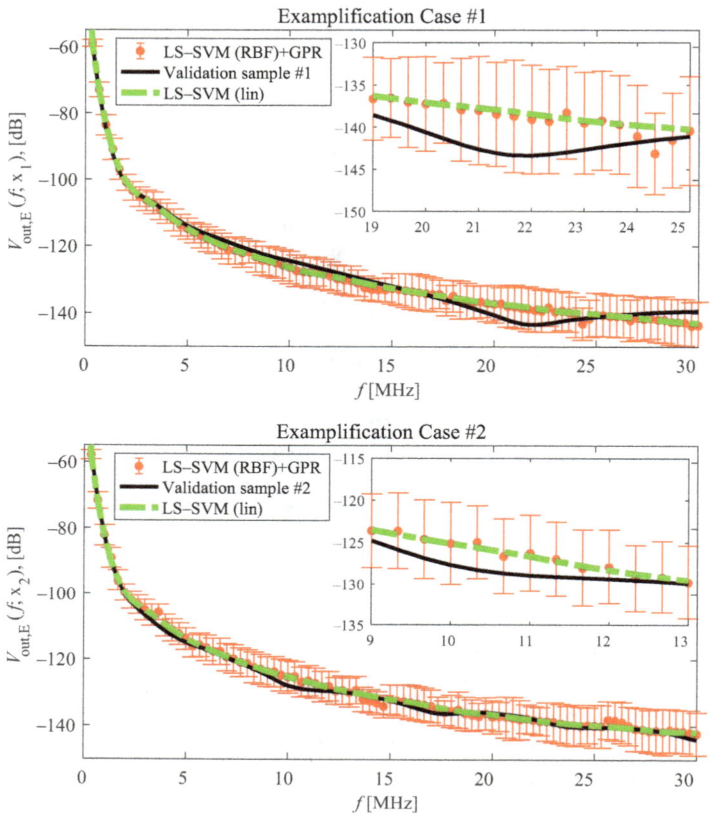

Figure 7.19 Comparison between the envelope spectra for two different configurations of the converter parameters (black curve) randomly chosen among the 10,000 realizations of the MC simulation with the corresponding predictions of the deterministic model based on the LS-SVM with RBF kernel (dashed green curve) and the mean values and 99% CI estimated by the probabilistic models built via the LS-SVM (linear)+GPR (red vertical bars)

7.11.3 Summary

We presented three ML regression techniques for the construction of accurate and fast-to-evaluate surrogate models for the UQ. All the proposed techniques allow constructing non-parametric surrogate models, and thus mitigating the curse of dimensionality issue. Specifically, the tolerance region provided by the ε-insensitive loss function used by the quadratic optimization for SVM regression allows building a surrogate model which is extremely robust to noise. On the other hand,

Table 7.2 Comparison of the computational costs needed to generate the surrogate models and to evaluate 10,000 samples. Reference is the MC simulation

Method	L=*50*		L = **150**		L = **300**	
	t_{model}	t_{pred}	t_{model}	t_{pred}	t_{model}	t_{pred}
MC	–	4 h 43 min	–	4 h 43 min	–	4 h 43 min
LS-SVM (linear)	42 s	<1 s	45 s	<1 s	115 s	<1 s
LS-SVM (RBF)	51 s	<1 s	61 s	2 s	165 s	4 s
LS-SVM (linear)+GPR	52 s	2.5 s	129 s	10 s	414 s	23 s
LS-SVM (RBF)+GPR	59 s	3 s	131 s	12 s	465 s	26 s
GPR (constant trend)	13 s	2.2 s	73 s	10 s	304 s	19 s
GPR (linear trend)	10 s	2.3 s	73 s	10 s	294 s	19 s
GPR (poly trend order 2)	11 s	2.3 s	73 s	10 s	307 s	19 s

the LS-SVM regression allows constructing accurate non-parametric surrogates models via a simple matrix inversion, without requiring to solve nonlinear optimization. Also, the GPR can be adopted to construct a probabilist non-parametric metamodel able to provide, for any configuration of the input parameters, a prediction of the system output along with a statistical interpretation of its reliability in terms of its CI.

7.12 From single-output static systems to multi output dynamic systems

The results presented in Section 7.11 highlight the capability of advanced ML regression techniques of building compact and accurate surrogate models for the UQ of nonlinear static system with few output variables, by mitigating the curse of dimensionality issue. On the other hand, the direct application of the above techniques to the modeling of the stochastic behavior of multiple output nonlinear dynamical system turns out to be rather cumbersome and complex.

Usually, there are two different approaches to tackle the above issue. One solution is to include the effect of both system parameters and transient dynamics into a single recursive model [33], for example, via a neural network [34–36]. Such models also include the realizations at adjacent points of the independent variables (e.g., time, frequency, temperature, etc.) as additional input parameters. Despite its compactness, the resulting model is so highly complex that a very large number of training samples (in the order of thousands) is needed, since a huge number of model parameters must be simultaneously tuned. This approach might be a viable solution for the cases in which the parametric responses can be represented as a low-order linear systems, but the model complexity grows exponentially for high-order nonlinear systems.

A reasonable alternative is to build a different surrogate model for each output variable and time instant of interest (see, e.g., [23,37,38]). Whilst this reduces the complexity of the single model, the overall number of models to be created is potentially huge, especially if a large number of values of the independent variable must be accounted to capture the actual evolution of the system, since such modeling scheme would require to solve a large set of optimization problem for tuning of their hyperparameters. The principal component analysis (PCA) can be seen as a promising solution, which can be used to overcome the above limitation [32,39]. Specifically, the PCA is used to reduce the number of metamodels to be created, by exploiting and removing the inherent correlation existing among the realizations of the system outputs at different value of the independent variable. The resulting compressed model turns out to be fully compatible with any surrogate model and allows reducing the model building cost.

7.12.1 Problem statement

Let us now consider the problem of constructing a compact and accurate metamodel $\widetilde{M}$ able of approximating the transient responses of the outputs of a stochastic dynamic nonlinear system:

$$\mathbf{y}(t) = \mathbf{M}(t; \mathbf{x}), \tag{7.70}$$

where $\mathbf{x} = [x_1, \ldots, x_d]^T \in \mathbb{R}^d$ is a set of (uncertain) input design parameters, $\mathbf{y}(t) = [y_1(t), \ldots, y_P(t)]^T \in \mathbb{R}^P$ are the system outputs, and t is an independent sweep time variable for which the outputs depend on. It is important to remark that the proposed approach is general and could be applied to any other sweep variable like frequency, input power, temperature, etc.

Also in this case, the surrogate model must be trained via a set of training pairs:

$$\mathcal{D} = \{(\mathbf{x}_i, \mathbf{y}_i(t_k))\}_{i,k=1}^{L,K} \tag{7.71}$$

where $\mathbf{y}_i(t_k)$ are vectors collecting the P system outputs, computed from (7.73) for a specific configuration $\mathbf{x}_i$ of the input parameters at K distinct and increasing time points $\{t_k\}_{k=1}^{K}$. For the sake of notation compactness, the set of training pairs is rewritten as a union of subsets

$$\mathcal{D} = \bigcup_{k,p=1}^{K,P} \mathcal{D}_{k,p}, \tag{7.72}$$

where each subset is defined as

$$\mathcal{D}_{k,p} = \left\{(\mathbf{x}_i, y_{i,k,p})\right\}_{i=1}^{L}, \tag{7.73}$$

and $y_{i,k,p}$ denotes the pth output evaluated at the kth time instant for the ith configuration of the input parameters, that is, $y_{i,k,p} = M_p(t_k; \mathbf{x}_i)$, with subscript p denoting the specific system output considered.

Starting from a set of training pairs $\mathcal{D}_{k,p}$, we seek for the best configurations of parameters $\{\mathbf{w}_{k,p}\}_{k,p=1}^{K,P}$, each defining a surrogate model $\widetilde{M}_p(t_k;\mathbf{x},\mathbf{w}_{k,p})$ that minimizes the following optimization problem (see Section 6 for additional details)

$$\min_{\mathbf{w}_{k,p}} \frac{1}{L}\sum_{l=1}^{L} \mathcal{L}(\widetilde{M}_p(t_k;\mathbf{x}_l,\mathbf{w}_{k,p}), y_{l,k,p}) + \lambda r(\mathbf{w}_{k,p}) \tag{7.74}$$

for $k = 1\ldots, K$ and $p = 1,\ldots, P$.

It is important to remark, that the solution of the above optimization problem for the case of system consisting of a large number of output variables P and/or time points K could be extremely expensive in terms of computational time, since it would require to train and construct $K \times P$ different metamodels.

7.13 PCA-compressed surrogate modeling

To overcome the above limitation, we introduce a compression strategy based on PCA [32,39]. To this aim, the complete dataset of training responses in $\mathcal{D}$, let us denote it with $Y = \{y_{l,k,p}\}_{l,k,p=1}^{L,K,P}$, can be interpreted as a $L \times K \times P$ three-way tensor. The data in tensor Y exhibit some inherent correlation, since they come from the same system. Specifically, data for different responses, at different time points, are not completely unrelated to each other, but rather exhibit a large amount of interdependency. If tensor Y is reshaped by stacking data for all outputs and time points row-wise, a matrix $\mathbf{Y} \in \mathbb{R}^{KP\times L}$ is obtained. This new dataset $\mathbf{Y}$ can be interpreted as a columnwise collection of L realizations $\{\zeta_l\}_{l=1}^{L}$ of a KP-variate stochastic variable ζ.

We calculate the experimental covariance matrix of ζ, that is,

$$\Sigma_\zeta = \frac{\widetilde{\mathbf{Y}}\widetilde{\mathbf{Y}}^{\mathrm{T}}}{L-1}, \tag{7.75}$$

where $\Sigma_\zeta \in \mathbb{R}^{KP\times KP}$ and

$$\widetilde{\mathbf{Y}} = \mathbf{Y} - \mu_\zeta \tag{7.76}$$

is the dataset with the samples epurated from the mean

$$\mu_\zeta = \frac{1}{L}\sum_{l=1}^{L} \zeta_l, \tag{7.77}$$

which is subtracted columnwise.

Each of the ζ-realizations can be expressed as [39]

$$\zeta_l = \mu_\zeta + \sum_{n=1}^{KP} Z_{l,n}\mathbf{u}_n, \tag{7.78}$$

for $l = 1, \ldots, L$, where

$$Z_{l,n} = \mathbf{u}_n^{\mathrm{T}}(\zeta_l - \mu_\zeta) \tag{7.79}$$

and $\{\mathbf{u}_n\}_{n=1}^{KP}$ are the left singular vectors, obtained by calculating the SVD

$$\widetilde{\mathbf{Y}} = \mathbf{USV}^{\mathrm{T}} \tag{7.80}$$

and taking the columns of $\mathbf{U}$.

We can now truncate (7.78) to retain only the $\bar{n}$ "principal components":

$$\zeta_l \approx \widehat{\zeta}_l = \mu_\zeta + \sum_{n=1}^{\bar{n}} Z_{l,n}\mathbf{u}_n, \tag{7.81}$$

which can be identified by setting a threshold ε on the relative magnitude of the singular values of $\widetilde{\mathbf{Y}}$, collected in descending order into the diagonal matrix $\mathbf{S}$. The singular values of $\widetilde{\mathbf{Y}}$ are proportional to the square root of the eigenvalues of its covariance matrix $\mathbf{\Sigma}_\zeta$. If we define $\Sigma_{\widehat{\zeta}}$ as the covariance matrix of the approximated data (91), using matrix properties we can conclude that

$$\frac{||\Sigma_{\widehat{\zeta}} - \Sigma_\zeta||_2}{||\Sigma_\zeta||_2} = \left(\frac{\sigma_{\bar{n}} + 1}{\sigma_1}\right)^2 \leq \varepsilon^2 \tag{7.82}$$

where σ_1 is the first singular value of $\widetilde{\mathbf{Y}}$ and $\sigma_{\bar{n}+}1$ is the first singular value that is *discarded* by the PCA truncation. Hence, setting a threshold on the singular values allows a rigorous control on the approximation in terms of variance. Here after, we truncate when the magnitude falls below 1% of the first singular value (i.e., $\varepsilon = 10^{-2}$).

The key achievement is that the set $\{Z_{l,n}\}_{l,n=1}^{L,\bar{n}}$ of PCA coefficients (7.79) can be interpreted as a collection of L samples of $\bar{n}$ new output system variables $Z_n(\mathbf{x})$, with $Z_{l,n} = Z_n(\mathbf{x}_l)$, describing the information pertaining to the entire set of original time-dependent outputs $\mathbf{y}(t)$. Since typically $\bar{n} \lll KP$, the PCA truncation leads to a substantial compression of the number of variables to be modeled. Each of these reduced variables can be approximated using any surrogate model. Once a model is available for the compressed variables Z_n, new samples for the original time-dependent output variables can be recovered via (7.84). It is important to remark that no specific assumption on the nature of the correlation among the available data is required by the PCA algorithm. Moreover, the PCA compression can be suitably combined with any regression or interpolation techniques, with the aim of constructing accurate and compact surrogate model of the responses of complex dynamic systems.

7.14 Application example: 16-bit flash memory bus

The feasibility and performance of the proposed compression strategy based on the PCA are assessed on an application examples consisting of a 16-bit channel with 54 uncertain parameters driven by Flash memory chips.

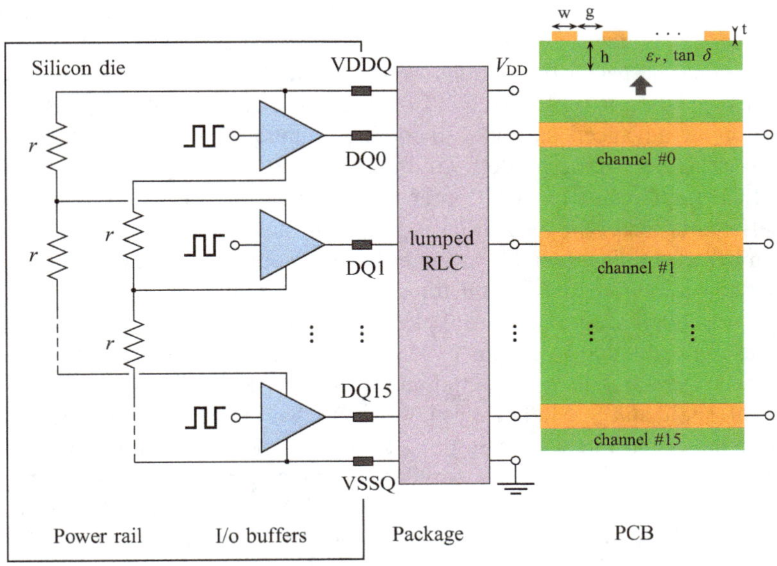

Figure 7.20 Illustration of the 16-bit digital channel for the Flash memory chip

Specifically, the proposed methodology is applied to the link depicted in Figure 7.20, which represents a 16-bit transmission channel of a memory chip [32]. For the I/O buffers, we use behavioral macromodels compatible with the drivers of a Flash technology operating at 66 MHz. The structure includes a resistive rail for the power supply of the buffers, a RLC network with package parasitics, and a transmission channel consisting of 16 coupled microstrip traces. The far-end terminations are floating to mimic connection to high-impedance receivers. For the drivers, we use behavioral macromodels constructed using the method in [40], but any other model, either behavioral or transistor-level, may be used. The even-bit drivers transmit slightly asynchronous pulses with a duration of 15 ns and rise/fall times of 0.1 ns, whereas odd-bit drivers remain quiet in the "low" state.

There are $d = 54$ uncertain design parameters, each with an independent Gaussian distribution and a 10% standard deviation from the nominal value. Namely: the resistance of the power rail (nominal value: $r = 1.14\ \Omega$), the package parasitics (nominal values: resistance $R = 50$ m Ω, inductance $L = 2$ nH, capacitance $C = 5.5$ pF), the width and thickness of each microstrip trace (nominal values: $w = 150\ \mu$m and $t = 30\ \mu$m), the gap between the traces (nominal value: $g = 150\ \mu$m), and the substrate parameters (nominal values: thickness $h = 460\ \mu$m, relative dielectric permittivity $\varepsilon_r = 4.1$, loss tangent $tan\delta = 0.02$). The values of package parasitics are here assumed to be independent, even though some correlation usually exists between them. The power supply voltage is $V_{DD} = 1.8$ V and the microstrip trace resistivity is $\rho = 1.72\ \Omega$m. The bus is 18-cm long.

The outputs of interest are the terminal voltages at the receiver side of each line and the supply voltages of each driver. These $M = 32$ outputs are evaluated at $K = 901$ equally spaced time points between 0 and 45 ns. The time step of the transient simulation is set to one half of the risetime of the digital signals. A naive application of advanced surrogate modeling techniques would require the construction of $KP = 28,832$ models. On the other hand, the use of a standard, non-sparse PCE implementation with order $p = 3$ would require the determination of the coefficients for $|\mathcal{K}| = 29,260$ basis functions, and hence the use of an exorbitant number of training samples to solve the corresponding regression problem.

We consider instead $L = 300$ training configurations of the uncertain parameters, generated according to a Latin hypercube scheme. The corresponding responses are evaluated by means of HSPICE simulations. Figure 7.21 shows the normalized singular values of the training dataset. The singular values drop below the 1% threshold at $\bar{n} = 51$. Therefore, we retain the first 51 terms in the PCA (88), with a compression rate amounting to less than 0.2% of the original variables. It should be noted that lowering the simulation time step only adds redundant information to the data, which is then cut off by the PCA compression. Moreover, if the PCA is applied separately to each of the 32 outputs (thus only exploiting the correlation of data over time), the overall number of models to be computed becomes 624. This demonstrates that modeling all outputs concurrently allows to effectively take advantage of their interdependency and leads to further compression.

We construct surrogate models for the PCA coefficients using both a sparse third-order PCE by using toolbox [41] and a LS-SVM regression with RBF kernel calculated via the LS-SVMlab toolbox [20]. In the former case, the cardinality of the subsets $\breve{\mathcal{K}}$ for the sparse models of the various PCA coefficients varies between 29 and 132, meaning that the number of non-negligible PCE coefficients is between 0.1% and 0.5% of the total. On the other hand, the hyperparameters of the LS-SVM regression are tuned based on the available set of training samples using the LOO CV score [20].

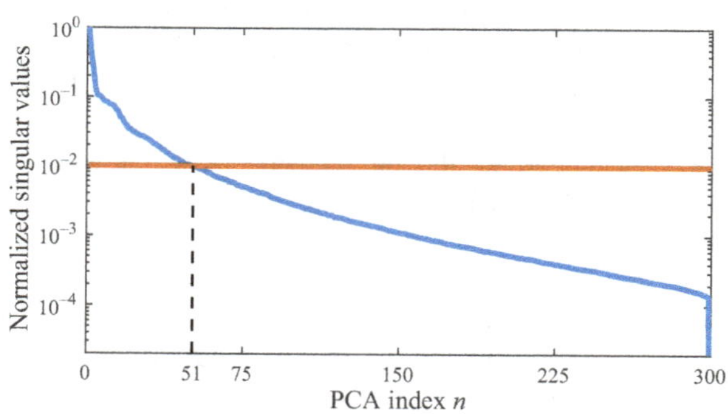

Figure 7.21 Normalized singular value plot of the 16-bit bus dataset (blue line). The horizontal line indicates the 1% threshold for the PCA truncation

Figure 7.22 shows the variability and the standard deviation of a selection of four outputs, namely the terminal voltages on channels #0, #1, and #11, and the supply voltage of the buffer of channel #15. The green lines are a set of 500 responses from a reference MC analysis, which help visualize the voltage variation due to the parameter uncertainty. It should be noted that the second and the third plot refer to crosstalk voltages, as the drivers of channels #1 and #11 are not transmitting. Moreover, from the fourth plot, a large fluctuation of the supply voltage is observed, resulting from the commutation of the buffers. The solid blue, dashed red and dotted yellow curves are the standard deviation estimated from 10,000 MC samples, with the sparse PCE, and with the LS-SVM regression, respectively. The results provided by the two surrogate models compare well, and they are also in fairly good agreement with the reference MC results. An exception is the case of the crosstalk voltage on channel #1, for which a rather large discrepancy is observed. The reason could be ascribed to the large amount of "outlying" responses that can be observed in Figure 7.22, which particularly affects the estimation of the standard deviation, especially around 20 ns.

To further assess the accuracy, we calculate the PDF of the quantities in Figure 7.22 at time points exhibiting large variability. The results are shown in Figure 7.23. A remarkable accuracy is established for both the PCE and the

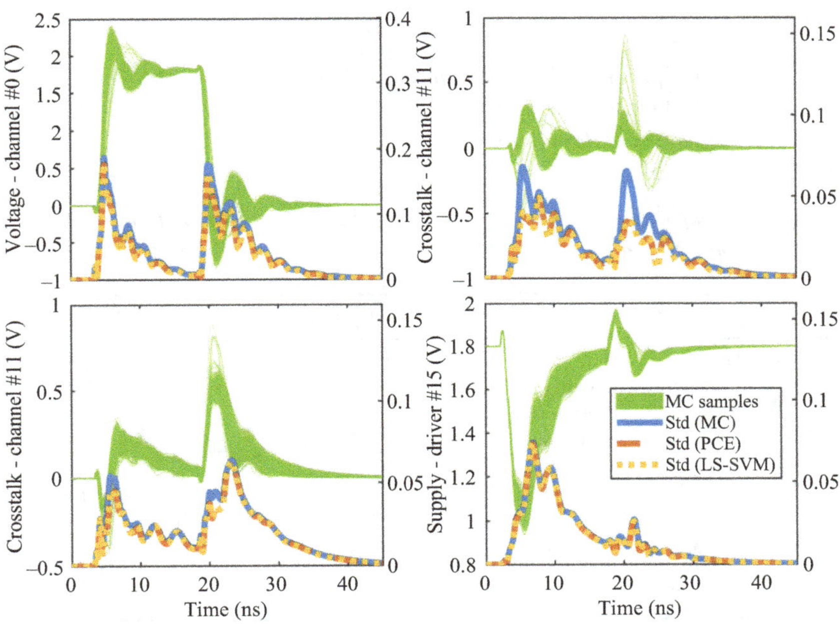

Figure 7.22 Collection of four of the 32 output voltages for the 16-bit bus of Figure 7.20. Green lines: MC samples; solid blue, dashed red, and dotted yellow lines: standard deviation computed with MC, sparse PCE and LS-SVM regression methods, respectively

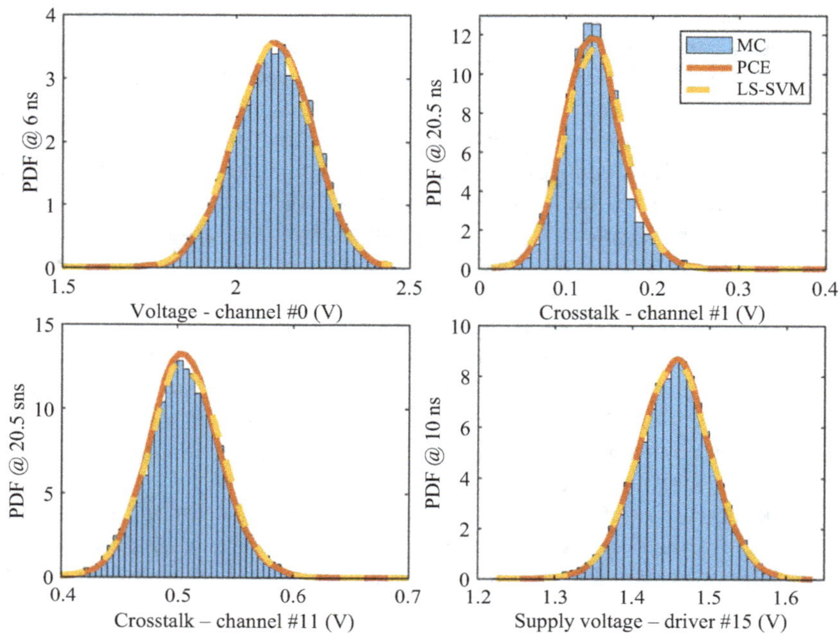

Figure 7.23 PDFs of the output variables in Figure 7.22 computed at selected time points of large variability using the MC samples (blue histograms), the sparse PCE (solid red line), and the LS-SVM model (dashed yellow line)

Table 7.3 Performance of the sparse PCE and LS-SVM surrogate models in conjunction with PCA compression for the 16-bit bus of Figure 7.20

		Sparse PCE	**LS-SVM**
RMS error (V)	Minimum	5.8095×10^{-7}	2.4235×10^{-7}
	Average	4.8885×10^{-3}	5.2712×10^{-3}
	Maximum	7.0072×10^{-2}	7.1321×10^{-2}
Time (s)	Training	4,798.4	
	Compression	1.6	
	Building	84.8	87.4
	Evaluation	4.4	3.1
	Recovery	8.8	
	Total	4,898.0	4,899.3
	Speed-up	32.7 ×	32.7 ×

LS-SVM models, even for the crosstalk on channel #1. Similar or better results are found for the remaining outputs.

Table 7.3 provides the main figures concerning the accuracy and efficiency of the sparse PCE and LS-SVM regression methods in conjunction with PCA compression. The accuracy is assessed in terms of the RMS error over the 10,000 samples of the reference MC analysis. The minimum, average and maximum error over the time points and the various outputs are provided. As far as the computational times are concerned, the table lists the time required by the HSPICE simulation of the training samples, by the PCA compression, for building and evaluating the surrogate models of the PCA coefficients, and by the recovery of the corresponding samples of the original variables, from which the statistical information is eventually extracted. For comparison, the HSPICE simulation of the reference MC responses takes about 160,000 s (1 day 20 h 27 min). All the simulations have been performed on a Dell Precision 5820 workstation with an Intel(R) Core(TM) i9-7900X, CPU running at 3.30 GHz, and 32 GB of RAM. The processing time of both the sparse PCE and LS-SVM regression is minor compared to the simulation of the training samples, and the two techniques exhibit similar performance in terms of accuracy and computational efficiency for this test case.

References

[1] Fang KT, Li R, *Sudjianto A. Design and Modeling for Computer Experiments (Computer Science & Data Analysis)*. Boca Raton, FL: Chapman & Hall/CRC; 2005.

[2] Bourinet JM. Reliability analysis and optimal design under uncertainty – Focus on adaptive surrogate-based approaches [Habilitation à diriger des recherches]. *Université Clermont Auvergne*; 2018. Available from: https://tel.archives-ouvertes.fr/tel-01737299

[3] Sudret B, Marelli S, Wiart J. Surrogate models for uncertainty quantification: An overview. In: *2017 11th European Conference on Antennas and Propagation (EUCAP)*; 2017. pp. 793–797.

[4] Shahriari B, Swersky K, Wang Z, *et al.* Taking the human out of the loop: A review of Bayesian optimization. *Proceedings of the IEEE*. 2016; 104(1): 148–175.

[5] Vapnik VN. *The Nature of Statistical Learning Theory*. New York, NY: Springer-Verlag, Inc.; 1995.

[6] Ochoa JS, Cangellaris AC. Random-space dimensionality reduction for expedient yield estimation of passive microwave structures. *IEEE Transactions on Microwave Theory and Techniques*. 2013; 61(12): 4313–4321.

[7] Weinberger K. *Lecture Notes in Machine Learning for Intelligent Systems*. Ithaca, NY: Cornell University; 2018.

[8] McKay MD, Beckman RJ, Conover WJ. Comparison of three methods for selecting values of input variables in the analysis of output from a computer code. *Technometrics*. 1979; 21(2): 239–245.

[9] Ghojogh B, Crowley M. The Theory Behind Overfitting, Cross Validation, Regularization, Bagging, and Boosting: Tutorial; 2019.

[10] Hastie T, Tibshirani R, Friedman J. *The Elements of Statistical Learning: Data Mining, Inference and Prediction*. 2nd ed. New York, NY: Springer; 2009.

[11] Russell S, Norvig P. *Artificial Intelligence: A Modern Approach*. 3rd ed. Hoboken, NJ: Prentice Hall Press; 2009.

[12] Bellman RE. Adaptive Control Processes: A Guided Tour; 1961. First introduction of curse of dimensionality.

[13] Kuhn M, Johnson K. *Applied Predictive Modeling*. New York, NY: Springer; 2013.

[14] Blatman G, Sudret B. Adaptive sparse polynomial chaos expansion based on least angle regression. *Journal of Computational Physics*. 2011; 230(6): 2345–2367. Available from: http://www.sciencedirect.com/science/article/pii/S0021999110006856

[15] Vapnik VN. *Statistical Learning Theory*. Hoboken, NJ: Wiley-Interscience; 1998.

[16] Cherkassky V, Ma Y. Practical selection of SVM parameters and noise estimation for SVM regression. *Neural Networks*. 2004; 17(1): 113–126.

[17] Smola AJ, Schölkopf B. A tutorial on support vector regression. *Statistics and Computing*. 2004; 14(3): 199–222.

[18] Trinchero R, Manfredi P, Stievano IS, *et al.* Machine learning for the performance assessment of high-speed links. *IEEE Transactions on Electromagnetic Compatibility*. 2018; 60(6): 1627–1634.

[19] Suykens JAK, Gestel TV, Brabanter JD, *et al. Least Squares Support Vector Machines*. Singapore: World Scientific; 2002.

[20] LS-SVMlab v. Leuven: Department of Electrical Engineering (ESAT), Katholieke Universiteit Leuven; 2011. Available from: http://www.esat.kuleuven.be/sista/lssvmlab/

[21] Rasmussen CE, Williams CKI. *Gaussian Processes for Machine Learning*. Cambridge, MA: MIT Press; 2006.

[22] Ebden M. Gaussian Processes: A Quick Introduction; 2015.

[23] Trinchero R, Canavero FG. Combining LS-SVM and GP regression for the uncertainty quantification of the EMI of power converters affected by several uncertain parameters. *IEEE Transactions on Electromagnetic Compatibility*. 2020; 1–8.

[24] Trinchero R, Larbi M, Swaminathan M, *et al.* Statistical analysis of the efficiency of an integrated voltage regulator by means of a machine learning model coupled with kriging regression. In: *2019 IEEE 23rd Workshop on Signal and Power Integrity (SPI)*; 2019. pp. 1–4.

[25] Larbi M, Trinchero R, Canavero FG, *et al.* Analysis of parameter variability in an integrated wireless power transfer system via partial least squares regression. *IEEE Transactions on Components, Packaging and Manufacturing Technology*. 2020; pp. 1–1.

[26] Statistics, Machine Learning Toolbox v. Natick, MA: The MathWorks, Inc.; 2020.

[27] Paul CR. *Analysis of Multiconductor Transmission Lines/Clayton R. Paul*. 2nd ed. Hoboken, NJ: Wiley-Interscience; 2007.

[28] Spina D, De Jonghe D, Deschrijver D, *et al.* Stochastic macromodeling of nonlinear systems via polynomial chaos expansion and transfer function trajectories. *IEEE Transactions on Microwave Theory and Techniques*. 2014; 62(7): 1454–1460.

[29] Muyshondt R, Krein PT. 20 W benchmark converters for simulation and control comparisons. In: *COM.P.EL.98. Record 6th Workshop on Computer in Power Electronics (Cat. No. 98TH8358)*; 1998. pp. 201–212.

[30] Neal RM. *Bayesian Learning for Neural Networks*. Berlin: Springer-Verlag; 1996.

[31] Trinchero R, Larbi M, Torun HM, *et al.* Machine learning and uncertainty quantification for surrogate models of integrated devices with a large number of parameters. *IEEE Access*. 2019; 7: 4056–4066.

[32] Manfredi P, Trinchero R. A data compression strategy for the efficient uncertainty quantification of time-domain circuit responses. *IEEE Access*. 2020; 8: 92019–92027.

[33] Ljung L. *System Identification: Theory for the User*. Hoboken, NJ: Prentice Hall, Inc.; 1986.

[34] Haykin SS. *Neural Networks and Learning Machines*. 3rd ed. Upper Saddle River, NJ: Pearson Education; 2009.

[35] Torun HM, Yu H, Dasari N, *et al.* A spectral convolutional net for co-optimization of integrated voltage regulators and embedded inductors. In: *2019 IEEE/ACM International Conference on Computer-Aided Design (ICCAD)*; 2019. pp. 1–8.

[36] Yu H, Michalka T, Larbi M, *et al.* Behavioral modeling of tunable I/O drivers with preemphasis including power supply noise. *IEEE Transactions on Very Large Scale Integration (VLSI) Systems*. 2020; 28(1): 233–242.

[37] Trinchero R, Stievano IS, Canavero FG. Black-box modeling of the maximum currents induced in harnesses during automotive radiated immunity tests. *IEEE Transactions on Electromagnetic Compatibility*. 2020; 62(2): 627–630.

[38] Larbi M, Stievano IS, Canavero FG, *et al.* Variability impact of many design parameters: The case of a realistic electronic link. *IEEE Transactions on Electromagnetic Compatibility*. 2018; 60(1): 34–41.

[39] Yaghoubi V, Marelli S, Sudret B, *et al.* Sparse polynomial chaos expansions of frequency response functions using stochastic frequency transformation. *Probabilistic Engineering Mechanics*. 2017; 48: 39–58. Available from: http://www.sciencedirect.com/science/article/pii/S0266892016301825

[40] Stievano IS, Rigazio L, Maio IA, *et al.* Behavioral modeling of IC core power-delivery networks from measured data. *IEEE Transactions on Components, Packaging and Manufacturing Technology*. 2011; 1(3): 367–373.

[41] Marelli S, Sudret B. UQLab: A Framework for Uncertainty Quantification in Matlab; 2014.

[42] Kohestani V, Hassanlourad M. Modeling the mechanical behavior of carbonate sands using artificial neural networks and support vector machines. *International Journal of Geomechanics*. 2015; 06(16): 04015038.
[43] Haifeng Wang, Dejin Hu. Comparison of SVM and LS-SVM for regression. In: *2005 International Conference on Neural Networks and Brain*. vol. 1; 2005. pp. 279–283.

Chapter 8

Artificial neural network-based yield optimization with uncertainties in EM structural parameters*

Feng Feng[1], Jianan Zhang[2] and Qi-Jun Zhang[2]

This chapter provides an overview of recent advances in artificial neural network-based yield optimization with uncertainties in electromagnetic (EM) structure parameters. We first give an introduction to EM-based yield estimation and optimization. Then, we review the formulation for performing yield-driven EM optimization with the traditional Monte Carlo (MC) approach. Following that, we review the space mapping (SM)-based yield optimization where the computationally expensive EM simulations are replaced by the evaluations of SM surrogates. Next, we describe the polynominal chaos (PC)-based approach to EM-based yield estimation and optimization. The PC approach can capture the stochastic properties of EM responses accurately with much fewer EM samples than the MC approach. A waveguide bandpass filter example is provided to demonstrate the advantages of the PC-based approach. Lastly, we discuss the pure artifical neural network-based yield optimization method and its advances, that is, the adaptively weighted yield optimization incorporating neuro-transfer function surrogate. This method is illustrated by a four-pole waveguide filter example.

8.1 Introduction

Design optimization with uncertainties, a.k.a. yield-driven design, or yield optimization, is an indispensable step in microwave design due to the presence of process variations. Such variations usually lead to uncertainties on the physical dimensions and material properties of the fabricated microwave components. This can further make the performance of a certain number of fabricated components fail to meet the design specifications. Yield refers to the ratio of the number of "good" circuits over the total number of circuit outcomes. Yield-driven design aims

*Feng Feng and Jianan Zhang have equal contribution.
[1]School of Microelectronics, Tianjin University, Tianjin, China
[2]Department of Electronics, Carleton University, Ottawa, ON, Canada

to find the optimal nominal point with the maximum yield through an optimization procedure.

In the circuit community, the history of yield-driven design can be traced back to 1980s, when various geometrical approaches [1,2] and Monte Carlo (MC)-based approaches [3–6] were developed. Yield-driven EM optimization (also known as EM-based yield optimization) have attracted great attentions from the research community since the 1990s when EM simulations start being popularly used in microwave design [7–10]. EM-based yield optimization is much more challenging than its circuit-based counterpart in that EM simulations are typically much more computationally extensive than circuit simulations. Not to mention that a huge number of EM simulations are usually required to perform EM-based yield optimization. As a recognized engineering methodology in the microwave area, space mapping (SM) [11] has been introduced to the field of yield-driven EM optimization to mitigate the computation burden [12–17]. Most of existing SM-based yield optimization methods train a surrogate to represent the EM behavior of the fine model and combine the SM concept with the trust region algorithm to find the optimal yield solution iteratively. The computational burden is significantly reduced since evaluating the surrogate model responses is much cheaper than calculating the fine model responses. For example, in [12], an SM neuro-model has been first developed then used as an computationally efficient surrogate to find the optimal yield solution of microwave structures. In [15], the ellipsoidal technique has been combined with SM to expedite EM-based yield optimization. More recently, parallel computation has been incorporated into SM to speed up the data generation and model training processes for SM-based yield optimization [17]. Consequently, the SM surrogate can be trained over multiple points and therefore can represent the EM behavior in a relatively larger neighborhood. The overall SM iterations during yield optimization is thus reduced and the total CPU time is greatly shortened.

The aforementioned SM-based yield optimization approaches require an equivalent circuit model to be available. When this requirement cannot be satisfied, the SM-based approaches cannot be applied in the classical way. To alleviate this difficulty, the features of EM responses have been exploited to achieve low-cost yield estimation and optimization of microwave structures [18–20]. In [18], the cognition-driven formulation of SM has been extended from nominal design to yield-driven design of microwave filters. The feature parameters of EM responses have been first extracted and then utilized to conduct yield estimation and optimization of microwave filters. Another feature-based method has been presented in [19,20], where the constructed feature model has been used in the pattern search algorithm to find the optimal yield solution.

The last decade has witnessed the development of various polynomial chaos (PC) approaches to uncertainty quantification of microwave structures [21–29]. In the PC approach, the relationship between a stochastic quantity and the random parameters is expressed as the sum of weighted orthogonal basis functions [30]. Each basis function is associated with a coefficient (called the PC coefficient), which can be found by multidimensional integration using a numerical quadrature [21]. It is well-known that by using PC, the stochastic properties of the EM responses, such as mean and variance, can be obtained analytically from the PC coefficients [30]. It has been shown that the number of EM samples used to evaluate the PC coefficients is typically much fewer than that in an MC analysis [31]. Because of this, the PC approach finds many applications in statistical analysis in the microwave area, including uncertainty quantification of the time-domain electric and magnetic fields [22], variability analysis of high-speed interconnects [24], and uncertainty quantification for the indoor radio channels [21]. More recently, the PC approach has been elevated from EM-based yield estimation to EM-based yield optimization in [32]. The PC-based yield optimization approach does not require the availability of a coarse model and provides a systematic formulation to yield-driven design of microwave structures.

The neuro-transfer function (neuro-TF) method is another recognized tool to deal with EM parametric modeling/nominal optimization/yield optimization problems when equivalent circuit coarse models are difficult to obtain [33–38]. In neuro-TF methods, the relationship between the EM response of microwave structures and frequency are expressed by transfer functions. The transfer function can be in a rational [34] format, a pole-residue format [38], or a pole-zero-gain format [35]. The vector fitting method [39] is typically needed to extract the coefficients (or poles/residues, poles/zeros/gain) in the transfer functions corresponding to a given set of EM responses. The relationship between the coefficients and geometrical parameters are learned by neural networks, while the EM responses (e.g., *S*-parameters) are utilized as training data to refine the overall model. Once trained properly, the neuro-TF model can serve as an efficient surrogate to provide fast and accurate predictions of the EM behavior of the microwave structure, accelerating the subsequent nominal design and yield-driven design processes. Recently, an adaptively weighted yield optimization approach incorporating neuro-TF surrogate for the yield-driven design of microwave structures has been reported [40]. By incorporating adaptive weighting factors into the formulation of the objective function, the adaptively weighted neuro-TF-based yield optimization reduces the number of iterations required in the neuro-TF-assisted yield optimization approach, thereby expediting the overall yield-driven design procedure.

This chapter provides an overview of some recent advances in surrogate-assisted yield-driven design of microwave structures. The MC-based approach to yield-driven EM optimization is first introduced, which simply replaces the circuit simulator with an EM simulator and accumulates the EM responses at multiple random samples to calculate the objective function [6]. The computational overhead is in general very high when the MC-based approach is used. To mitigate this issue, various surrogate-assisted yield optimization approaches have been developed in recent years. We first review the SM-based approaches which replace the computationally expensive EM-based fine model by a computationally cheaper coarse model. Following that, we describe the PC-based approach which takes advantages of the PC coefficients to reduce the overall CPU time for yield-driven EM optimization [32]. The PC coefficients have been incorporated into the formulation of the objective function, which greatly reduces the number of EM simulations required to achieve a considerable yield improvement. Lastly, we discuss the pure artifical neural network (ANN)-based approach to yield-driven design of microwave structures. To enhance the capability and generality of pure ANN models, neuro-TF modeling methods are then introduced where the relationship between the EM response of microwave structures and versus frequency are expressed by transfer functions, and the neural networks are trained to learn the relationship between the coefficients in the transfer functions and geometrical parameters.

8.2 The MC-based approach to yield-driven EM optimization

Let $\boldsymbol{x}$ be a n-dimensional vector which contains the n design parameters such as geometrical parameters of the considered microwave structure. Let $\boldsymbol{x}^0$ represent the nominal point of $\boldsymbol{x}$. When we consider yield-driven design of microwave structures, the actual values of $\boldsymbol{x}$ are typically assumed to be distributed around $\boldsymbol{x}^0$ following a certain type of distribution such as uniform or Gaussian distribution. Let these parameter values be denoted by

$$\boldsymbol{x}^m = \boldsymbol{x}^0 + \Delta\boldsymbol{x}^m, \ m = 1, 2, \dots, N_{\mathrm{mc}} \tag{8.1}$$

where N_{mc} represents the total number of random outcomes and $\Delta\boldsymbol{x}^m$ is the deviation from the nominal point $\boldsymbol{x}^0$. The goals of microwave design are typically expressed by a number of design specifications imposed on the EM responses of microwave structures. The design specifications are usually sampled at a number of frequency points, which are uniformly distributed in the whole frequency range of interest. Let N_s denote the total number of design specification samples, which can consist of the upper and the lower ones. Let the jth design specification sample be denoted by S_j, where $j = 1, \dots, N_s$. Let N_s^u and N_s^l ($N_s^l = N_s - N_s^u$) be the number of upper specification samples and the number of lower specification samples, respectively. Without the loss of generality, we assume that the first N_s^u samples, $S_1, \dots, S_{N_s^u}$, represent the upper design specification samples, and that the remaining N_s^l samples, $S_{N_s^u+1}, \dots, S_{N_s}$, are the lower design specification samples. Let $R_j^{EM}(\boldsymbol{x})$ represent the EM response of the microwave structure at the frequency of interest corresponding to S_j. Let $\boldsymbol{e}(\boldsymbol{x})$ be an error vector used to indicate how well the EM response vector of the microwave structure satisfies the design specifications. Specifically, $\boldsymbol{e}(\boldsymbol{x})$ is defined as follows:

$$\boldsymbol{e}(\boldsymbol{x}) = [e_1 \, e_2 \cdots e_{N_s^u} \cdots e_{N_s}]^T \tag{8.2}$$

where the jth element in the above vector, $e_j(\boldsymbol{x})$, is given by

$$e_j(\mathbf{x}) = \begin{cases} R_j^{\mathrm{EM}}(\boldsymbol{x}) - S_j, & \text{if } 1 \le j \le N_s^u \\ S_j - R_j^{\mathrm{EM}}(\boldsymbol{x}), & \text{if } N_s^u < j \le N_s \end{cases} \tag{8.3}$$

Due to the uncertainties introduced by the manufacturing process, the EM responses at some outcomes may violate the design specifications. Yield is defined as the ratio of "good" designs over the total number of fabricated design outcomes. Let the yield at nominal point $\boldsymbol{x}^0$ be denoted as $Y(\boldsymbol{x}^0)$. Then, $Y(\boldsymbol{x}^0)$ is usually approximated by

$$Y(\boldsymbol{x}^0) \approx N_{\text{good}}/N_{\text{mc}} \tag{8.4}$$

where N_{good} represents the total number of "good" designs. Let $H_p(\cdot)$ represent the one-sided least pth function. According to [6], the MC-based approach utilizes the following objective function $U(\boldsymbol{x}^0)$ to solve the EM-based yield optimization problem

$$U(\boldsymbol{x}^0) = H_p(\boldsymbol{u}(\boldsymbol{x}^0)) \tag{8.5}$$

where $\boldsymbol{u} = \left[u_1, u_2, \dots, u_{N_{mc}}\right]^T$, and the mth element in $\boldsymbol{u}$ is found from

$$u_m = \alpha_m H_q(\boldsymbol{e}(\boldsymbol{x}^m)), \quad m = 1, \dots, N_{mc} \tag{8.6}$$

where q and p are the two parameters indicating the norms used for $\boldsymbol{e}$ and $\boldsymbol{u}$, respectively. Following the suggestions in [6], the two parameters are usually set as $p = q = 1$ and the weighting factor is set by $\alpha_m = 1$. These settings results in an objective function used in the MC-based approach in the following form [32]:

$$U(\boldsymbol{x}^0) = \sum_{m \in M} \sum_{j \in J(\boldsymbol{x}^m)} e_j(\boldsymbol{x}^m). \tag{8.7}$$

$$J(\boldsymbol{x}^m) = \{j | e_j(\boldsymbol{x}^m) > 0\} \tag{8.8}$$

$$M = \{m | J(\boldsymbol{x}^m) \neq \varnothing\} \tag{8.9}$$

A smaller value of $U(\boldsymbol{x}^0)$ indicates a smaller degree of the EM responses violating the design specifications at the random outcomes, which implies a higher yield. Therefore, the yield of the microwave structure can be optimized by minimizing $U(\boldsymbol{x}^0)$ using a certain optimization (such as a gradient-based) algorithm. However, in most cases, a reasonably large number of random outcomes are needed to achieve an effective minimization of the objective function defined in (8.2)–(8.9). If the responses are obtained from circuit simulations, as in [6], this MC-based approach will work efficiently. When the responses need to be solved by EM simulators, the MC-based approach cannot be directly applied due to the huge number of EM simulations, which is computationally prohibitive.

8.3 The SM-based approach to yield-driven EM optimization

In the SM-based yield optimization approaches, we assume the availability of a computationally fast but not so accurate coarse model (e.g., equivalent circuit model) and an accurate but computationally expensive fine model (e.g., fine-mesh EM simulation) [41]. Let $\boldsymbol{R}^c(\boldsymbol{x})$ and $\boldsymbol{R}^f(\boldsymbol{x})$ denote the response vectors of the coarse and the fine models corresponding to $\boldsymbol{x}$, respectively. The SM-based approach establishes a surrogate model which combines the coarse model with a linear input mapping as follows:

$$\boldsymbol{x}_c = \boldsymbol{B}\boldsymbol{x} + \boldsymbol{c} \tag{8.10}$$

$$\boldsymbol{R}^s(\boldsymbol{x}, \boldsymbol{w}) = \boldsymbol{R}^c(\boldsymbol{x}_c) = \boldsymbol{R}^c(\boldsymbol{B}\boldsymbol{x} + \boldsymbol{c}) \tag{8.11}$$

where $\boldsymbol{x}_c$ and $\boldsymbol{x}$ are two vectors containing all the design variables of the coarse model and the fine model, respectively; $\boldsymbol{R}^c(\boldsymbol{x}_c)$ is the response vector of the coarse model corresponding to $\boldsymbol{x}_c$, while $\boldsymbol{R}^s(\boldsymbol{x}, \boldsymbol{w})$ represents the response vector of the SM surrogate; $\boldsymbol{w}$ is a vector of mapping parameters containing all the elements in $\boldsymbol{B}$ and $\boldsymbol{c}$, where $\boldsymbol{B}$ and $\boldsymbol{c}$ represent the coefficients in the linear mapping function.

In the traditional sequential SM-based approach, the SM surrogate is trained with the EM responses solved at a single geometrical sample. Let the nominal point at the kth ($k = 1, 2, \ldots$) iteration during yield optimization be denoted by $\boldsymbol{x}^{0,k}$, where $\boldsymbol{x}^{0,1}$ is typically initialized as the optimal solution from nominal optimization. In the kth iteration, the EM responses (e.g., S-parameters) solved at the nominal point $\boldsymbol{x}^{0,k}$ by EM simulation are used to train the surrogate model. The error function for training the surrogate model is defined as

$$E(\boldsymbol{w}) = \|\boldsymbol{R}^f(\boldsymbol{x}^{0,k}) - \boldsymbol{R}^s(\boldsymbol{x}^{0,k}, \boldsymbol{w})\|^2 \tag{8.12}$$

where $\boldsymbol{R}^f(\boldsymbol{x}^{0,k})$ and $\boldsymbol{R}^s(\boldsymbol{x}^{0,k})$ denote the response vectors of the fine model and the surrogate model corresponding to $\boldsymbol{x}^{0,k}$, respectively. The training process of the surrogate model can be formulated as

$$\boldsymbol{w}^* = \arg\min_{w} E(\boldsymbol{w}) \tag{8.13}$$

where $\boldsymbol{w}^*$ contains the optimal parameters of the mapping function after training.

Next, we perform yield optimization using the trained surrogate to obtain the next optimal yield solution $\boldsymbol{x}^{0,k+1}$. The one-sided l_1 design centering algorithm [12] is typically used to optimize the yield of the trained surrogate. Let $\boldsymbol{e}^s(\boldsymbol{x})$ represent the error vector used to indicate how well the response vector of the surrogate model satisfies the design specifications. Similar to (8.2), we have

$$\boldsymbol{e}^s(\boldsymbol{x}) = \left[e_1^s \; e_2^s \cdots e_{N_s^u}^s \cdots e_{N_s}^s\right]^T. \tag{8.14}$$

For each random outcome $\boldsymbol{x}^m$, we have a corresponding error vector $\boldsymbol{e}^s(\boldsymbol{x}^m)$. The value of $e_j^s(\boldsymbol{x}^m)$, where $j = 1, \ldots, N_s$, indicates the degree of the surrogate model response violating the jth design specification sample S_j. A larger value of $e_j^s(\boldsymbol{x}^m)$ implies that S_j is violated to a larger degree. To optimize the yield of the microwave structure at $\boldsymbol{x}^0$, the l_1 design centering algorithm minimizes the following objective function using a certain gradient-based optimization algorithm:

$$U^s(\boldsymbol{x}^0) = \sum_{m \in M} \sum_{j \in J(\boldsymbol{x}^m)} e_j^s(\boldsymbol{x}^m) \tag{8.15}$$

$$J(\boldsymbol{x}^m) = \{j | e_j^s(\boldsymbol{x}^m) > 0\}$$

$$M = \{m | J(\boldsymbol{x}^m) \neq \varnothing\}$$

where $e_j^s(\boldsymbol{x}^m)$ is defined in the same way as that in (8.3), but this time the responses are evaluated from the surrogate model instead of the fine model,

$$e_j^s(\boldsymbol{x}) = \begin{cases} R_j^s(\boldsymbol{x}, \boldsymbol{w}) - S_j, & \text{if } 1 \leq j \leq N_s^u, \\ S_j - R_j^s(\boldsymbol{x}, \boldsymbol{w}), & \text{if } N_s^u < j \leq N_s. \end{cases} \tag{8.16}$$

It is seen that the yield objective function defined in (8.15) is closely related to the number of failed designs. By minimizing (8.15), the yield of the fine model is expected to be increased. Therefore, the new nominal point used for the next iteration of yield optimization, $\boldsymbol{x}^{0,k+1}$, is obtained as follows:

$$\boldsymbol{x}^{0,k+1} = \arg\min_{\boldsymbol{x}^0} U^s(\boldsymbol{x}^0). \tag{8.17}$$

Note that by replacing the EM-based fine model by a computationally efficient SM surrogate model, and applying (8.15)–(8.17) on the SM surrogate, we avoid the large number of computationally expensive EM simulations and thus the EM-based yield optimization procedure can be accelerated.

8.4 The PC-based approach to yield-driven EM optimization

In this section, we review the recently reported PC-based approach to accelerate yield-driven EM optimization when equivalent circuit models are not available [32]. A waveguide bandpass filter example is provided to illustrate the advantages of the PC-based approach.

8.4.1 Formulations of the PC approach

The PC approach transforms the original random parameters $\boldsymbol{x}$ to independent standard random parameters $\boldsymbol{\xi}$ and then applies the stochastic expansion in the "$\boldsymbol{\xi}$-space" [42]. Let the vector of the independent standard random parameters be denoted by $\boldsymbol{\xi} = [\xi_1, \ldots, \xi_n]^T$. In yield optimization, the nominal point $\boldsymbol{x}^0$ is a variable and the relationship between $\boldsymbol{\xi}$ and $\boldsymbol{x}$ can be denoted by [32]

$$\boldsymbol{x} = T^{-1}(\boldsymbol{x}^0, \boldsymbol{\xi}) \tag{8.18}$$

where T denotes the transformation function used to transform $\boldsymbol{x}$ to be $\boldsymbol{\xi}$. The functional form between $R_j^{\mathrm{EM}}(\boldsymbol{x})$ and $\boldsymbol{\xi}$ is approximated by the sum of weighted orthogonal basis polynomials in terms of $\boldsymbol{\xi}$ [30]

$$R_j^{\mathrm{EM}}(\boldsymbol{x}) = R_j^{\mathrm{EM}}(T^{-1}(\boldsymbol{x}^0, \boldsymbol{\xi})) = \sum_{i=0}^{\mathrm{P}} a_{ij}(\boldsymbol{x}^0)\Phi_i(\boldsymbol{\xi}) \tag{8.19}$$

where $\Phi_i(\cdot)$ is the generalized PC basis function. Different probability distribution types of the design parameters have different optimal bases to construct the multivariate basis function [42]. The coefficient $a_{ij}(\boldsymbol{x}^0)$ is the weighting coefficient for the ith basis function Φ_i of the jth EM response $R_j^{\mathrm{EM}}(\boldsymbol{x})$. P determines the total number of terms [30].

In the non-intrusive PC approaches, the PC coefficients $a_{ij}(\boldsymbol{x}^0)$ can be found by [21]

$$a_{ij}(\boldsymbol{x}^0) = \frac{\int_{\Omega^n} R_j^{\mathrm{EM}}(T^{-1}(\boldsymbol{x}^0, \boldsymbol{\xi}))\Phi_i(\boldsymbol{\xi})\rho(\boldsymbol{\xi})d\boldsymbol{\xi}}{\int_{\Omega^n} \Phi_i^2(\boldsymbol{\xi})\rho(\boldsymbol{\xi})d\boldsymbol{\xi}} \tag{8.20}$$

where Ω^n is the n-dimensional space covering all the possible values of $\boldsymbol{\xi}$; $\rho(\boldsymbol{\xi})$ is the joint probability density function (PDF) of the standard random parameters $\boldsymbol{\xi}$. Note that the PC coefficients a_{ij} need to be re-computed from one iteration to another since $\boldsymbol{x}^0$ is a variable during yield optimization. The multidimensional integration in (8.20) can be evaluated using numerical quadrature, for example, [21]

$$\int_{\Omega^n} R_j^{\mathrm{EM}}(T^{-1}(\boldsymbol{x}^0, \boldsymbol{\xi}))\Phi_i(\boldsymbol{\xi})\rho(\boldsymbol{\xi})d\boldsymbol{\xi} \approx \sum_{l=1}^{M} R_j^{\mathrm{EM}}(T^{-1}(\boldsymbol{x}^0, \boldsymbol{\xi}^{(l)}))\Phi_i(\boldsymbol{\xi}^{(l)})w^{(l)} \tag{8.21}$$

$$\int_{\Omega^n} \Phi_i^2(\boldsymbol{\xi})\rho(\boldsymbol{\xi})d\boldsymbol{\xi} \approx \sum_{l=1}^{M} \Phi_i^2(\boldsymbol{\xi}^{(l)})w^{(l)} \tag{8.22}$$

where $\boldsymbol{\xi}^{(l)}$ and $w^{(l)}$ are the integration quadrature points (also called "nodes") and weights in the "$\boldsymbol{\xi}$-space," respectively. M is the total number of integration quadrature points. $R_j^{\mathrm{EM}}(T^{-1}(\boldsymbol{x}^0, \boldsymbol{\xi}^{(l)}))$ is the EM response evaluated at the lth sampling point in the original random space. The sparse grid techniques based on the Smolyak algorithm [43] are typically applied to reduce the computational costs for multidimensional numerical integration in (8.21) and (8.22).

Since the result of the numerical quadrature in (8.22) does not depend on the change of $\boldsymbol{x}^0$, for notational convenience, we define b_i as follows

$$b_i = \sum_{l=1}^{M} \Phi_i^2(\boldsymbol{\xi}^{(l)})w^{(l)}. \tag{8.23}$$

It is noticeable that the value of b_i only depends on M and the distribution types of standard random parameters $\boldsymbol{\xi}$. Therefore, we can determine the value b_i before the yield optimization process begins. Substituting (8.20)–(8.22) and (8.23) into (8.20) yields

$$a_{ij}(\boldsymbol{x}^0) = \frac{1}{b_i}\sum_{l=1}^{M} R_j^{\mathrm{EM}}(T^{-1}(\boldsymbol{x}^0, \boldsymbol{\xi}^{(l)}))\Phi_i(\boldsymbol{\xi}^{(l)})w^{(l)} \tag{8.24}$$

where $\Phi_i(\boldsymbol{\xi}^{(l)})$ and $w^{(l)}$ are constants during yield optimization.

One of the most attractive features of the PC approach is that the statistical properties (such as mean $\mu_j(\boldsymbol{x}^0)$ and variance $\sigma_j^2(\boldsymbol{x}^0)$) of the stochastic quantity $R_j^{\mathrm{EM}}(\boldsymbol{x})$ can be obtained analytically through the PC coefficients $a_{ij}(\boldsymbol{x}^0)$, that is, [21]

$$\mu_j(\boldsymbol{x}^0) = E(R_j^{\mathrm{EM}}) = a_{0j}(\boldsymbol{x}^0) \tag{8.25}$$

$$\sigma_j^2(\boldsymbol{x}^0) = E\left[(R_j^{\mathrm{EM}} - \mu_j)^2\right] = \sum_{i=1}^{\mathrm{P}} a_{ij}^2(\boldsymbol{x}^0)b_i. \tag{8.26}$$

8.4.2 *Formulation of the optimization objective function incorporating PC coefficients*

In order to allow the incorporation of the PC coefficients into EM-based yield optimization, we re-organize the objective function defined in (8.7) into a new form so that the statistical properties of the EM responses offered by the PC coefficients can be exploited. Specifically, we swap the order of the summations and at the same time rearrange the elements in J and K to rewrite the original objective function $U(\boldsymbol{x}^0)$ in (8.7) as [32]

$$U(\boldsymbol{x}^0) = \sum_{j\in\bar{J}}\sum_{k\in\bar{K}_j} e_j(T^{-1}(\boldsymbol{x}^0, \boldsymbol{\xi}^k)) \tag{8.27}$$

$$\bar{K}_j = \{k|e_j(T^{-1}(\boldsymbol{x}^0, \boldsymbol{\xi}^k)) \quad > 0\} \tag{8.28}$$

$$\bar{J} = \{j|\bar{K}_j \neq \varnothing\} \tag{8.29}$$

where $\bar{K}_j$ is a set which contains all the indices of the design outcomes whose EM responses violate the design specification sample S_j; $\bar{J}$ is a set which contains all the indices of the design specification samples which are violated by at least one random outcome. For ease of description, the re-organized objective function (8.27) is divided into two parts according to the design specification type as follows:

$$U(\boldsymbol{x}^0) = \sum_{j\in\bar{J}_u}\sum_{k\in\bar{K}_j} e_j(T^{-1}(\boldsymbol{x}^0, \boldsymbol{\xi}^k)) + \sum_{j\in\bar{J}_l}\sum_{k\in\bar{K}_j} e_j(T^{-1}(\boldsymbol{x}^0, \boldsymbol{\xi}^k)) \tag{8.30}$$

where $\bar{J}_u$ and $\bar{J}_l$ contain the indices of the upper and lower specification samples which are violated by at least one design outcome, respectively,

$$\bar{J}_u = \{j|j \in \bar{J}, 1 \leq j \leq m_u\} \tag{8.31}$$

$$\bar{J}_l = \{j|j \in \bar{J}, m_u < j \leq m\}. \tag{8.32}$$

Define $\bar{u}_j(\boldsymbol{x}^0)$ as a yield indicator with respect to the specification sample S_j as follows [32]:

$$\bar{u}_j(\boldsymbol{x}^0)\begin{cases} -S_j\int_{S_j}^{\infty} f(R_j^{\mathrm{EM}})\mathrm{d}R_j^{\mathrm{EM}} + \int_{S_j}^{\infty} R_j^{\mathrm{EM}} f(R_j^{\mathrm{EM}})\mathrm{d}R_j^{\mathrm{EM}},\ j \in \bar{J}_u. \\ S_j\int_{-\infty}^{S_j} f(R_j^{\mathrm{EM}})\mathrm{d}R_j^{\mathrm{EM}} - \int_{-\infty}^{S_j} R_j^{\mathrm{EM}} f(R_j^{\mathrm{EM}})\mathrm{d}R_j^{\mathrm{EM}},\ j \in \bar{J}_l. \end{cases} \tag{8.33}$$

Through some mathematical derivaitons, we can prove that the re-organized objective function in (8.30) is mathematically related to $\bar{u}_j(\boldsymbol{x}^0)$ [32]. Loosely speaking, a smaller value of $\bar{u}_j(\boldsymbol{x}^0)$ implies that there are fewer outcomes whose responses violate the specification sample S_j, indicating a higher yield. Therefore, reducing the value of $\bar{u}_j(\boldsymbol{x}^0)$ is expected to lead to an increase of the yield.

According to (8.25) and (8.26), we can relate the first term of $\bar{u}_j(\boldsymbol{x}^0), j \in \bar{J}_u$, to the nominal point $\boldsymbol{x}^0$ via PC coefficients as follows [32]:

$$-S_j\int_{S_j}^{\infty} f(R_j^{\mathrm{EM}})\mathrm{d}R_j^{\mathrm{EM}} = S_j\left(\phi\left(\frac{S_j - a_{0j}(\boldsymbol{x}^0)}{\sqrt{\sum_{i=1}^{\mathrm{P}} a_{ij}^2(\boldsymbol{x}^0)b_i}}\right) - 1\right) \tag{8.34}$$

where $\phi(\cdot)$ is the cumulative distribution function of the standard normal distribution; a_{0j} and a_{ij} represent the PC coefficients as functions of the yield optimization variable $\boldsymbol{x}^0$. For notational convenience, we define $\gamma_j(\boldsymbol{x}^0)$ as

$$\gamma_j(\boldsymbol{x}^0) = \frac{S_j - a_{0j}(\boldsymbol{x}^0)}{\sqrt{\sum_{i=1}^{P} a_{ij}^2(\boldsymbol{x}^0)b_i}}, \quad j \in \bar{J}_u \cup \bar{J}_l. \tag{8.35}$$

Here, $\phi(\gamma_j(\boldsymbol{x}^0))$ represents the probability that R_j^{EM} satisfies the specification S_j when $j \in \bar{J}_u$ or that R_j^{EM} violates S_j when $j \in \bar{J}_l$.

To analytically relate the second term of $\bar{u}_j(\boldsymbol{x}^0), j \in \bar{J}_u$, to $\boldsymbol{x}^0$ via PC coefficients, we need to find the mean value of the EM response R_j^{EM} under the condition that R_j^{EM} violates the upper specification S_j, that is, the value of $E(R_j^{\mathrm{EM}}|R_j^{\mathrm{EM}} > S_j)$. Then $\int_{S_j}^{\infty} R_j^{\mathrm{EM}} f(R_j^{\mathrm{EM}})\mathrm{d}R_j^{\mathrm{EM}}$ can be analytically related to the nominal point $\boldsymbol{x}^0$ as follows [44]:

$$\int_{S_j}^{\infty} R_j^{\mathrm{EM}} f(R_j^{\mathrm{EM}})\mathrm{d}R_j^{\mathrm{EM}} = \mu_j(1 - \phi(\gamma_j(\boldsymbol{x}^0))) + \frac{1}{\sqrt{2\pi}}\sigma_j \cdot e^{-\frac{1}{2}\gamma_j^2(\boldsymbol{x}^0)} \tag{8.36}$$

where μ_j and σ_j can be computed from the PC coefficients as shown in (8.25) and (8.26).

By substituting (8.25)–(8.26) and (8.34)–(8.36) into $\bar{u}_j(\boldsymbol{x}^0)$, we incoporate the PC coefficients into $\bar{u}_j(\boldsymbol{x}^0)$ for the upper specifications, $j \in \bar{J}_u$, as follows [32]:

$$\bar{u}_j(\boldsymbol{x}^0) = \sqrt{\sum_{i=1}^{P} a_{ij}^2(\boldsymbol{x}^0) b_i \cdot [(\phi(\gamma_j(\boldsymbol{x}^0)) - 1)\gamma_j(\boldsymbol{x}^0) + \frac{1}{\sqrt{2\pi}} e^{-\frac{1}{2}\gamma_j^2(\boldsymbol{x}^0)}]}. \tag{8.37}$$

In a similar manner, the PC coefficients are incorporated into $\bar{u}_j(\boldsymbol{x}^0)$ for the lower specifications, $j \in \bar{J}_l$, as follows:

$$\bar{u}_j(\boldsymbol{x}^0) = \sqrt{\sum_{i=1}^{P} a_{ij}^2(\boldsymbol{x}^0) b_i \cdot [\phi(\gamma_j(\boldsymbol{x}^0))\gamma_j(\boldsymbol{x}^0) + \frac{1}{\sqrt{2\pi}} e^{-\frac{1}{2}\gamma_j^2(\boldsymbol{x}^0)}]}. \tag{8.38}$$

Accordingly, the PC coefficients-incorporated objective function $\bar{U}(\boldsymbol{x}^0)$ for yield-driven EM optimization including both upper and lower specifications is formulated as

$$\begin{aligned} \bar{U}(\boldsymbol{x}^0) &= \sum_{j \in \bar{J}_u} \bar{u}_j(\boldsymbol{x}^0) + \sum_{j \in \bar{J}_l} \bar{u}_j(\boldsymbol{x}^0) \\ &= \sum_{j \in \bar{J}} V_j(\boldsymbol{x}^0) + \sum_{j \in \bar{J}_u} (a_{0j}(\boldsymbol{x}^0) - S_j) \end{aligned} \tag{8.39}$$

where $V_j(\boldsymbol{x}^0)$ is computed from the PC coefficients a_{ij} and the constant b_i as follows [32]:

$$V_j(\boldsymbol{x}^0) = \sqrt{\sum_{i=1}^{P} a_{ij}^2(\boldsymbol{x}^0) b_i \cdot [\phi(\gamma_j(\boldsymbol{x}^0))\gamma_j(\boldsymbol{x}^0) + \frac{1}{\sqrt{2\pi}} e^{-\frac{1}{2}\gamma_j^2(\boldsymbol{x}^0)}]}. \tag{8.40}$$

It can be concluded from (8.27) to (8.40) that we have reformulated the original objective function defined in (8.7) into a new format where the PC coefficients are incorporated. The PC coefficients are able to provide the statistical properties and approximate the PDF of the EM response with fewer EM samples than that required by MC analysis. Therefore, the new PC-based formulation requires fewer EM simulations to obtain accurate yield representation than the original formulation (which is based on assembling the responses evaluated at the MC samples). This advantage allows the PC coefficients-incorporated objective function to facilitate the overall EM-based yield optimization process effectively.

In order to apply the PC coefficients-incorporated objective function in yield-driven EM optimization, we need to deduce the derivatives of $\bar{U}(\boldsymbol{x}^0)$ with respect to the varying nominal point $\boldsymbol{x}^0$ so that a gradient-based optimization algorithm (e.g., the quasi-Newton method) can be employed. As shown in Figure 8.1, as the nominal point changes from $\boldsymbol{x}^0$ to $\boldsymbol{x}^0_{\text{new}}$, the sparse grid samples in the original parameter space change accordingly, to form a new set of sparse grid samples, denoted by $\mathscr{K}_{\text{new}}$. That is to say, the change of the nominal point affects the locations of the sparse grid points in the original parameter space, which in turn affects the value of each PC coefficient a_{ij}. Therefore, the derivative information can be obtained by first deriving the sensitivity formulas of PC coefficients w.r.t. $\boldsymbol{x}^0$ and then deducing the sensitivities for the objective function based on the

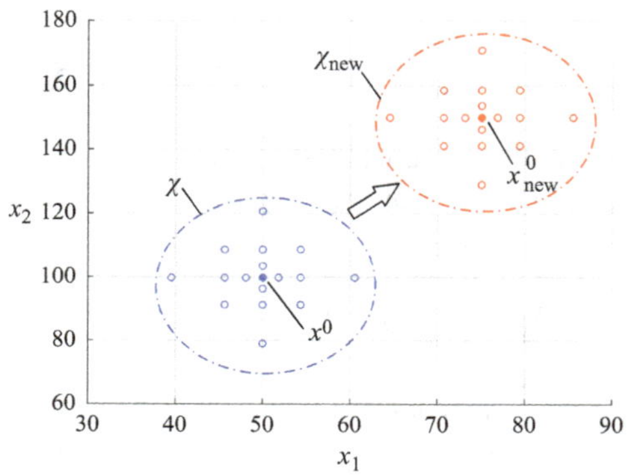

Figure 8.1 Illustration of the change of the locations of sparse grid samples in the PC-based approach between two consecutive iterations during yield optimization. The two black dots represent the current and the new nominal points. The hollow circles represent the integration samples generated by the sparse grid technique (based on [32])

sensitivity formulas of the PC coefficients [32]. The overall flow chart of the PC-based yield optimization approach is shown in Figure 8.2.

8.4.3 Yield optimization of a waveguide bandpass filter with FSI using the PC-based approach

Consider a waveguide bandpass filter with FSI, as shown in Figure 8.3(a) [45]. The design parameters of the filter are $\boldsymbol{x} = [d_1\ d_2\ d_3]^T$ (mm). The detailed geometrical dimensions of one of the FSI are shown in Figure 8.3(b). The section of the waveguide is $a = 22.86$ mm, $b = 10.16$ mm (WR-90). The design specification for this filter is given by $|S_{11}| \leq -22$ dB, in the frequency range from 9.2 to 9.8 GHz. The cognition formulation of SM [46] is used to obtain the optimal nominal solution $\boldsymbol{x}^0_{\text{ini}} = [9.3444\ 4.9203\ \ 3.7423]^T$ (mm).

All the design parameters are assumed to have independent normal distributions to allow yield estimation and optimization. The standard deviation for each design parameter is assumed to be 20 μm. The proposed approach and the MC-based yield optimization approach are both used to optimize the yield. For the proposed approach, Hermite polynomials are used and the integration accuracy level required is 3 for the sparse grid technique. This results in the number of sparse grid samples $M = 19$. For the MC-based yield optimization approach, it is found that $N = 100$ is suitable to represent the statistics of all the possible outcomes. For comparison purpose, different number of random samples ($N = 19, 50, 100$) per iteration are used for the MC-based yield optimization approach. To allow both optimization approaches to converge, the maximum iteration count is set to be a large number.

The comparison between the two approaches in terms of the final yield value and the total number of EM simulations needed during optimization is given by Table 8.1. We can conclude from the table that the proposed method achieves the

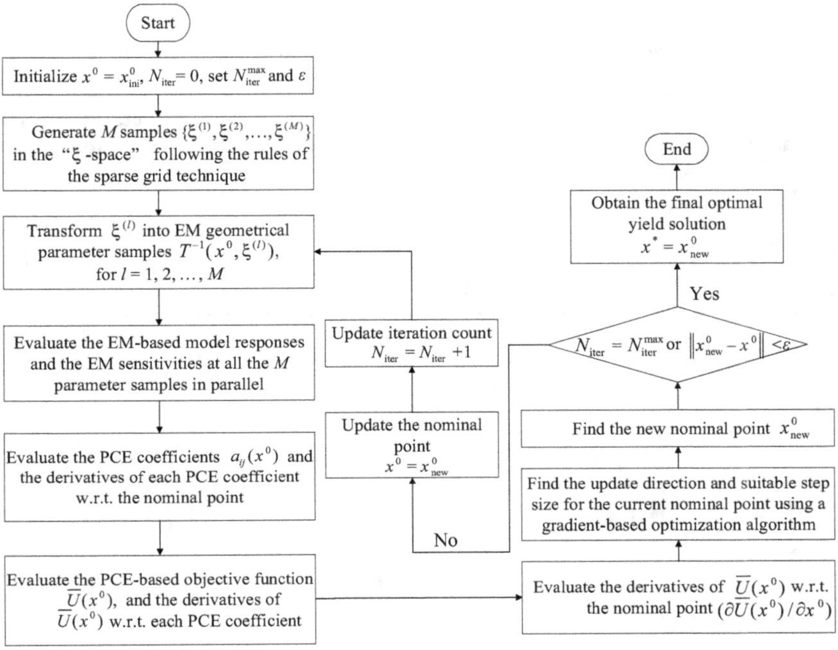

Figure 8.2 Overall flow chart of the PC-based yield optimization approach (based on [32])

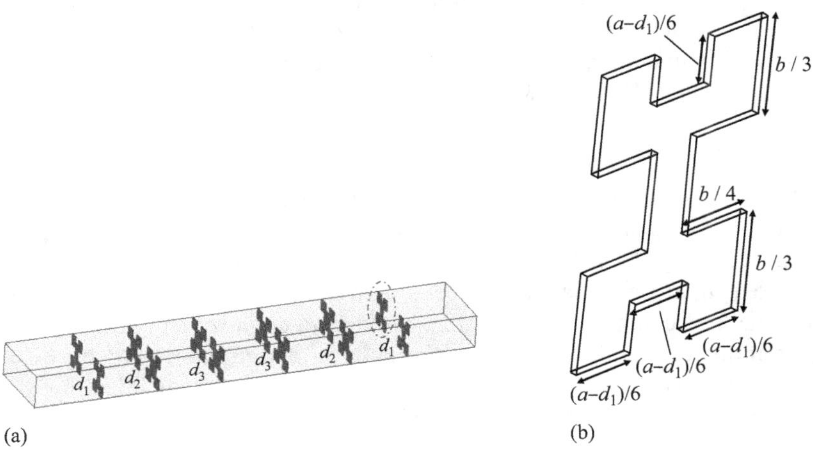

Figure 8.3 (a) The structure of the waveguide bandpass filter with fractal shaped irises (FSI): defined for EM simulation and yield optimization. (b) Detailed geometrical dimensions of the FSI as circled in (a) (based on [32])

Table 8.1 Comparison of yield optimization results for the waveguide FSI filter (based on [32])

Yield Optimization Approach	Number of EM Simulations Per iteration	Initial yield	Final yield	Total number of EM simulations
Monte Carlo-based	19	67%	69%	304
	50	67%	72%	900
	100	67%	79%	1,100
PC-based [32]	19	67%	84%	475

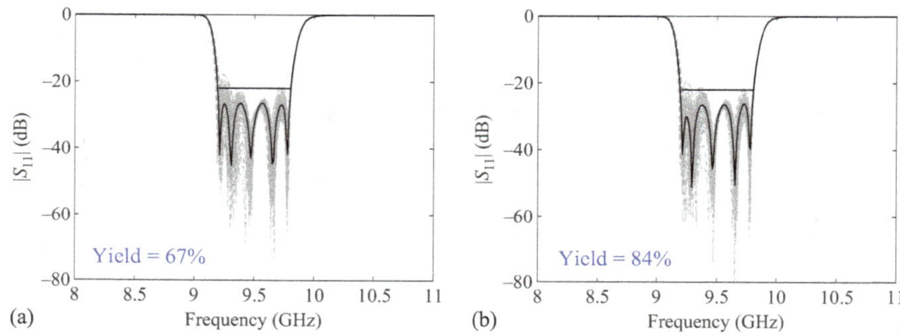

Figure 8.4 Yield optimization results of the waveguide bandpass filter with FSI: (a) before yield optimization and (b) after yield optimization. The black solid line indicates the response evaluated at the nominal point, while the gray dashed lines indicate the responses evaluated at the 100 MC samples (based on [32])

greatest yield increase among four cases of optimizations compared. It can also be observed that to achieve a similar yield increase, the total number of EM simulations required by the PC-based approach is less than half of that required by the MC-based yield optimization approach. This reduction allows us to achieve a substantial speed up for the overall EM-based yield optimization process. The yield values of the waveguide FSI filter before and after optimization using the PC-based approach are shown in Figure 8.4. The optimal yield solution found by the PC-based approach is $[9.3682\ 4.9542\ 3.7585]^T$ (mm).

8.5 The ANN-based approach to yield-driven EM optimization

In this section, we first review the pure ANN-based approach to yield-driven EM optimization. Then, we describe the neuro-TF approach and its recent advance to improve the capability and generality of pure ANN models. A four-pole waveguide filter example is used to illustrate the adaptively weighted neuro-TF approach.

8.5.1 Formulations of pure ANN-based approach to yield-driven EM optimization

In the pure ANN-based approach to yield-driven design of microwave structures, we first train an ANN model to represent the EM behavior of the microwave structure in a certain region in the design parameter space. Then, we use the trained ANN to replace the EM-based fine model to perform yield optimization [47]. Let the input and output vectors of the ANN model be denoted by $\boldsymbol{x}$ and $\boldsymbol{y}$, respectively. Let ω represent the frequency, which is an addtional input to the ANN model. Let $\boldsymbol{w}$ be a vector containing all the neural network weights in the ANN model. The model output $\boldsymbol{y}$ is a function of the geometrical parameters, the neural network weights, and frequency, that is,

$$\boldsymbol{y} = \boldsymbol{y}(\boldsymbol{w}, \boldsymbol{x}, \omega_i) \tag{8.41}$$

where ω_i represents the ith frequency sample among all the frequency samples of interest. Let $\boldsymbol{d}$ represents the EM responses obtained from EM simulation. Let the training data for training the ANN model be denoted by $\{\boldsymbol{x}^{(l)}, \boldsymbol{d}^{(l)}\}$, where $n = 1, 2, \ldots, n_s$, and n_s is the total number of training samples. The value of n_s depends on the specific sampling method (e.g., random distribution and star distribution) used to generate the training samples. Let $E_{\text{ANN}}(\boldsymbol{w})$ be the error function for training the ANN model, formulated as [47]

$$E_{\text{ANN}}(\boldsymbol{w}) = \frac{1}{2n_s} \sum_{l=1}^{n_s} \sum_{i=1}^{F} ||\boldsymbol{y}(\boldsymbol{w}, \boldsymbol{x}^{(l)}, \omega_i) - d_i^{(l)}||^2 \tag{8.42}$$

where $d_i^{(l)}$ represents the EM response corresponding to $\boldsymbol{x}^{(l)}$ and ω_i; the symbol F denotes the total number of frequency samples of interest. In practice, the ANN structure (e.g., number of hidden neurons) and the amount of training data may need to be adjusted in order to achieve the desired model accuracy.

Once the training process is finished, the trained ANN can be used as an efficient surrogate to provide fast and accurate predictions of the EM behavior of the considered microwave structure. Subsequently, it can be utilized in the high-level yield-driven design of microwave structures exploiting any existing statistical approaches [3–6] or geometrical approaches [1,2]. However, this pure ANN-based approach suffers from several limitaions in practice. Firstly, yield optimization approaches typically address the situation where the range of design space that covers the optimal yield solution is not known in advance. This necessitates accurate models in a large range in the design parameter space, which could defy the modeling capability of pure ANNs. Moreover, it is well-known that the generality of pure ANN models is not satisfactory when applied to design optimization of complex microwave structures. A huge amount of training data are usually needed and a highly complex ANN structure is typically used to achieve the desired model accuracy. One of the solutions to this issue is to leverage the neuro-TF modeling method to improve the capability and generality of the pure ANN modeling method. In the following, we describe the neuro-TF approach to yield-driven EM optimization.

8.5.2 Neuro-TF approach to yield-driven EM optimization

The neuro-TF approach to yield-driven EM optimization finds the optimal yield solution iteratively. Let k $(k = 1, 2, \ldots,)$ denote the iteration counter. In each iteration, a pole-residue-based neuro-TF model [38] is first developed around the current nominal point through a two-stage training process. Then, the trained neuro-TF is utilized to perform yield optimization exploiting the generalized l_1 design centering algorithm [12].

The pole-residue-based neuro-TF model mainly consists of two components, that is, pole-residue-based transfer functions and neural networks. The pole-residue-based transfer function is expressed as follows [38]:

$$R_{\text{TF}}(\boldsymbol{x}, s) = \sum_{j=1}^{q} \frac{r_j(\boldsymbol{x})}{s - p_j(\boldsymbol{x})} \tag{8.43}$$

where R_{TF} denotes the response of the neuro-TF model; s represents the frequency in Laplace domain; q is the order of the transfer function; r_j and p_j $(j = 1, \ldots, q)$ are the jth residue and the jth pole for the transfer function, respectively. Both r_j and p_j are functions of the geometrical parameters $\boldsymbol{x}$. As geometrical parameters change, the poles and residues in the transfer function change accordingly. The vector fitting technique [39] is used to extract the poles and residues in transfer function for a given set of EM responses. Neural networks are utilized to learn the unknown and nonlinear relationships between the poles/residues and geometrical parameters. The training of the neuro-TF model consists of two stages: (1) preliminary training of the neural networks; and (2) refinement training of the overall neuro-TF model.

In the first stage, the neural networks are preliminarily trained to learn the relationships between the poles/residues in the transfer function and the geometrical parameters. The details of this training stage are provided by [38]. In the second stage, the overall neuro-TF model is further refined with initial neural network weights found in the first stage. The purpose is to further align the outputs of the neuro-TF model with those of the fine model by adjusting the internal neural network weights using a certain optimization algorithm. The error function used in the refinement training stage can be formulated as:

$$\begin{aligned} E_{tr}(\boldsymbol{w}_p, \boldsymbol{w}_r) = \sum_{l=1}^{n_s} \sum_{i=1}^{F} || R_{\text{TF}}(p_{NN}(\boldsymbol{x}^{(l),k}, \boldsymbol{w}_p), r_{NN}(\boldsymbol{x}^{(l),k}, \boldsymbol{w}_r), \omega_i) \\ - R_{\text{EM}}(\boldsymbol{x}^{(l),k}, \omega_i)||^2 \end{aligned} \tag{8.44}$$

where $\boldsymbol{x}^{(l),k}$ is the lth geometrical sample in the kth iteration, generated according to a certain sampling method; $R_{\text{EM}}(\boldsymbol{x}^{(l),k}, \omega_i)$ denotes the EM response corresponding to $\boldsymbol{x}^{(l),k}$ at frequency ω_i; n_s represents the total number of training samples; F represents the total number of frequency samples; p_{NN} and r_{NN} represent the neural network outputs for poles and residues, respectively; $\boldsymbol{w}_p$ and $\boldsymbol{w}_r$ represent the internal neural network weights corresponding to p_{NN} and r_{NN}, respectively. Once the refinement training of the overall model is finished, the pole-residue-based neuro-TF surrogate is used in the subsequent yield optimization process.

The nominal point $\boldsymbol{x}^{0,k}$ is taken as the yield optimization variable at each iteration during optimization. Let X_{mc}^k be a set containing all the random outcomes around the nominal point $\boldsymbol{x}^{0,k}$, that is,

$$X_{mc}^k = \{\boldsymbol{x}^{m,k} | m = 1, \ldots, N_{\mathrm{mc}}\}. \tag{8.45}$$

where $\boldsymbol{x}^{m,k}$ represents the mth random outcome in the kth iteration. Let $\mathscr{S}_u$ and $\mathscr{S}_l$ be two sets which contain the upper and lower design specification samples, respectively. Let $X_{\mathrm{TF}}^k(\omega_i)$ be a set which contains all the random outcomes for which the neuro-TF model response at ω_i violates the corresponding design specification, that is,

$$X_{\mathrm{TF}}^k(\omega_i) = \{\boldsymbol{x}^{m,k} | e^{\mathrm{TF}}(\boldsymbol{x}^{m,k}, \omega_i) > 0, m = 1, \ldots, N_{mc}\}. \tag{8.46}$$

where

$$e^{\mathrm{TF}}(\boldsymbol{x}^{m,k}, \omega_i) = \begin{cases} R_{\mathrm{TF}}(\boldsymbol{x}^{m,k}, \omega_i) - S(\omega_i), & \text{if} S(\omega_i) \in \mathscr{S}_u \\ S(\omega_i) - R_{\mathrm{TF}}(\boldsymbol{x}^{m,k}, \omega_i), & \text{if} S(\omega_i) \in \mathscr{S}_l \end{cases} \tag{8.47}$$

where $R_{\mathrm{TF}}(\boldsymbol{x}^{m,k}, \omega_i)$ denotes the response of neuro-TF model for the random outcome $\boldsymbol{x}^{m,k}$ at frequency ω_i, and $S(\omega_i)$ is the design specification sample corresponding to frequency ω_i.

The yield optimization is performed by minimizing the following objective function with a certain gradient-based optimization algorithm (e.g., the quasi-Newton method):

$$U^{\mathrm{TF}}(\boldsymbol{x}^{0,k}) = \sum_{i \in F} \sum_{\boldsymbol{x}^{m,k} \in X_{\mathrm{TF}}^k(\omega_i)} e^{\mathrm{TF}}(\boldsymbol{x}^{m,k}, \omega_i). \tag{8.48}$$

Let $\boldsymbol{x}^{0,k+1}$ denote the new nominal point found by the gradient-based optimization algorithm from the nominal point $\boldsymbol{x}^{0,k}$ in the present iteration, that is, the kth iteration. Formally, $\boldsymbol{x}^{0,k+1}$ is obtained as follows:

$$\boldsymbol{x}^{0,k+1} = \arg \min_{\boldsymbol{x}^{0,k}} U^{\mathrm{TF}}(\boldsymbol{x}^{0,k}). \tag{8.49}$$

It is noticeable that the replacement of the EM-based fine model by the neuro-TF surrogate allows us to avoid the huge number of computationally expensive EM simulations, thereby accelerating the overall EM-based yield optimization procedure significantly.

8.5.3 Adaptively weighted yield-driven EM optimization incorporating neuro-TF surrogate

In this section, we review the recently reported adaptively weighted yield optimization approach incorporating neuro-TF surrogate [40]. In the first part, the definitions of key frequency points, ordinary frequency points, and adaptive weighting factors are presented. Then, a two-stage training process is presented for the development of the adaptively weighted neuro-TF model. In the second part, we review the formulation of a novel objective function for yield-driven design incorporating the adaptive weighting factors. The formulated objective function is then used to iteratively find the optimal yield by employing a gradient-based optimization algorithm.

Let $\mathscr{F} = \{1, \ldots, F\}$ be the index set of all the frequency points of interest, where F denotes the total number of frequencies. The adaptively weighted neuro-TF approach distinguishes *Key Frequency Points* from *Ordinary Frequency Points* in $\mathscr{F}$, defined as follows:

Key Frequency Points: The key frequency points are the frequencies at which at least one outcome from X_{mc}^k violates the corresponding design specification sample. Let $\mathscr{F}_{\text{key}}^k \subset \mathscr{F}$ represent the index set of the key frequency points in iteration k. Then, $\mathscr{F}_{\text{key}}^k$ is obtained as follows:

$$\mathscr{F}_{\text{key}}^k = \{i | X^k(\omega_i) \neq \varnothing, i = 1, \ldots, \mathscr{F}\} \tag{8.50}$$

where ω_i represents the ith frequency point of interest; $X^k(\omega_i)$ is a set which contains all the random outcomes for which the EM response at ω_i violates the corresponding design specification, that is,

$$X^k(\omega_i) = \{\boldsymbol{x}^{m,k} | e(\boldsymbol{x}^{m,k}, \omega_i) > 0, m = 1, \ldots, N_{\text{mc}}\}. \tag{8.51}$$

Here, $e(\boldsymbol{x}^{m,k}, \omega_i)$ is an error defined to measure the violation degree of the EM response relative to the design specification for random outcome $\boldsymbol{x}^{m,k}$ at frequency ω_i, that is,

$$e(\boldsymbol{x}^{m,k}, \omega_i) = \begin{cases} R_{\text{EM}}(\boldsymbol{x}^{m,k}, \omega_i) - S(\omega_i), & \text{if } S(\omega_i) \in \mathscr{S}_u \\ S(\omega_i) - R_{\text{EM}}(\boldsymbol{x}^{m,k}, \omega_i), & \text{if } S(\omega_i) \in \mathscr{S}_l \end{cases} \tag{8.52}$$

where $R_{\text{EM}}(\boldsymbol{x}^{m,k}, \omega_i)$ denotes the EM response for random outcome $\boldsymbol{x}^{m,k}$ at frequency ω_i, and $S(\omega_i)$ is the design specification sample corresponding to frequency ω_i. Clearly, a positive (negative) value of $e(\boldsymbol{x}^{m,k}, \omega_i)$ indicates that the design specification is violated (satisfied) by $R_{\text{EM}}(\boldsymbol{x}^{m,k}, \omega_i)$.

Ordinary Frequency Points: The ordinary frequency points are the frequencies onto which no design specification is imposed, and those at which no outcome from X_{mc}^k violates the corresponding design specification. Let $\mathscr{F}_{\text{ord}}^k \subset \mathscr{F}$ represent the index set of ordinary frequency points in iteration k. Clearly, we have $\mathscr{F}_{\text{key}}^k \cup \mathscr{F}_{\text{ord}}^k = \mathscr{F}$ and therefore,

$$\mathscr{F}_{\text{ord}}^k = \mathscr{F} \backslash \mathscr{F}_{\text{key}}^k \tag{8.53}$$

where the symbol \ is used to denote the difference between two sets.

Let α_i^k be defined as a factor indicating the degree to which the EM response may violate the design specification sample at each frequency point in iteration k. The adaptively weighted neuro-TF computes α_i^k as follows:

$$\alpha_i^k = \begin{cases} K \cdot |X^k(\omega_i)|, & \text{if} i \in \mathscr{F}_{\text{key}}^k \\ 1.0, & \text{if} i \in \mathscr{F}_{\text{ord}}^k \end{cases} \tag{8.54}$$

where $|X^k(\omega_i)|$ denotes the number of elements in the set $X^k(\omega_i)$, and K is a coefficient used to control the dominance of the key frequency points over the ordinary frequency points in training the surrogate. The weighting factor α_i^k $(i \in \mathscr{F})$ intuitively indicates the sensitive degree of the EM response relative to the design specification at ω_i. A larger value of α_i^k implies a more sensitive EM response to the design specification and thus should be assigned with a larger weighting factor in the surrogate model training process.

The expression of the pole-residue-based transfer function, and the first training stage of the neuro-TF model are the same as that in the neuro-TF approach. In the second training stage, the overall neuro-TF model is further refined using a adaptive weighting factors incorporated training error function, formulated as [40]:

$$\begin{aligned} \bar{E}_{tr}(\boldsymbol{w}_p, \boldsymbol{w}_r) = \sum_{l=1}^{n_s} \sum_{i \in \mathscr{F}} \alpha_i^k || R_{\text{TF}}(p_{\text{NN}}(\boldsymbol{x}^{(l),k}, \boldsymbol{w}_p), r_{\text{NN}}(\boldsymbol{x}^{(l),k}, \boldsymbol{w}_r), \omega_i) \\ - R_{\text{EM}}(\boldsymbol{x}^{(l),k}, \omega_i)||^2 \end{aligned} \tag{8.55}$$

where α_i^k is the weighting factor for the ith frequency point calculated by (8.54), which is used to enhance the surrogate yield estimation accuracy at the key frequency points; R_{TF} and R_{EM} denote the response of the neuro-TF model and that of the EM-based fine model, respectively. Once the adaptively weighted refinement training of the overall model is finished, the pole-residue-based neuro-TF surrogate is used in the subsequent yield optimization process.

The nominal point $\boldsymbol{x}^{0,k}$ is taken as the yield optimization variables at each iteration during optimization. The yield optimization is performed by minimizing the following objective function with a certain gradient-based optimization algorithm:

$$\bar{U}^{\text{TF}}(\boldsymbol{x}^{0,k}) = \sum_{i \in \mathscr{F}} \sum_{\boldsymbol{x}^{m,k} \in X^k(\omega_i)} \alpha_i^k e^{\text{TF}}(\boldsymbol{x}^{m,k}, \omega_i). \tag{8.56}$$

Let $\boldsymbol{x}^{0,k+1}$ denote the new nominal point found by the gradient-based optimization algorithm from the nominal point $\boldsymbol{x}^{0,k}$ in the present iteration, that is, the kth iteration. Formally, $\boldsymbol{x}^{0,k+1}$ is obtained as follows:

$$\boldsymbol{x}^{0,k+1} = \arg\min_{\boldsymbol{x}^{0,k}} \bar{U}^{\text{TF}}(\boldsymbol{x}^{0,k}). \tag{8.57}$$

Note that by incorporating the adaptive weighting factor α_i^k to the optimization objective function (8.56), more attentions are paid to the key frequency points in the optimization. This can further help accelerate the convergence of the overall yield optimization process [40]. After the surrogate training is finished, the neuro-TF surrogate learns the EM behavior of the EM-based fine model in a certain region in the design parameter space. Therefore, it is necessary to define a trust region for the trained surrogate in each iteration. For example, the yield-specific trust region framework presented in [17] can be adopted to update the trust radius of each design parameter in each iteration during yield optimization.

The adaptively weighted yield optimization approach terminates whenever one of the following three conditions is satisfied: (1) the difference between the current nominal point $\boldsymbol{x}^{0,k}$ and the new nominal point $\boldsymbol{x}^{0,k+1}$ is sufficiently small, that is, smaller than a user-defined threshold ε_1; (2) the difference between the yield values of the EM-based fine model in the two successive iterations is less than another user-defined threshold ε_2; and (3) the iteration counter k exceeds the maximum iteration count $k_{\max}$. These three conditions can be expressed as:

$$\|\boldsymbol{x}^{0,k+1} - \boldsymbol{x}^{0,k}\| \leq \varepsilon_1 \tag{8.58}$$

$$\|Y_{\text{EM}}(\boldsymbol{x}^{0,k+1}) - Y_{\text{EM}}(\boldsymbol{x}^{0,k})\| \leq \varepsilon_2 \tag{8.59}$$

$$k > k_{\max} \tag{8.60}$$

where $Y_{\text{EM}}(\boldsymbol{x}^{0,k})$ and $Y_{\text{EM}}(\boldsymbol{x}^{0,k+1})$ denote the yield values of the EM-based fine model at iterations k and $k+1$, respectively, which can be obtained by the traditional MC approach; ε_1, ε_2, and $k_{\max}$ are all user-defined criteria. The overall flow chart of the adaptively weighted yield optimization approach incorporating neuro-TF surrogate is shown in Figure 8.5.

8.5.4 *Yield optimization of a four-pole waveguide filter using the adaptively weighted yield optimization approach incorporating neuro-transfer function surrogate*

The adaptively weighted neuro-TF-based yield optimization approach is applied to the yield optimization of a four-pole waveguide filter [46], as shown in Figure 8.6. Four design parameters are considered, that is, $\boldsymbol{x} = [h_1\ h_2\ h_3\ h_{c1}]^T$. The design specification is given by $|S_{11}| \leq -17$ dB, in the frequency range from 10.85 to 11.15 GHz. The optimal nominal solution is $\boldsymbol{x}^{0,1} = [3.4248\ 4.3379\ 4.0401\ 3.2511]^T$ (mm). We assume independent normal distribution for all the design parameters with 10 μm standard deviation. The MC analysis is used to estimate the yield at the initial nominal point. A convergence on the yield value of the EM-based fine model is observed (20%) when $N_{\text{mc}} = 200$ by varying the number of random MC samples in the range of 50–300. The pole-residue-based neuro-TF model has five inputs (i.e., h_1, h_2, h_3, h_{c1}, and frequency) and one output (i.e., $|S_{11}|$). The order of the transfer function is $q = 12$. The initial trust radius is set to be 80 μm ($\boldsymbol{\delta}^1 = 80\,\mu\text{m}$) for each design

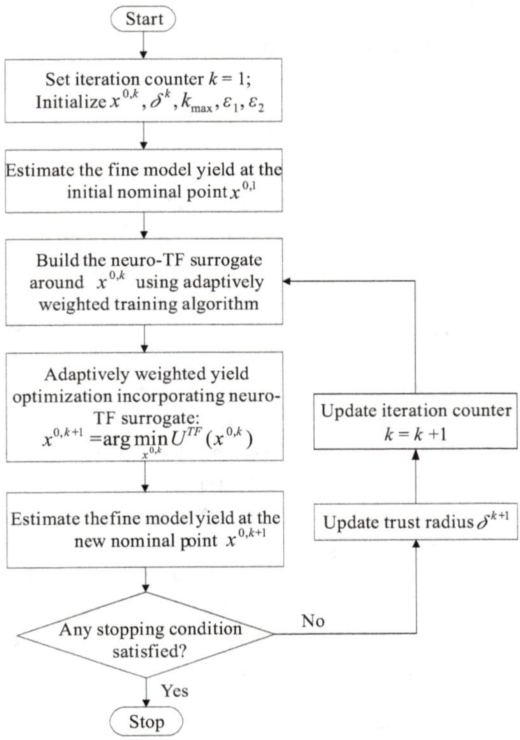

Figure 8.5 Flow chart of the adaptively weighted neuro-TF approach for efficient yield-driven EM optimization (based on [40])

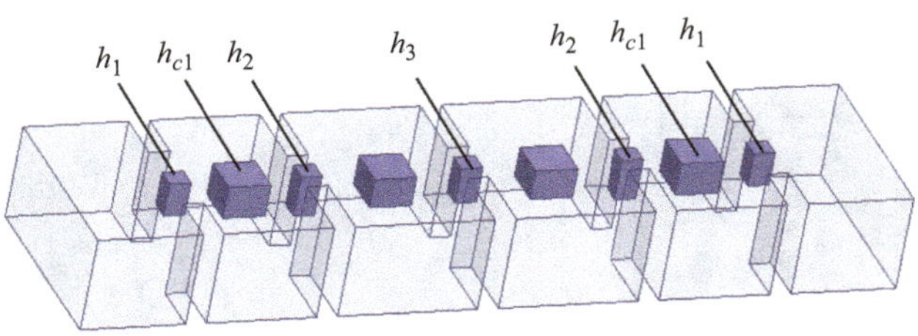

Figure 8.6 The structure of the four-pole waveguide filter example to be used for EM simulation and yield optimization (based on [40])

parameter. Nine levels of design of experiments (DOE) are used to generate the training data, resulting in a total of 81 samples ($n_s = 81$) of training data. The number of hidden neurons in the neural network models for the poles and residues is 10. The coefficient in computing the adaptive weighting factors is set to be 1.0 ($K = 1.0$). The termination conditions are set as $\varepsilon_1 = 1.0e^{-5}$, $\varepsilon_2 = 1\%$, and $k_{max} = 10$.

The adaptively weighted yield optimization terminates after three iterations, meeting the second termination condition. The final yield optimal solution obtained from the adaptively weighted neuro-TF approach is $\boldsymbol{x}^*= [3.5811\ 4.3949\ 3.9523\ 3.2005]^T$ (mm) with final yield value $Y^* = 97\%$. Figure 8.7(a) and (b) shows the initial yield and the final yield for the four-pole waveguide filter, respectively. It can be observed that a substantial improvement of the yield is achieved by using the adaptively weighted neuro-TF approach. In particular, in the first iteration, among 101 frequency points, 29 frequency points are identified as the key frequency points ($\mathscr{F}^1_{\text{key}}$) at which the weighting factors are assigned according to the value of $|X^1(\omega_i)|$. The remaining 72 frequency points are identified as the ordinary frequency points ($\mathscr{F}^1_{\text{ord}}$) at which the weighting factors are assigned to be constant ($\alpha_i^1 = 1.0$). In the second iteration, 31 frequency points are identified as the key frequency points ($\mathscr{F}^2_{\text{key}}$) while 70 frequency points are identified as the ordinary frequency points ($\mathscr{F}^2_{\text{ord}}$).

For the purpose of comparison, we also apply three existing approaches to optimize the yield of this filter. For the MC-based approach, 200 random samples are used per iteration. For the PC-based approach, the number of sparse grid samples is 33 in each iteration. For the neuro-TF approach, all the settings are the same with the proposed technique except that constant weighting factors are used for surrogate training and surrogate-assisted yield optimization. Table 8.2 compares the yield optimization results for all the four approaches in comparison. From Table 8.2 it can be observed that the proposed technique is able to find the optimal yield solution with much shorter CPU time than the other three methods. Compared with the neuro-TF approach, the adaptively weighted neuro-TF approach requires two less iterations to reach the desired yield threshold, leading to nearly 35% less CPU time for the whole yield optimization process.

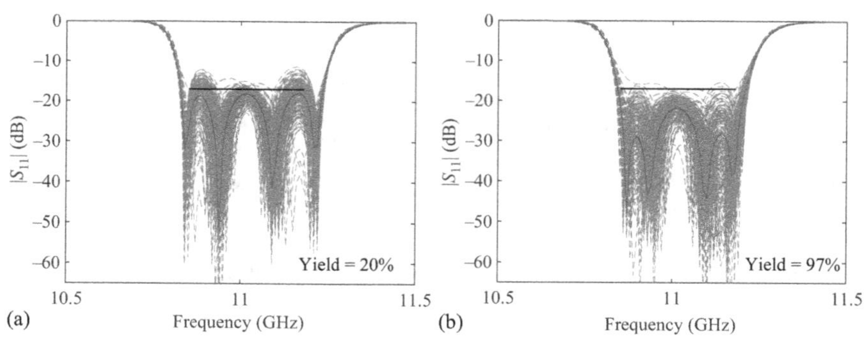

Figure 8.7 Yield optimization results of the four-pole waveguide filter: (a) initial yield before optimization and (b) final yield after optimization using the adaptively weighted neuro-TF approach. Gray dashed lines indicate the EM responses solved at the 200 MC samples, while the black solid line represents the EM response solved at the nominal point (based on [40])

Table 8.2 Comparison of yield optimization results for the four-pole waveguide filter (based on [40])

Yield optimization approach	MC-based	PC-based [32]	Neuro-TF [38]	Adaptively weighted neuro-TF [40]
Initial yield	20%	20%	20%	20%
Number of iterations	–	–	5	3
EM data generation	–	–	23 min × 5	23 min × 3
EM-based yield verification	–	–	56 min × 6	56 min × 4
Total surrogate training time	–	–	11 min × 5	11 min × 3
Total yield optimization time on surrogate	–	–	150 s	50 s
Total CPU time	72 h 48 min	12 h 35 min	8 h 28 min	5 h 27 min
Final yield	93%	92%	95%	97%

8.6 Discussion

This chapter has presented some recent advances in surrogate-assisted yield-driven design of microwave structures. Here, we discuss the proper usage of different approaches in different practical situations. When an equivalent circuit coarse model is available, the SM-based approaches such as that in [17] are more convenient to use. Compared with the PC-based approach [32] and the adaptively weighted neuro-TF approach [40], the SM-based approach usually requires a smaller amount of EM training data in each iteration in order to train the SM surrogate accurately. This is common when the behavior of the coarse model is close to that of the fine model. However, when the quality of the behavior of the coarse model is far from that of the fine model, we typically need a large amount of training data and a highly complicated mapping structure to train the SM surrogate accurately. This will result in a much longer CPU time for due to data generation and model training, which could potentially make the SM-based approach less efficient than the other two yield optimization approaches.

When an equivalent circuit coarse model is difficult to obtain and the microwave structure is to be designed using *S*-parameters, the adaptively weighted neuro-TF approach [40] is usually more efficient to use than the other two approaches. The adaptively weighted neuro-TF approach expresses the *S*-parameters by transfer function in a certain format and uses neural networks to learn the relationship between the coefficients in the transfer function and geometrical parameters. The use of the transfer function greatly reduce the nonlinearity of the neural network structure. Thus, the neuro-TF approach requires fewer EM training samples to develop the surrogate, thereby having great potential to shorten the total time for yield optimization of microwave structures. However, the neuro-TF approach is usually accompanied with the "order-changing" problem or the issue of "mismatch of poles/zeros," which requires specific algorithms to obtain continuous poles/zeros/residues among different geometrical parameter values during the development process of neuro-TF models.

When an equivalent circuit coarse model is not available, or the microwave structure is to be designed using other metrics such as characteristic impedance and gain, the PC-based approach [32] is probably easier to use. In the PC-based approach, the stochastic quantity (e.g., characteristic impedance and gain) is expressed as a function of the random parameters (e.g., geometrical dimensions) and the function is approximated by a weighted orthogonal basis polynomials. The construction of the PC model does not involve any training and testing processes. However, since the evaluation of PC coefficients in each iteration requires the EM responses solved at sparse grid samples, the total CPU time of the PC-based yield optimization is in general higher than the SM-based and neuro-TF-based approaches [40]. Therefore, in typical cases, the PC-based approach should only be used when the former two approaches cannot be applied conveniently.

8.7 Conclusion

A review of recent advances in surrogate-assisted yield-driven design of microwave structures has been provided in this chapter. By replacing the EM-based fine model with computationally efficient surrogate model in yield-driven design, the total yield optimization process can be siginificantly expedited. We have first introduced the MC-based approach to yield-driven design, which originates from the circuit community. Then, we have described the SM-based yield optimization approach which can mitigate the computational burden of MC-based yield optimization. Following that, we have illustrated the PC approach to uncertainty quntification and reviewed the PC-based yield optimization approach. Lastly, we have discussed the adaptively weighted yield optimization approach incorporating neuro-TF surrogate. The surrogate-assisted yield-driven EM optimization approaches help to shorten the total time in seeking for the optimal design of microwave structures with uncertainties in EM structural parameters.

References

[1] H. L. Abdel-Malek, A. S. O. Hassan, and M. H. Bakr, "A boundary gradient search technique and its application in design centering," *IEEE Trans. Comput.-Aided Design Integr. Circuits Syst.*, vol. 18, no. 11, pp. 1654–1661, 1999.

[2] H. L. Abdel-Malek, A. S. O. Hassan, and M. H. Bakr, "Statistical circuit design with the use of a modified ellipsoidal technique," *Int. J. Microw. Millim.-Wave Comput.-Aided Eng.*, vol. 7, no. 1, pp. 117–129, 1997.

[3] K. Singhal and J. F. Pinel, "Statistical design centering and tolerancing using parametric sampling," *IEEE Trans. Circuits Syst.*, vol. CAS-28, no. 7, pp. 692–702, 1981.

[4] A. H. Zaabab, Q. J. Zhang, and M. Nakhla, "A neural network modeling approach to circuit optimization and statistical design," *IEEE Trans. Microw. Theory Techn.*, vol. 43, no. 6, pp. 1349–1358, 1995.

[5] D. E. Hocevar, M. R. Lightner, and T. N. Trick, "An extrapolated yield approximation for use in yield maximization," *IEEE Trans. Comput.-Aided Design Integr. Circuits Syst.*, vol. CAD-3, no. 4, pp. 279–287, 1984.

[6] J. W. Bandler and S. H. Chen, "Circuit optimization: The state of the art," *IEEE Trans. Microw. Theory Techn.*, vol. 36, no. 2, pp. 424–443, 1988.

[7] N. K. Nikolova, X. Zhu, Y. Song, A. Hasib, and M. H. Bakr, "*S*-parameter sensitivities for electromagnetic optimization based on volume field solutions," *IEEE Trans. Microw. Theory Techn.*, vol. 57, no. 6, pp. 1526–1538, 2009.

[8] J. V. Morro, P. S. Pacheco, H. E. Gonzalez *et al.*, "Fast automated design of waveguide filters using aggressive space mapping with a new segmentation strategy and a hybrid optimization algorithm," *IEEE Trans. Microw. Theory Techn.*, vol. 53, no. 4, pp. 1130–1142, 2005.

[9] M. A. Ismail, D. Smith, A. Panariello, Y. Wang, and M. Yu, "EM-based design of large-scale dielectric-resonator filters and multiplexers by space mapping," *IEEE Trans. Microw. Theory Techn.*, vol. 52, no. 1, pp. 386–392, 2004.

[10] J. E. Rayas-Sanchez, "EM-based optimization of microwave circuits using artificial neural networks: The state-of-the-art," *IEEE Trans. Microw. Theory Techn.*, vol. 52, no. 1, pp. 420–435, 2004.

[11] J. W. Bandler, R. M. Biernacki, S. H. Chen, P. A. Grobelny, and R. H. Hemmers, "Space mapping technique for electromagnetic optimization," *IEEE Trans. Microwave Theory Tech.*, vol. 42, no. 12, pp. 2536–2544, 1994.

[12] J. W. Bandler, J. E. Rayas-Sanchez, and Q. J. Zhang, "Yield-driven electromagnetic optimization via space mapping-based neuromodels," *Int. J. RF Microw. Comput.-Aided Eng.*, vol. 12, no. 1, pp. 79–89, 2002.

[13] A. S. O. Hassan, A. S. A. Mohamed, and A. Y. El-Sharabasy, "EM-based yield optimization exploiting trust-region optimization and space mapping technology," *Int. J. RF Microwave CAE*, vol. 25, no. 6, pp. 474–484, 2015.

[14] Q. S. Cheng, J. W. Bandler, and S. Koziel, "Response corrected tuning space mapping for yield estimation and design centering," *IEEE MTT-S Int. Microw. Symp. Dig.*, Anaheim, CA, 2010, pp. 193–196.

[15] H. L. Abdel-Malek, A. S. O. Hassan, E. A. Soliman, and S. A. Dakroury, "The ellipsoidal technique for design centering of microwave circuits exploiting space-mapping interpolating surrogates," *IEEE Trans. Microw. Theory Techn.*, vol. 54, no. 10, pp. 3731–3738, 2006.

[16] A. S. Hassan, A. S. Mohamed, and A. Y. El-Sharabasy, "Statistical microwave circuit optimization via a non-derivative trust region approach and space mapping surrogates," *IEEE MTT-S Int. Microwave Symp. Dig.*, Baltimore, MD, 2011, pp. 1–4.

[17] J. Zhang, F. Feng, W. Na, S. Yan, and Q. J. Zhang, "Parallel space mapping based yield-driven EM optimization incorporating trust region algorithm and polynomial chaos expansion," *IEEE Access*, vol. 7, no. 1, pp. 143673–143683, 2019.

[18] C. Zhang, W. Na, Q. J. Zhang, and J. W. Bandler, "Fast yield estimation and optimization of microwave filters using a cognition-driven formulation of space mapping," *IEEE MTT-S Int. Microw. Symp. Dig.*, San Francisco, CA, 2016, pp. 1–4.

[19] S. Koziel and A. Bekasiewicz, "Low-cost surrogate-assisted statistical analysis of miniaturized microstrip couplers," *J. Electromagnetic Waves Appl.*, vol. 30, no. 10, pp. 1345–1353, 2016.

[20] S. Koziel and J. W. Bandler, "Rapid yield estimation and optimization of microwave structures exploiting feature-based statistical analysis," *IEEE Trans. Microw. Theory Techn.*, vol. 63, no. 1, pp. 107–114, 2015.

[21] A. C. M. Austin, N. Sood, J. Siu, and C. D. Sarris, "Application of polynomial chaos to quantify uncertainty in deterministic channel models," *IEEE Trans. Antennas Propag.*, vol. 61, no. 11, pp. 5754–5761, 2013.

[22] A. C. M. Austin, and C. D. Sarris, "Efficient analysis of geometrical uncertainty in the FDTD method using polynomial chaos with application to microwave circuits," *IEEE Trans. Microw. Theory Techn.*, vol. 61, no. 12, pp. 4293–4301, 2013.

[23] P. Sumant, H. Wu, A. Cangellaris, and N. Aluru, "Reduced-order models of finite element approximations of electromagnetic devices exhibiting statistical variability," *IEEE Trans. Antennas Propag.*, vol. 60, no. 1, pp. 301–309, 2012.

[24] T.-A. Pham, E. Gad, M. S. Nakhla, and R. Achar, "Decoupled polynomial chaos and its applications to statistical analysis of high-speed interconnects," *IEEE Trans. Compon. Packag., Manuf. Technol.*, vol. 4, no. 10, pp. 1634–1647, 2014.

[25] A. K. Prasad, M. Ahadi, and S. Roy, "Multidimensional uncertainty quantification of microwave/RF networks using linear regression and optimal design of experiments," *IEEE Trans. Microw. Theory Techn.*, vol. 64, no. 8, pp. 2433–2446, 2016.

[26] D. V. Ginste, D. De Zutter, D. Deschrijver, T. Dhaene, P. Manfredi, and F. Canavero, "Stochastic modeling-based variability analysis of on-chip interconnects," *IEEE Trans. Compon. Packag., Manuf. Technol.*, vol. 2, no. 7, pp. 1182–1192, 2012.

[27] A. K. Prasad and S. Roy, "Accurate reduced dimensional polynomial chaos for efficient uncertainty quantification of microwave/RF networks," *IEEE Trans. Microw. Theory Techn.*, vol. 65, no. 10, pp. 3697–3708, 2017.

[28] M. Ahadi and S. Roy, "Sparse linear regression (SPLINER) approach for efficient multidimensional uncertainty quantification of high-speed circuits," *IEEE Trans. Comput.-Aided Design Integr. Circuits Syst.*, vol. 35, no. 10, pp. 1640–1652, 2016.

[29] M. R. Rufuie, E. Gad, M. Nakhla, and R. Achar, "Generalized hermite polynomial chaos for variability analysis of macromodels embedded in nonlinear circuits," *IEEE Trans. Compon. Packag., Manuf. Technol.*, vol. 4, no. 4, pp. 673–684, 2014.

[30] D. Xiu and G. Karniadakis, "The Wiener-Askey polynomial chaos for stochastic differential equations," *SIAM J. Sci. Comput.*, vol. 24, no. 2, pp. 619–644, 2002.

[31] J. Zhang, C. Zhang, F. Feng, W. Zhang, and Q. J. Zhang, "Recent advances in EM-based yield estimation," *IEEE Int. Conf. on Numerical Electromagnetic and Multiphysics Modeling and Optimization for RF, Microwave, and Terahertz Applications (NEMO)*, Reykjavik, 2018.

[32] J. Zhang, C. Zhang, F. Feng, W. Zhang, J. Ma, and Q. J. Zhang, "Polynomial chaos-based approach to yield-driven EM optimization," *IEEE Trans. Microw. Theory Techn.*, vol. 66, no. 7, pp. 3186–3199, 2018.

[33] V. M. R. Gongal-Reddy, F. Feng, C. Zhang, S. Zhang, and Q. J. Zhang, "Parallel decomposition approach to gradient based EM optimization," *IEEE Trans. Microw. Theory Techn.*, vol. 64, no. 11, pp. 3380–3399, 2016.

[34] Y. Cao, G. Wang, and Q. J. Zhang, "A new training approach for parametric modeling of microwave passive components using combined neural networks and transfer functions," *IEEE Trans. Microw. Theory Techn.*, vol. 57, no. 11, pp. 2727–2742, 2009.

[35] J. Zhang, F. Feng, W. Zhang, J. Jin, and Q. J. Zhang, "Parametric modeling of microwave passive components using combined neural networks and reduced-order models," *IEEE Trans. Microw. Theory Techn.*, vol. 68, no. 6, pp. 2215–2233, 2020.

[36] V. Gongal-Reddy, S. Zhang, C. Zhang, and Q. J. Zhang, "Parallel computational approach to gradient based EM optimization of passive microwave circuits," *IEEE Trans. Microw. Theory Techn.*, vol. 64, no. 1, pp. 44–59, 2016.

[37] L.-Y. Xiao, W. Shao, X. Ding, and B.-Z. Wang, "Dynamic adjustment kernel extreme learning machine for microwave component design," *IEEE Trans. Microw. Theory Techn.*, vol. 66, no. 10, pp. 4452–4461, 2018.

[38] F. Feng, C. Zhang, J. Ma, and Q. J. Zhang, "Parametric modeling of EM behavior of microwave components using combined neural networks and pole-residue-based transfer functions," *IEEE Trans. Microw. Theory Techn.*, vol. 64, no. 1, pp. 60–77, 2016.

[39] B. Gustavsen and A. Semlyen, "Rational approximation of frequency domain responses by vector fitting," *IEEE Trans. Power Del.*, vol. 14, no. 3, pp. 1052–1061, 1999.

[40] J. Zhang, F. Feng, J. Jin, W. Zhang, Z. Zhao, and Q. J. Zhang, "Adaptively weighted yield-driven EM optimization incorporating neuro-transfer function surrogate with applications to microwave filters," *IEEE Trans. Microw. Theory Techn.*, vol. 69, no. 1, pp. 518–528, 2021.

[41] J. W. Bandler, Q. S. Cheng, S. A. Dakroury *et al.*, "Space mapping: The state of the art," *IEEE Trans. Microw. Theory Techn.*, vol. 52, no. 1, pp. 337–361, 2004.

[42] M. S. Eldred, "Recent advances in non-intrusive polynomial chaos and stochastic collocation methods for uncertainty analysis and design," in *Proceedings of Structures, Structural Dynamics, and Materials Conference*, Palm Springs, CA, 2009, pp. 1–20.

[43] F. Heiss and V. Winschel, "Likelihood approximation by numerical integration on sparse grids," *J. Econometr.*, vol. 144, no. 1, pp. 62–80, 2008.

[44] A. Papoulis, *Probability, Random Variables and Stochastic Processes*. New York, NY: McGraw-Hill, 1991.

[45] D. Oloumi, A. Kordzadeh, and A. A. Lotfi Neyestanak, "Size reduction and bandwidth enhancement of a waveguide bandpass filter using fractal-shaped irises," *IEEE Antennas Wirel. Propag. Lett.*, vol. 8, pp. 1214–1217, 2009.

[46] C. Zhang, F. Feng, V. Gongal-Reddy, Q. J. Zhang, and J. W. Bandler, "Cognition-driven formulation of space mapping for equal-ripple optimization of microwave filters," *IEEE Trans. Microw. Theory Techn.*, vol. 63, no. 7, pp. 2154–2165, 2015.

[47] A. H. Zaabab, Q. J. Zhang, and M. S. Nakhla, "A neural network modeling approach to circuit optimization and statistical design," *IEEE Trans. Microwave Theory Tech.*, vol. 43, pp. 1349–1358, 1995.

Chapter 9

Exploring order reduction clustering methods for uncertainty quantification of electromagnetic composite structures

Sébastien Lallèchère[1,2]

9.1 Introduction

For at least two decades, the use of composite materials has been generalized in a variety of industrial sectors including (but not restricted to): aerospace, automotive, communication, and defense. Indeed, various industrial actors are demanding efficient and realistic tools to predict the behavior of such materials, with increasing frequency needs (from direct current to tens of GHz).

Moreover, due to their increasing spread in different industrial fields (such as transport: e.g., automotive and/or aerospace), composite materials have experienced a constant interest for electromagnetic (EM) purposes from early 2000s (e.g., trade-off between mass and electrical properties grounding, shielding, conductivity, etc.). Thus, different domains including wireless propagation [1] and electromagnetic compatibility (EMC) [2] have shown a growing interest for theoretical [3], numerical [1,4–6], and/or experimental [7] developments in this framework.

The future deployment of fifth generation communication systems and the growing share by communicating objects (internet of things) complicate the analysis of EMC electronic systems [8]. Indeed, the modifications imposed on the nature of sources of EM radiation (types, levels, and uses), coupling paths and victims' potential disturbances require a detailed analysis of the EMC of these systems [2].

As part of the protection of victims, the use of carbon fibers and nanotubes (CNTs) for materials composites has seen increasing interest over the past 15 years [9]. The work in [10] gives an example of MC generation and EM characterization of such materials. Thus, the work initiated in [1] has allowed to show the importance of a multi-scale and multi-physical characterization for applications of broadband EM shielding.

The interest of MC modeling for the statistical characterization of electrical conductivity of composite material is illustrated in [11].

[1]Université Clermont Auvergne, Institut Pascal, Clermont-Ferrand, France
[2]AFSCET, Systemic Scientific Society, Paris, France

Previous studies in [10,11] have highlighted composite materials are, by nature, characterized by uncertain input factors. The latter statement is mainly justified by the intrinsic drifts surrounding the definition of composite structures, including but not restricted to drifts due to:

- electrical properties (conductivity, complex permittivity, ...) of composite matrix and inclusions;
- geometrical description of conductive inclusions (e.g., shapes, sizes, locations, etc.); and
- or pure ignorance of the deterministic distributions of inputs, see for instance the work in [10] for inclusions, electric contacts distribution in [11].

The works in [2,9,12] have demonstrated the interest of equivalent electrical model and electromagnetic mixture theory (EMT), allowing efficient modeling of composite structures. In this chapter, the numerical simulations will rely on the homogenization principle and EMT developed in [2,9,12] to provide accurate and efficient assessment of material shielding effectiveness (SE).

Following this physical context introduction, the rest of the chapter is organized through three distinct sections:

- Section 9.2 is devoted to the description of the problem under consideration, here the EM assessment of composite material shielding (e.g., through SE characteristic). This is achieved considering various steps including: need for random description of composite structures, propagation of uncertain inputs jointly with electromagnetic mixing theory (EMT) techniques and Monte Carlo (MC) strategy, theoretical description of stochastic reduced order method (SROM) as an alternative to brute-force sampling.
- The validation of the physical and stochastic assumptions previously proposed are detailed in Section 9.3 through EMT modeling tools and SROM realizations. The method SROM is also challenged with full-wave 3D time domain MC simulations (relying on CST simulation tool) to highlight the interest of the proposed methodology, coupling computational resources saves (e.g., limiting the number of simulations required) and consistency with respect to the reference brute-force MC strategy (i.e., comparing the accuracy of SE statistical assessments).
- Section 9.4 provides concluding words and synthetizes the different achievements proposed when facing high frequency composite modeling with random constraints (electrical and geometrical ones for instance).

9.2 Statement of the problem

This chapter focuses on the realization of a mixture between carbon fibers and a matrix made of Polymethyl methacrylate (PMMA, based on a Debye model). For demonstration purposes, the choice of matrices (PMMA) and inclusions (carbon fibers) is based on the deterministic case explained in [12]. The recent study of the properties for an application in the protection of electronic systems shows different composites and the use of polyethylene terephthalate and CNTs [13], more in phase with the usage properties (material level) for the production of EM protective braids [14]. In the following, it will be assumed that the carbon fibers have micrometric dimensions with generic cylindrical shapes. In the following, it is considered that the produced mixtures are subjected to uncertain constraints, here the distribution of classes of fibers with variable dimensions, see for instance results in [15] and equivalent R-L-C model of thin nanorods in [9].

9.2.1 Random composition of composite

As pointed out in the literature (see for instance [9,15]), when dealing with composite material manufacturing (including non-conductive matrix and conductive particles for instance), the chemical processing faces a natural diversity of parameters considering for instance the intrinsic electrical and geometrical drifts. Examples are given in [10,11] from the electrical and EM characterization of the composite material.

The work in [2] has carefully pointed out the effect of conductive particles shapes for EMC or electromagnetic interface issues.

The works in [9,15] offer interesting overviews of the distributions of inclusions regarding their natural variations through geometries (length and height of carbon nanotube/epoxy mixtures). Thus, the different models developed in [9,15] offer interesting prospects both for:

- the uncertainty quantification (UQ) of initial random assumptions needed to better match with the dispersion of the shapes of inclusions; and
- the uncertainty propagation of these random assumptions through a given mapping and/or transfer function, expected to be representative with respect to the problem under consideration (SE of composite materials).

The two latter points are crucial ones when dealing with uncertain configurations and stochastic systems.

The distributions of geometrical input parameters are varying following various statistical laws (log-Normal, Normal, and quasi-Uniform ones, see for instance [15]). This demonstrates the importance of trustworthy statistical definition rules for input parameters when modeling composite materials.

Figures 9.1 and 9.2 illustrate the proposed stochastic definition of conductive fibers relying on cylindrical nanorods (see examples given in [9]):

- Composite matrix material is considered as PMMA through equivalent Debye model.
- Conductive fibers with: constant conductivity, random sizes of length L and diameter 2R, and various bins of conductive fibers as illustrated in Figure 9.1. Random parameters are given through length and diameter of cylindrical inclusions (assumed uniformly distributed), jointly with constituting bins of conductive fibers.

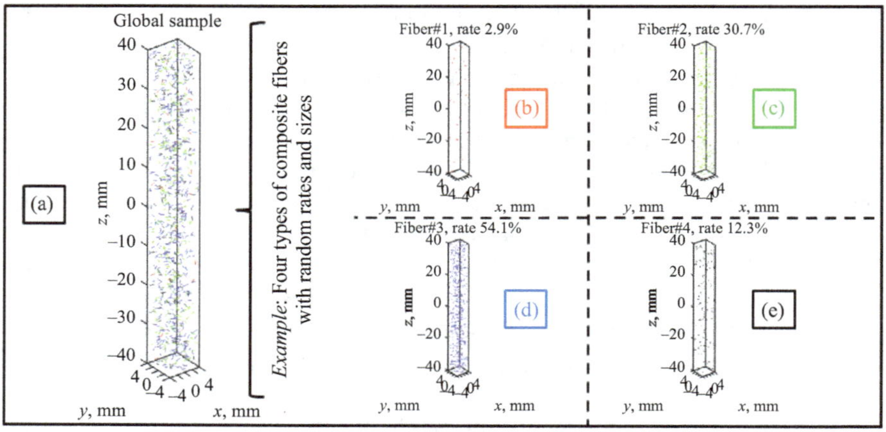

Figure 9.1 (a) Variability of nanorods in composite material made of thin cylindrical conductive inclusions: illustration with four distinct bins of fibers (b)–(e) (here example randomly generating the rate of each class of fibers). In the following, the composite material will be modeled with 20 bins of conductive fibers, and the reference dataset will be given from 10,000 MC simulations

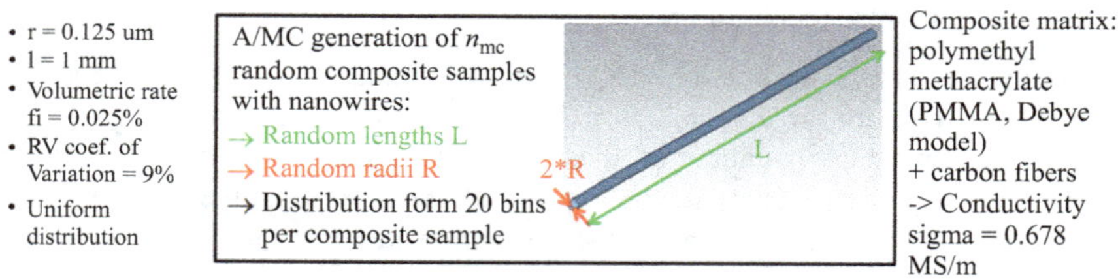

Figure 9.2 First step (Step A) of the proposed methodology. Description of the numerical test case including: sizes of nanorods conductive structure (conductivity given here), volumetric rate, variation intensity, and given distribution law. It is to be noticed 20 different bins of nanorods are considered in this work

- Volumetric rates of conductive fibers fi = 0.025%.
- MC random generation of conductive fibers is arbitrarily chosen through 20 distinct bins, randomly choosing the rate of each fiber bin: for the illustrative purpose, an example is given in Figure 9.1 where four classes are given, see Figure 9.1(b)–(e).

The methodology followed in this study is summarized according to the four levels represented in Figures 9.2 and 9.3:

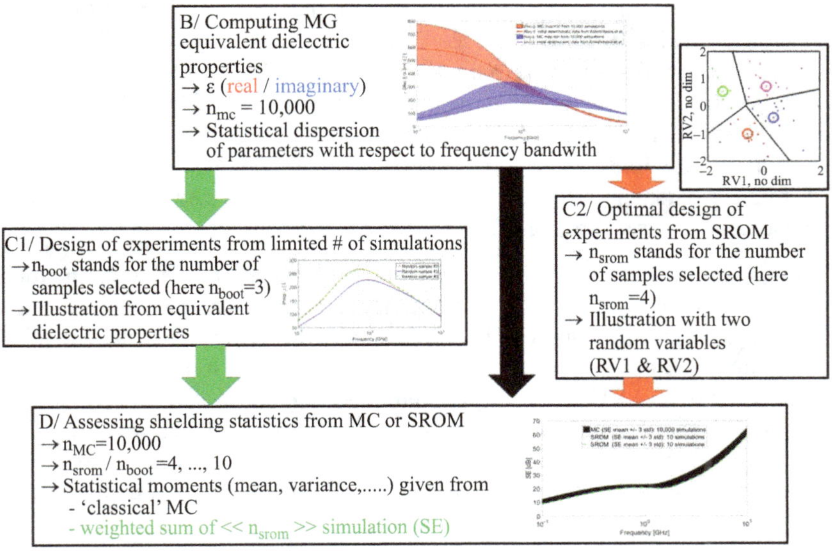

Figure 9.3 Flowchart of the proposed methodology including: equivalent dielectric properties computation through MG EMT (Step B), variations when defining design of experiments (DOE) (randomly choosing a limited number nboot of inputs/Step C1, SROM/Step C2), characterizing SE through classical MC or SROM (weighted sum of nsrom limited number of simulations, see details in the following for SROM explanations)

- *Step A*: Generation of sample collections (several thousand, tens of thousands: nmc) including classes of carbon fibers whose parameters are defined randomly (see Figure 9.2). Figure 9.1 gives a representation of a sample of composite material containing four classes of fibers with random lengths and diameters within the limits mentioned above. The effect on the distribution of the complex permittivity is illustrated in Figure 9.3 (2) with the real (A) and imaginary (B) parts.
- *Step B*: Evaluation of the dielectric properties of composites according to work given in [12].
- *Step C*: Definition of classical experimental designs (Step C1 in Figure 9.3; random choices of nboot realizations among the nmc of the initial population (step A)) or optimum (step C2 in Figure 9.3; SROM, nsrom).
- *Step D*: Final post-treatment (estimation of the shielding efficiency of the material subjected to a wave plane at normal incidence: mean and standard deviation).

As previously mentioned, one may notice that the various random constraints (size of the carbon fibers, definition of classes of fibers of different sizes for the same sample) lead to significant variations in the dielectric properties of the material as a function of frequency (see Figure 9.3, block (B) above). We thus see, for three distinct frequencies (namely f1 = 100 MHz, f2 = 1 GHz, and f3 = 2.51 GHz, see upcoming Figure 9.5) a distribution of the real and imaginary parts of the very variable dielectric permittivity including Weibul, normal, and/or log-normal distributions.

The latter statement requires to properly deals with the UQ issue, considering two drastically different strategies:

- *Frequency by frequency*: this requires the identification of the parameters of the distribution laws complex permittivity (real and imaginary parts) for each frequency; strategies can then be implemented including or not the use of MC. This methodology is well suited if the calculation of the EMC shielding properties (e.g., SE) includes a frequency dependence.
- Taking into account all the frequency data. In this case, the use of time domain tools is possible by integrating the dependence on the dielectric properties of the material from one frequency to another. In this work, this methodology is favored (see in the following).

9.2.2 Uncertainty propagation: MC as golden standard and optimal sampling using SROM

This section will detail the theoretical foundations of SROM modeling, complementary to classical MC method. As pointed out in [16,17], MC technique is often considered as the reference technique to tackle UQ of measurement and computational experiments (here with application in EMC). However, as previously pointed out, the statistical dispersion of complex dielectric permittivity with respect to frequency (sizes of inclusions considered as random parameters) is a serious bottleneck whether each real/imaginary relative permittivity per frequency had to be considered as a random variable (RV) by itself.

9.2.2.1 Theoretical foundations

The previous findings (Figure 9.3/Step B) justify the interest of proposing an alternative strategy to the MC method allowing to naturally take into account random parameters over a wide range of frequencies (the distributions evolving as a function of the frequency, see upcoming Figure 9.5). Various alternative techniques were listed in [16–19], most of them are facing the curse of dimensionality, which is well-known [16,17] as a serious bottleneck when increasing the number of RVs (due to the attached tensor product). One of the major aims in this work is to develop a priori strategy, meaning the proposed methodology is expected to take benefit from the MC relative independence to the number of RVs with a limited number of realizations, much lower than what is classically required for MC convergence (from thousands to hundreds of thousands of realizations when dealing with first statistical moments).

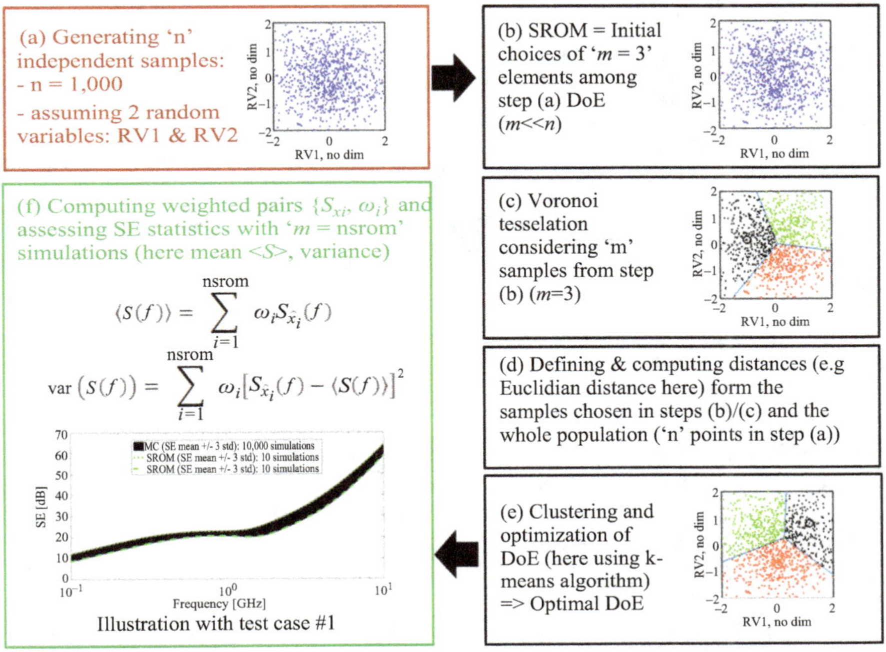

Figure 9.4 SROM step-by-step flowchart: illustration with two RVs, 1,000 initial MC realizations, optimal size nsrom of SROM DOE with $m = nsrom = 3$

The alternative strategy is based on the SROM approach for advanced sampling (see its use in the field of applied mathematics in [20]) and makes it possible to select the optimal experimental design for a number of reduced simulations (nsrom=10 at maximum here).

The methodology allows an automated choice of the data set ($\widehat{x}$i; w_i) with:

- index $i = (1, \ldots, \text{nsrom})$ is referring to the SRO indice; and
- $\widehat{xi}$ is the sample number from the initial set of size n (see Figure 9.4).

The choice is based on considering each value of the real and imaginary parts of the dielectric permittivity as a function of frequency (here 202 variables, 101 frequencies per real/imaginary complex permittivity component, each sampled in frequency between 100 MHz and 10 GHz).

In the literature, many clustering techniques and generating algorithms exist; the interested reader may find details for instance in [20,21]. The main idea relies on a minimization procedure when choosing a given set of points (optimized cluster) from an original population (considering input space constituted from RVs).

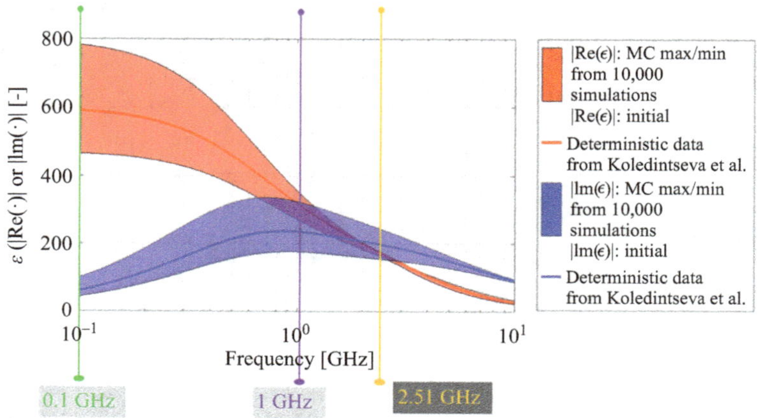

Figure 9.5 Statistical dispersion of equivalent dielectric properties: magnitudes of real and imaginary parts, respectively red and blue areas/curves. In the following (Figure. 9.6), a particular care is given to three distinct frequencies: f1=100 MHz (green), f2=1 GHz (purple), and f3=2.51 GHz (orange)

The governing principle is illustrated in Figure 9.4 with two RVs, 1,000 MC realization as initial dataset, m = nsrom = 3 as the expected optimal number of realizations, and a step-by-step procedure summarized as follows:

- *Step a*: generating n independent samples;
- *Step b*: randomly extracting m-size subset points ($m << n$);
- *Step c*: dividing the uncertain region into m Voronoi subregions (comprised of all the samples from *Step a* that are closest to points in *Step b*), known as tessellation process [22];
- *Step d*: computing the distances from all the samples in *Step b* to all the assumed samples in *Step a*, the optimized DoE $\{\widehat{x}i\}$ (this being the set of points, numbered from the initial dataset) with $i = (1, \ldots, m)$ will be given through a minimization procedure of the latter distance metric (e.g., Euclidean distance when designing initial samples in the random space).
- *Step e*: designing the optimal DoE from k-means technique, the interested reader may refer to [23] for details. Relying on initial random inputs and assumptions, this step provides the optimal set of required simulations, drastically saving the computing costs relatively to MC technique ($m << n$);
- *Step f*: counting the number of samples in each Voronoi regions relying on set $\{\widehat{x}_i\}$ will lead to the SROM weighted set of points ($\widehat{x}_i$; w_i).

Eventually, the average value of the quantity of interest (namely SE, noted S here) is calculated according to:

$$(S(f)) = \sum_{i=1}^{\text{nsrom}} \omega_i S_{\widehat{x}_i}(f) \tag{9.1}$$

where $\langle S(f) \rangle$ stands for the SE mean at frequency f, and $S_{\widehat{x}_i}(f)$ represents SE simulated from the realization number $\widehat{x}_{\mathrm{I}}$ from the initial input dataset (i.e. generated at *Step a* in Figure 9.4).

The assessment of SE variance also relies on the set of points $(\widehat{x}_{\mathrm{i}};\ w_i)$ used in the previous relation (1):

$$\mathrm{var}(S(f)) = \sum_{i=1}^{\mathrm{nsrom}} \omega_i \left[S_{\widehat{x}_i}(f) - \langle S(f) \rangle \right]^2. \tag{9.2}$$

The following subsection will depict the statistical dispersion of complex dielectric components assuming random geometrical constraints of conductive fibers as previously pointed out. In this framework, SROM technique will itself reveal as a useful tool to select the most influential MC realization of the composite samples (from the initial MC input dataset).

9.2.2.2 Application to equivalent mixed dielectric properties

As aforementioned, the proposed random assumptions characterizing the diversity of sizes of conductive inclusions led to high variability of relative complex permittivity.

Relying on the test case depicted in Figure 9.2, Figure 9.5 provides an overview of the maximum/minimum caliber both for real and imaginary components of complex permittivity simulated following formalism given in [12] through 10,000 MC realizations.

It is to be noticed the statistical distributions of real and imaginary components of complex dielectric permittivity are highly variable from one frequency to another. Figure 9.6 focuses on three chosen frequencies: $f1 = 100$ MHz, $f2 = 1$ GHz, and $f3 = 2.51$ GHz.

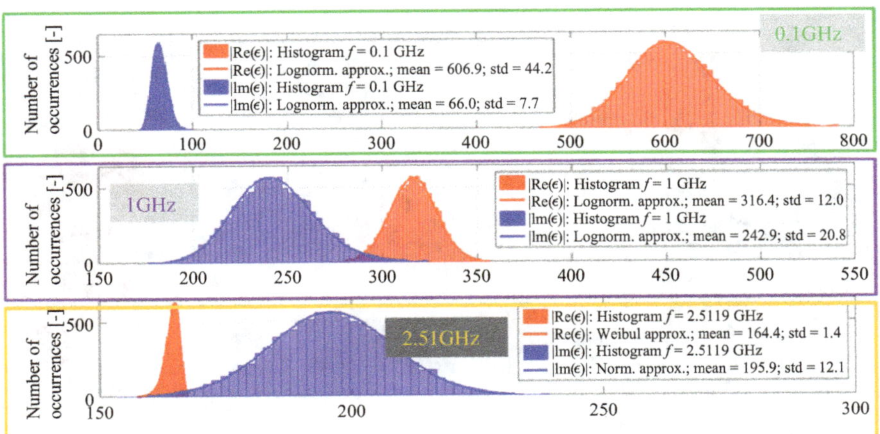

Figure 9.6 Number of occurrences of the magnitudes of real (red areas/curves) and imaginary (blue areas/curves) dielectric permittivity at: f1 = 100 MHz (green frame), f2 = 1 GHz (purple frame), and f3 = 2.51 GHz (orange frame)

Histograms and approximated statistical models (extracted from Matlab distribution tools) show in Figure 9.6 that the corresponding laws are varying from log-Normal/Normal to Weibul distributions.

The results obtained at that stage offer an initial dataset of 10,000 $\times$ 202 random parameters (see Figure 9.5 for the illustration). The proposed SROM methodology will consider the latter results as initial dataset in *Step a* in Figure 9.4. The following section will demonstrate the interest of SROM through different proof-of-concept (PoC), including two distinct numerical setups.

9.3 Assessing stochastic shielding properties with MC and SROM

9.3.1 PoC descriptions (test cases)

This subsection is dedicated to the presentation of the computational test cases defined to illustrate the interest of SROM methodology, comparatively to classical random choices and brute-force MC technique.

Figure 9.7 illustrates the two different configurations considered in this section (simulations from 100 MHz to 10 GHz):

- Computation of the composite SE considering semi-infinite layer (depth = 5 mm here) and normal incident plane wave stimulation (see Figure 9.7(a)). SE is obtained from the analytical expression (see reminder in [12]) providing reflection and transmission coefficients (and therefore SE) from the mismatch between free space and composite impedances (see Z0 and Zin in Figure 9.7(a)). For the interested reader, the SE formula is detailed in [24]. It straightforward depends on

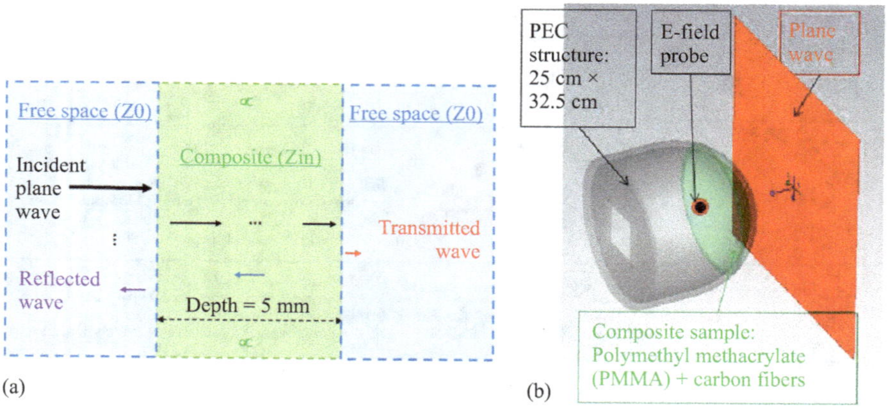

Figure 9.7 Numerical setups with: (a) test case#1 (plane wave with normal incidence in free space Z0-impedance, composite impedance Zin), and (b) 3D configuration for full-wave time domain CST simulation (plane wave illumination of PEC device with composite cap, see light green part)

the depth of the composite layer and its equivalent complex dielectric properties given from works in [12].

- 3D full-wave simulation of EM fields passing through composite structure in a generic perfect electric conductor (PEC) device with aperture. SE is characterized relating electric field (E-field probing) in absence and presence of the composite layer (see light green cep in Figure 9.7(b)).

The two latter cases are considered following the same composite material manufacturing. The main differences (out of numerical device under test) rely on the number of MC simulations achieved to provide golden standard (reference): 10,000 various equivalent dielectric properties (real and imaginary parts) with test case #1, and 1,000 simulations with test case #2. The decrease of the number of realizations from test case #1 to test case #2 is mainly due to the computation time needed to provide one given full-wave simulation of the configuration in Figure 9.7(b) (35 minutes with Quad-core Intel Xeon processor, RAM 12 Gb).

9.3.2 Test case #1: EM attenuation through semi-infinite composite material with normal incidence plane wave

Figure 9.8 illustrates the statistical trend (SE mean) obtained considering test case #1 (see Figure 9.7(a)):

- Randomly chosen simulations from the whole MC simulation set (10,000 simulations in total, see Figure 9.3 and *Step C1*); black areas in Figure 9.8.

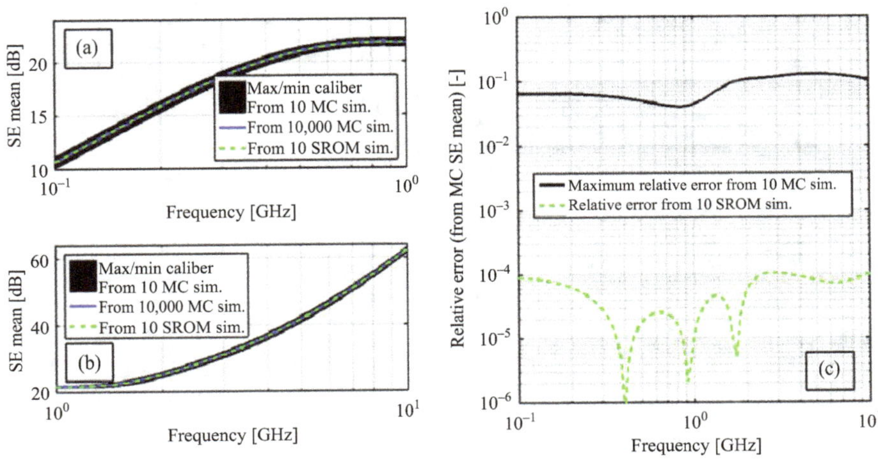

Figure 9.8 Assessment of the SE mean (reference data from 10,000 MC simulations with plain blue line, max./min. caliber from randomly choosing 10 simulations with black area, SROM result with dashed green line): (a) from 100 MHz to 1 GHz and (b) from 1 GHz to 10 GHz. (c) Maximum relative error from: random choice of 10 MC simulations (plain black line), and SROM dataset (dashed green line)

- Optimal SROM DOE (see Figure 9.3 and *Step C2*); dashed green line in Figure 9.8.
- Reference data considering the whole set of simulation (from MC brute-force modeling); plain blue line in Figure 9.8.

The results in Figure 9.8 shows the excellent agreement between EM shielding properties (SE) from nsrom = 10 well-chosen simulations (SROM) and 10,000 MC simulations reference (see Figure 9.8(a) from 100 MHz to 1 GHz and Figure 9.8(b) from 1 GHz to 10 GHz, respectively). Indeed, the accuracy of the SROM results are assessed considering as a reference the whole set of MC simulations (10,000 realizations), see in Figure 9.8(c). As depicted, the arbitrary choice of nboot = 10 simulations leads to linear relative error levels from 4% to 12% (from 100 MHz to 10 GHz), see plain black line. The optimal SROM design offers more accurate results for SE mean with linear errors lower than 0.01% over the given frequency bandwidth. Similar to SE mean, the results in Figure 9.9 clearly express the accuracy of SROM comparatively to reduced MC design of experiment, when dealing with SE standard deviation (std). Indeed, errors levels are respectively lower than 4% (SROM) and around 70% (10 MC simulations) over the whole frequency bandwidth (i.e., from 100 MHz to 10 GHz).

Figure 9.10 provides the distribution of SE with:

- maximum/minimum bounds due to reference results (10,000 MC simulations, light blue area); and

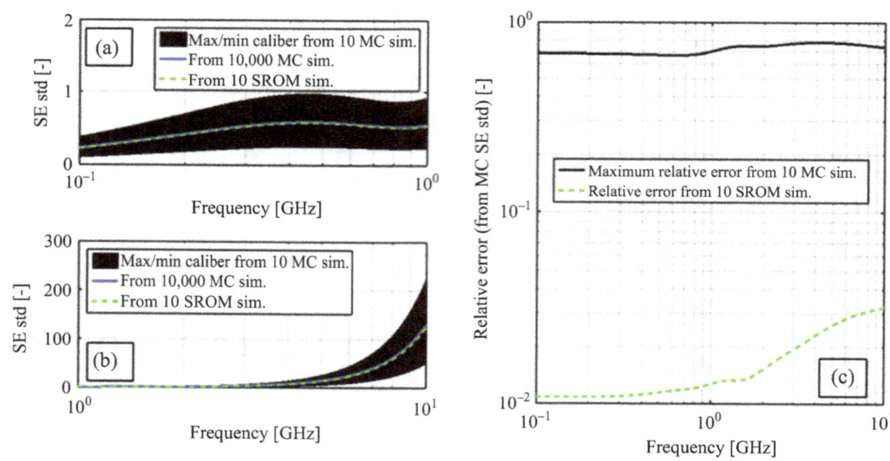

Figure 9.9 Assessment of the SE standard deviation (reference data from 10,000 MC simulations with plain blue line, max/min caliber from randomly choosing 10 simulations with black area, SROM result with dashed green line): (A) from 100 MHz to 1 GHz and (B) from 1 GHz to 10 GHz. (C) Maximum relative error from: random choice of 10 MC simulations (plain black line), and SROM dataset (dashed green line)

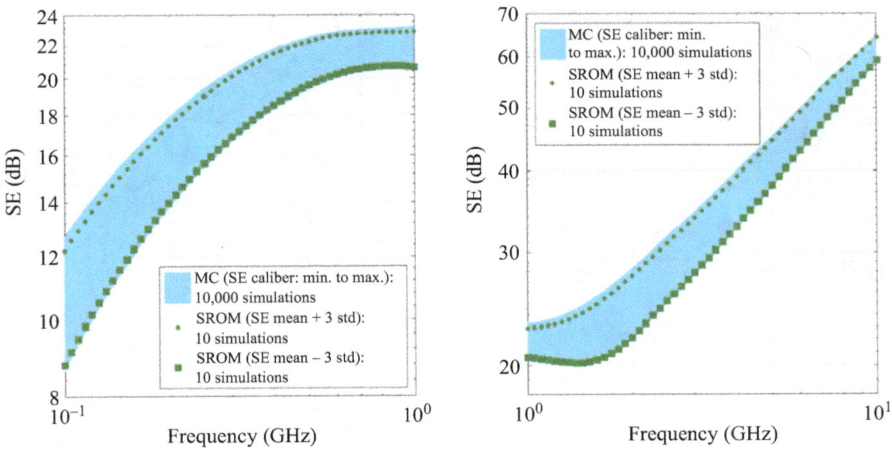

Figure 9.10 SE density (from 100 MHz to 1 GHz/from 1 GHz to 10 GHz, respectively left/right side) extracted from the whole reference dataset (10,000 MC simulations, light blue area), and extracted from SROM SE mean ± 3 standard deviation (dark green markers)

- assessment of SE caliber from SROM DOE (nsrom = 10 well-chosen realizations of initial dataset), computing SE mean ± 3 standard deviations (dark green markers).

The SROM results are in great accordance with MC 10,000 realizations reference, both on lower ([100 MHz; 1 GHz]) and higher ([1 GHz; 10 GHz]) frequency bandwidths.

The following subsection will highlight the ability of SROM technique to provide useful information regarding the most influential set of parameters (here equivalent dielectric EMT composite properties) with a restricted number of simulations (with nsrom = 10).

9.3.3 Test case #2: Composite material SE via full-wave time domain models

Test case #2 is dedicated to the full-wave simulation depicted in Figure 9.7(b). In this framework, Figure 9.11 presents the first moment (mean trend) obtained for a reference set of 1,000 MC simulations (considering equivalent dielectric permittivity including real and imaginary parts as previously given), see plain red line. As proposed in this work, the results given by 10 SROM simulations (dashed green line) are in excellent agreement with reference data (plain red line), whereas random choices of 10 MC simulations lead to dispersion around mean trend (see blue area).

Parallel to results from Test case #1, Figure 9.11 gives an overview of the relative linear error with respect to:

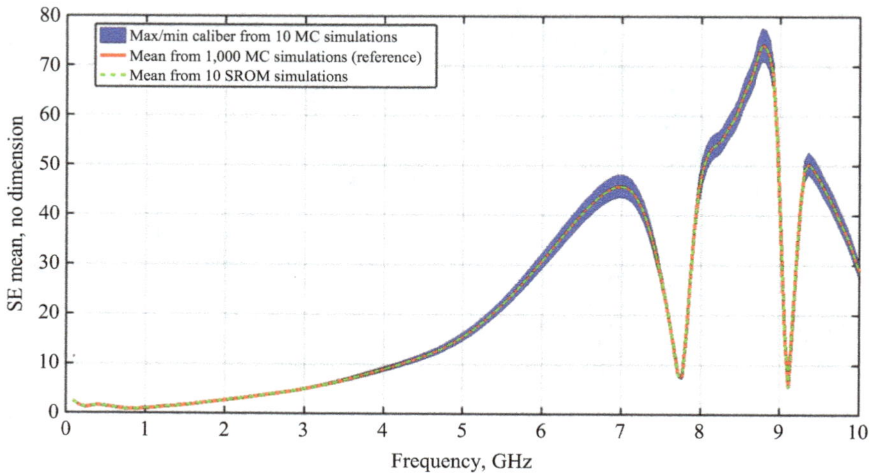

Figure 9.11 SE average trend with respect to frequency (from 100 MHz to 10 GHz): reference data (1,000 MC simulations) with plain red line, maximum/minimum caliber due to random choices of 10 simulations (blue area), and SROM mean SE with nsrom = 10 simulations (dashed green line)

- Randomly choosing 10 simulations (plain blue line, maximum error with respect to frequency).
- Averaging SE from SROM DOE (dashed green line).

Relying on Figure 9.12, the relative error levels between SE mean given by nsrom = 10 well-chosen simulations are lower than the one due to randomly choosing 10 simulations among reference dataset. Indeed, the accuracy when choosing SROM dataset (dashed green line) is approximately 1,500 times greater than the one obtained from random choices of 10 simulations (plain blue line). Error levels are bounded between 0.2% and 10% for random MC choices (plain blue line), whereas they are given between 0.0001% and 0.4% for SROM DOE (dashed green line).

Parallel to previous results assessing mean trends (SE mean, see Figures 9.11 and 9.12), Figures 9.13 and 9.14 are focusing on second-order moment (here SE variance). Again, SROM results are in excellent agreement with the reference dataset (see plain red and dashed green lines in Figure 9.13). Whereas the sensitivity is given with weak levels (see in Figure 9.13 SE variance close to 0 below 4G Hz), the SE variance drastically increases from 4 GHz to 10 GHz. The SROM prediction is fair comparatively to reference results as pointed out in Figure 9.14 (mostly between 2% and 10%, see dashed green line). Contrary to SROM results, randomly choosing 10 simulations among initial set may lead to extremely high drifts (comparatively to reference data) with maximum error levels greater than 200%. On the one hand, below 4 GHz, the sensitivity (i.e., SE variance) is rather

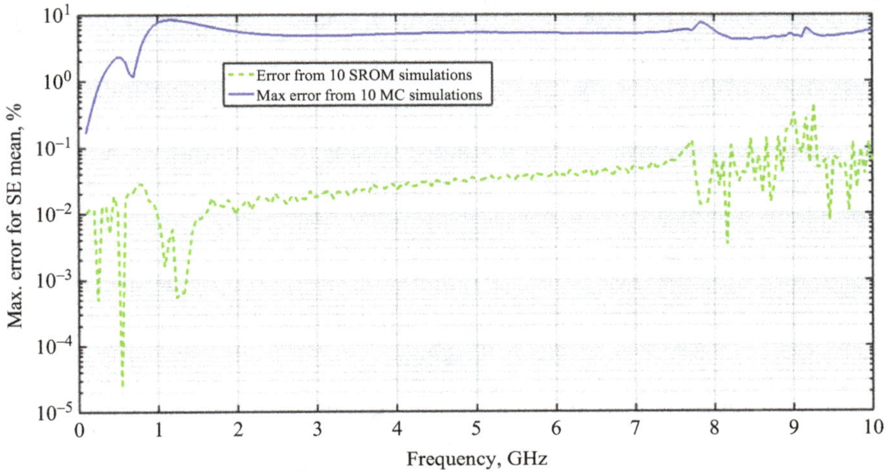

Figure 9.12 *Linear error of SE mean with limited number of simulations (nsrom = 10) from the reference dataset (1,000 MC simulations, see plain red line in Figure 9.11) considering: arbitrarily choosing nsrom = 10 simulations, SROM DOE (dashed green line)*

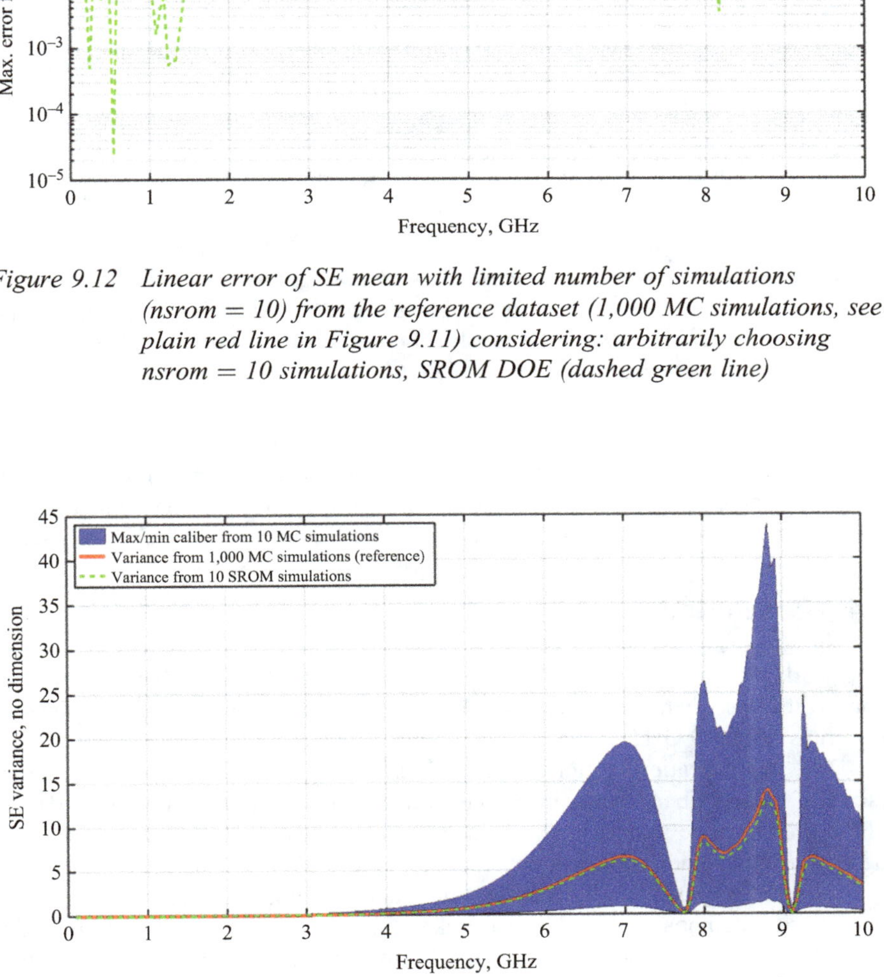

Figure 9.13 *SE variance (second order statistical moment) with respect to frequency (from 100 MHz to 10 GHz): reference data (1,000 MC simulations) with plain red line, max/min caliber due to random choices of 10 simulations (blue area), and SROM mean SE with nsrom = 10 simulations (dashed green line)*

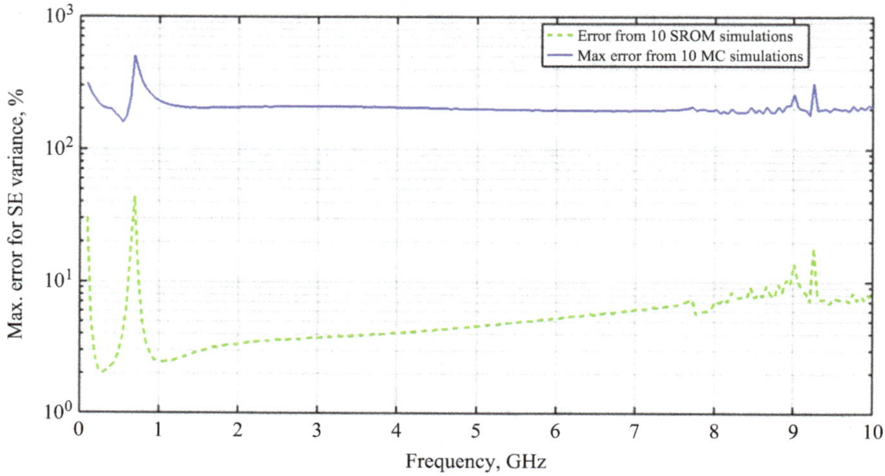

Figure 9.14 Linear error of SE variance with limited number of simulations (nsrom = 10) from the reference dataset (1,000 MC simulations, see plain red line in Figure 9.13) considering: arbitrarily choosing nsrom = 10 simulations, SROM DOE (dashed green line)

low and the consequences are limited regarding statistical dispersion of results. On the other hand, increasing frequency up to 10 GHz leads to very high level of dispersion, and potential drastic effects regarding SE assessment and EMC use.

9.4 Conclusion

The work presented in this chapter details the use of the SROM method in the context of the characterization of EM shielding levels induced by composite materials up to 10 GHz.

The quantification of uncertainties linked to geometric random variations (sizes of carbon fibers) was carried out over a frequency range from 100 MHz to 10 GHz. It is to be noticed the proposed model for test case #1 remains valuable below the appearance of skin effects (here up to tens of GHz, data not shown here).

The diversity of statistical distributions characterizing the equivalent dielectric permittivity composite samples was taken into account naturally through the SROM method. The natural ability of the method to integrate the dependence between the RVs offers interesting perspectives for the simulation of these materials in time domain simulation full-wave codes allowing joint access to results over a wide range of frequencies with an appreciable degree of realism compared to MC simulations. The saving in computing time (factor 1,000 here) and the estimation of the relative error (1–3 for variance estimation in Test case #1, around 10% in Test case #2) produced by the use of experimental designs SROM offers very interesting prospects for the integration of these models when dealing with sensitivity analysis of full-wave simulations, especially considering computational saves expected both in time and memory resources.

References

[1] Njoku, C., Whittow, W. G., and Vardaxoglou, J. C., 'Effective permittivity of heterogeneous substrates with cubes in a 3-D lattice', *IEEE Antennas Wirel. Propag. Lett.*, 2011, 10, pp. 1480–1483.

[2] Préault, V., Corcolle, R., Daniel, L., and Pichon, L., 'Shielding effectiveness of composite materials: effect of inclusion shape', *IEEE Trans. Magn.*, 2013, 49, pp. 1941–1944.

[3] Koledintseva, M., DuBroff, R. E., and Schwartz, R. W., 'Maxwell Garnett rule for dielectric mixtures with statistically distributed orientations of inclusions', *PIER*, 2009, 99, pp. 131–148.

[4] Liao, Y., Shi, G., Yuan, L., and Ying, X., *'Equivalent modeling of the microwave dielectric properties for fiber reinforced shielding composites'*, in *Proc. APEMC* 2016, Shenzhen, 2016.

[5] Boubekeur, M., Kameni, A., and Pichon, L., 'Modeling of thin heterogeneous sheets in the Discontinuous Galerkin method for 3D transient scattering problems', *Eur. Phys. J. App. Phys.*, 2016, 73, pp. 1–5.

[6] Lallechere, S., 'Advanced statistical 3D models of composite materials for microwave electromagnetic compatibility applications', *ACES J.*, 2017, 32(12), pp. 1113–1116.

[7] Bastianelli, L., Capra, S., Gradoni, G., *et al.*, *'Shielding effectiveness statistical evaluation of random concrete composites'*, in *Proc. IEEE Metrology for Aerospace* 2016, Florence, 2016.

[8] Vikstedt, J., *'The 5G New Radio – Implications for EMC and antenna testing'*, in *Proc. 2018 IEEE Symposium on EMC, SI & PI*, Aug. 2018.

[9] Koledintseva, M., DuBroff, R., and Schwartz, R., A Maxwell-Garnett model for dielectric mixtures containing conducting particle, PIER, 2006.

[10] Dawson, J. F., Flintoft, I. D., Austin, A. N., and Marvin, A. C., *Shielding effectiveness of non-woven carbon fibre sheets: modelling the micro-structure*, ESA workshop on EMC, 2016.

[11] Zhang, T. and Yi, Y., 'Monte Carlo simulations of effective electrical conductivity in short-fiber composites', *J. Appl. Phys.*, 2008.

[12] Koledintseva, M., *'Engineering of composite media for shields at microwave frequencies'*, in *Proc. IEEE International Conference on EMC*, 2005.

[13] M. Michel, Élaboration d'un fil nanocomposite PET – nanotubes de carbone pour le blindage électromagnétique. Suivi de la microstructure induite par le procédé d'étirage et caractérisations des basses fréquences aux hyperfréquences (in French), PhD Thesis, Université de Bourgogne, Dijon, Oct. 2019.

[14] Omerin, P., Omerin, G., and Clément, G., Gaine de blindage électromagnétique, à diamètre extensible (in French), Patent EP1348247B1, European Patent, 2009.

[15] Li, N., Huang, Y., Du, F., *et al.*, 'EM intefrerence (EMI), shielding of single-walled carbon nanotube epoxy composites', *China NanoLett.*, 2006.

[16] Carobbi, C. F. M., Lalléchère, S., and Arnaut, L. R., Review of uncertainty quantification of measurement and computational modeling in EMC Part I: measurement uncertainty', *IEEE Trans. Electromagn. Compat.*, Dec. 2019, 61(6), pp. 1690–1698, doi:10.1109/TEMC.2019.2904973

[17] Lalléchère, S., and Arnaut, L. R., 'Review of uncertainty quantification of measurement and computational modeling in EMC Part II: computational uncertainty', *IEEE Trans. Electromagn. Compat.*, Dec. 2019, 61(6), pp. 1699–1706. doi:10.1109/TEMC.2019.2904999

[18] Menezes, L., Ajayi, A., Christopoulos, C., Sewell, P., and Borges, G. A., 'Efficient computation of stochastic electromagnetic problems using unscented transforms', *IET Sci. Meas. Technol.*, 2008, 2(2), pp. 88–95.

[19] de Medeiros, J. E. G., Hadad, S. A. P., and de Menezes, L. R. A. X., 'Extended formulation for unscented transform and its application as Monte Carlo alternative', *Electron. Lett.*, 2016, 52(22), pp. 1842–1843.

[20] Grigoriu, M., 'Reduced order models for random functions. Application to stochastic problems', *Appl. Math. Model.*, 2009, 33(1), pp. 161–175.

[21] Zhuang, Y., Fei, Z., Chen, A., Huang, Y., Rabbi, K., and Zhou, J., 'Design of multioctave high-efficiency power amplifiers using stochastic reduced order models', *IEEE Trans. MTT*, 2018, 66(2), pp. 1015–1023.

[22] Shamos, M. and Hoey, D., 'Closest-point problems' in *Proceeding of 16th Annual IEEE Symposium on Foundations of Computer Science*, 1975.

[23] Spath, H., *Cluster Dissection and Analysis: Theory, FORTRAN Programs, Examples*. New York, NY: Halsted Press, 1985.

[24] Zheng, L., Zhao, Q., and Xing, X. J., 'Effect of plasma on electromagnetic wave propagation and THz communications for reentry flight', *ACES J.*, 2015, 30(11), pp. 1241–1245.

Chapter 10

Mixed epistemic-aleatory uncertainty using a new polynomial chaos formulation combined with machine learning

Domenico Spina[1], Tom Dhaene[1] and Flavia Grassi[2]

10.1 Introduction

Nowadays, it is common practice to adopt statistical techniques for uncertainty quantification (UQ) and propagation (UP) of radio frequency (RF) and microwave designs. Indeed, several sources of uncertainty affect the behavior of such circuits, including parasitic effects and tolerances of the manufacturing process, for example. As a result, several design parameters are inherently unknown and/or hard to control, even in relatively simple test setups.

UQ and UP problems are typically formulated by relying on probability theory. The main assumption is that a probability distribution can be assigned to every variable in the problem under study, representing the likelihood for a chosen parameter to assume a specific value close to its nominal one (aleatory uncertainty). In this framework, design parameters can be considered as random variables. Once the distribution of these random effects has been estimated (e.g., starting from a large set of experimental measurements), sampling-based methods can be adopted to solve UQ and UP problems. A typical example is the Monte Carlo (MC) method [1]: first, the circuit under study is simulated via suitable computer-aided design (CAD) programs for a large set of samples of the random parameters considered, chosen according to their pertinent distributions. Different CAD simulators can be used, depending on the type of system under study: from electromagnetic simulators, such as ADS Momentum (Momentum EEsof EDA; Agilent Technologies, Santa Rosa, CA), to circuit simulators, such as PSpice (Cadence Design Systems, San Jose, CA). Then, based on the obtained data, relevant statistic information can be computed, thus solving the UQ or UP problem at hand. However, the MC method requires a substantial number of samples to converge to an accurate solution [1]. Since simulations are often computationally expensive, due to the increased complexity of RF and microwave systems, the MC method has a high

[1]IDLab, Department of Information Technology, Ghent University-imec, Ghent, Belgium
[2]Department of Electronics, Information and Bioengineering, Politecnico di Milano, Milano, Italy

computational cost. Hence, several advanced statistical techniques have been proposed in the last decade [2–6] with the objective to outperform the MC method in terms of computational efficiency, while retaining comparable accuracy in predicting the variability of the output variables. However, a probabilistic framework is not always suitable to describe all types of UQ problems. For example, the geometry and material characteristics of textile antennas heavily depend on their operating conditions: while worn, a textile antenna will bend and the substrate will be compressed according to the movements of the wearer, changing the antenna radiation characteristics [7]. Instead of assuming specific distributions to describe the variability of certain geometrical and electrical parameters of these antennas, it is more realistic recognize that the uncertainty on those parameters stems from an inherent lack of knowledge about the antennas operating conditions (epistemic uncertainty).

Section 10.2 discusses how to solve UQ and UP problems involving epistemic uncertainty sources for microwave and RF systems. The proposed methodology is based on the evidence (or Dempster–Shafer) theory [8], summarized in Section 10.2.1. In particular, evidence theory is presented as a generalization of probability theory, where epistemic variables can be characterized as fuzzy variables (FVs). Studying epistemic UQ and UP problems via the proposed approach is described in Section 10.2.2, while Section 10.2.3 illustrates how machine learning can be used to drastically improve the computational efficiency when characterizing epistemic uncertainty sources [9]. In particular, the approach here proposed leverages on Bayesian optimization (BO) [10–12], a machine learning-based global optimization algorithm. Such approach is particularly useful when the system under study must be characterized via computationally expensive CAD simulations (or measurements). Hence, BO is particularly useful in electromagnetic problems, where numerical full wave simulations are often required. Finally, a relevant application example is presented in Section 10.2.4 to illustrate the characteristics of the proposed BO-based method.

However, when analyzing complex systems, it is possible that both aleatory and epistemic uncertainty sources are present. In this case, a hybrid approach is necessary to tackle UQ and UP problems. This scenario is discussed in Section 10.3, where the methodology presented in Section 10.2 is extended to characterize both aleatory and epistemic uncertainty sources. Note that, evidence theory provides a general framework to describe hybrid UQ and UP problems, since

- aleatory uncertainty can be described through the use of random variables, to which *probability* distribution functions (PDFs) are assigned; and
- epistemic uncertainty due to lack of information is tackled through the use of FVs, to which *possibility* distribution functions are assigned.

In particular, the hybrid algorithm [13] is discussed in Section 10.3.1. This approach relies on one main idea: quantifying the effect of the epistemic variables on the performance of the system under study for a large number of samples of the aleatory variables, obtained via MC sampling. Then, the overall UQ or UP problem can be solved by combining in a suitable way all the information obtained so far. Such hybrid approach has been adopted in [14] to estimate the radiated susceptibility of a non-ideally twisted wire pair (above ground) illuminated by a partially unknown impinging electromagnetic field. This application revealed the potential of the hybrid approach to provide a more general and exhaustive representation of the variability of the output variables, compared to a fully probabilistic solution of the problem (where all uncertain parameters were assigned specific PDFs). However, it also illustrates the computational inefficiency of such an algorithm: MC simulations are required to assess the influence of random variables, and for each MC run a computationally expensive problem needs to be solved to quantify the effect of the epistemic variables. This severely limits the applicability of the approach described in Section 10.3.1 to analyze RF and microwave systems, especially if simulations have an high computational cost (i.e., full wave simulations). A novel machine learning-based approach is presented in Section 10.3.2 to significantly increase the computational efficiency of solving hybrid epistemic-aleatory UQ and UP problems [15]. In particular, the BO framework described in Section 10.2.3 to characterize epistemic variability is combined with a state-of-the-art polynomial chaos (PC) [16–24] approach to efficiently include aleatory uncertainty sources. A relevant design example in Section 10.3.3 illustrates the computational efficiency and accuracy of the novel hybrid method.

10.2 UQ with epistemic variables

10.2.1 Basic elements of evidence theory

The evidence (or Dempster–Shafer) theory provides a general framework where possibility and probability distributions can be used to solve epistemic UQ and UP problems. Note that, evidence theory can also be used to characterize hybrid epistemic-aleatory uncertainty sources, as shown in Section 10.3. Hence, this section introduces how possibility and probability distributions are defined within evidence theory.

Let us assume that the set Σ contains all possible values x that a variable X may assume, and let us define A as a subset of Σ: $A \subseteq \Sigma$. Our goal is to evaluate the *evidence* of the statement "x may belong to A." First, let us consider the power set Ω formed by all the possible subsets of Σ, indicated as B_i. A probability weight $w(B_i)$ can be assigned to each subset, under the condition that their overall sum over Ω is equal to one. In particular, B_i is called *focal element* of Ω if its weight $w(B_i)$ is non-zero. Now, the evidence of the statement "x may belong to A" can be expressed by the Belief (Bel) and Plausibility (Pl) functions, defined as follows [8]:

$$\mathrm{Bel}(A) = \sum_{B_i \subseteq A} w(B_i) \tag{10.1}$$

$$\mathrm{Pl}(A) = \sum_{B_i \cap A \neq \varnothing} w(B_i) \tag{10.2}$$

Bel indicates the confidence we have in the statement "x may belong to A" based on the available information (represented by the weights $w(B_i)$). Note that, $B_i \cap A \neq \varnothing$ represents an event that contradicts the statement "x may belong to A." Hence, Pl in (10.2) expresses the confidence in the statement "x may belong to A" based on the confidence on the occurrence of events that contradict the statement itself. In particular, the plausibility function can be expressed as $\mathrm{Pl}(A) = 1 - \mathrm{Bel}(\overline{A})$.

The definition of Pl and Bel is fundamental to introduce possibility and probability theory in the framework of the Dempster–Shafer theory. Indeed, Pl and Bel represent a limiting upper and lower cumulative distribution function (CDF), respectively, that encompass all families of probability distributions P(A) [25], as follows:

$$\mathrm{Pl}(A) \leq \mathrm{P}(A) \leq \mathrm{Bel}(A) \tag{10.3}$$

Now, if the focal elements B_i are singletons, it is possible to assign weights to each element of Σ in a unique and precise way, which corresponds to the traditional probabilities. In this case, Bel and Pl degenerate into a single CDF P, leading to Bel = Pl = P. Hence, probability theory can be considered as a particular case of evidence theory.

Another case of interest is the following: let us assume that all focal elements are nested intervals $B_1 \subseteq B_2 \subseteq \ldots \subseteq B_N$. In this case, the Belief (10.1) and Plausibility (10.2) functions are called *necessity* (N) and *possibility* (Π) functions, respectively, and can be expressed as:

$$\mathrm{N}(A) = 1 - \sup_{x \in \overline{A}} \pi(x) \tag{10.4}$$

$$\Pi(A) = \sup_{x \in A} \pi(x) \tag{10.5}$$

where the function $\pi(x) : \Sigma \to [0, 1]$ is called possibility distribution (PD), and is defined as:

$$\pi(x) = \sum_{B_k \ni x} w(B_k), \quad \forall x \in \Sigma \tag{10.6}$$

While the probability distribution indicates the likelihood (frequency of occurrence) of an event, the PD represents the degree of confidence that a certain event can occur: $\pi(x) = 0$ corresponds to a *definitely impossible* event and $\pi(x) = 1$ to a *totally possible* one.

As an example, a rectangular PD is shown in Figure 10.1(a): all values x in the interval $A = [x_0, x_1]$ are perfectly possible, whereas all other values, in $\overline{A}$, are impossible. Compared to a uniform probability distribution, which assumes all values to be equally probable in $[x_0, x_1]$, the information provided by a rectangular PD is more vague. However, it is also more general, since it does not make any assumptions on the likelihood of the values in A. It is important to remark that possibility Π and necessity N functions correspond to Bel and Pl, when the focal elements are nested intervals. Hence, thanks to equation (10.3), it holds that all

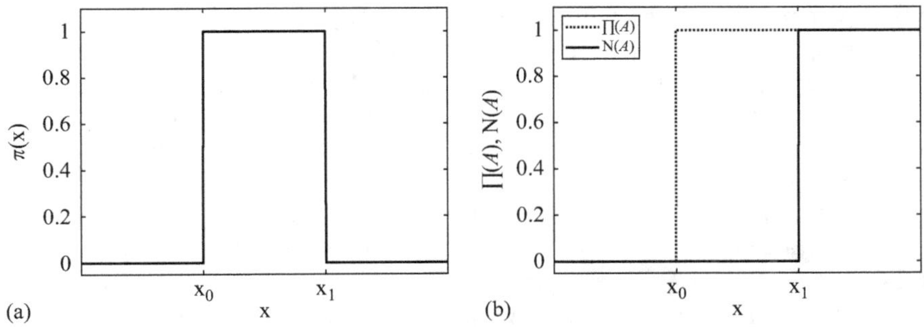

Figure 10.1 *(a) Example of rectangular PD defined in the interval* $A = [x_0, x_1]$. *(b) Corresponding possibility* $\Pi(A)$ *and necessity* $\mathrm{N}(A)$ *functions*

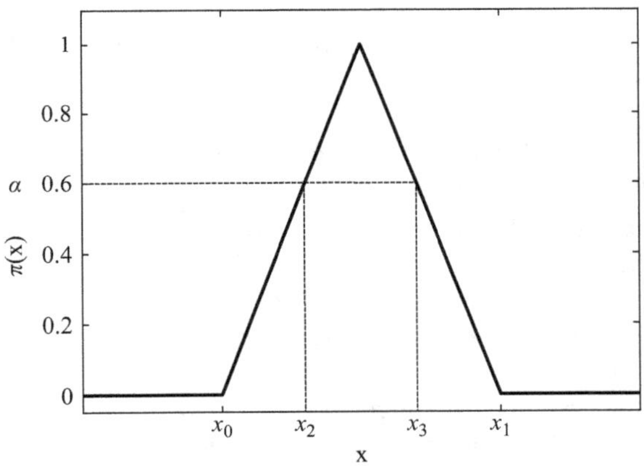

Figure 10.2 *Example of triangular PD defined in the interval* $A = [x_0, x_1]$. *Choosing a value of* α *defines a unique interval* A_α. *For example,* $\alpha = 0.6$ *identifies the interval* $A_\alpha = [x_2, x_3]$

possible CDFs $P(A)$ assigned to the interval $A = [x_0, x_1]$ (including the uniform CDF), are bounded by the possibility and necessity functions of the rectangular PD, shown in Figure 10.1(b). This example illustrates how it is possible to estimate the boundaries that delimit the distribution of a function $f(x)$ defined in an interval A in the framework of possibility theory; however, the specific distribution of $f(x)$ cannot be evaluated, since no probabilistic information about the variable x is available.

The theory of fuzzy sets [26] offers a rigorous mathematical framework to describe PDs. Indeed, the focal elements B_i are nested intervals obtained by "cutting" a PD at different levels $\alpha \in [0, 1]$. For example, let us consider the triangular PD in Figure 10.2. The value $\alpha = 0.6$ identifies the interval $[x_2, x_3]$. To each value of $\alpha \in [0, 1]$ corresponds a unique interval $A_\alpha = [\inf_\alpha, \sup_\alpha]$, called α-cut. It is important to remark that the entire PD is determined by the knowledge of the extreme values $\inf_\alpha$ and $\sup_\alpha$ of a sufficient amount of α-cuts: an epistemic variable x characterized by a PD $\pi(x)$ can be considered as a FV described by its α-cuts.

Note that the α-cuts are always nested, independently on the PD under study. For example, let us consider the triangular PD in Figure 10.2: for all the values of $\alpha \in (0.6, 1]$ it holds that $A_{\alpha \in (0.6,1]} \subset A_{\alpha=0.6}$. The opposite stands for $\alpha \in [0, 0.6)$: $A_{\alpha=0.6} \subset A_{\alpha \in [0,0.6)}$. A particular case is represented by the rectangular PD in Figure 10.1(a): all the α-cuts coincide with the interval $[x_0, x_1]$ where the PD is defined.

Additionally, it is possible to prove that each α-cut corresponds to a confidence interval with necessity $\mathrm{N}(A_\alpha)$, as follows [27]:

$$\mathrm{N}(A_\alpha) = 1 - \sup_{x \notin A_\alpha} \pi(x) = 1 - \alpha \tag{10.7}$$

The interested reader is referred to [26, 27] for a thorough discussion on the application of the FVs' mathematical formalism in the framework of possibility theory.

10.2.2 Characterization of systems in the presence of epistemic uncertainty

Now that the fundamental elements of possibility theory have been introduced in the previous section, their application to solve UQ problems is described here. Let us consider a system under uncertainty effects, such as a microwave filter or an antenna, where some of the parameters (e.g., the relative permittivity of the substrate) are epistemic variables. As a result of the variability of these parameters, the performance of the system will change. Our goal is to quantify the variability of the system's performance by adopting possibility theory.

More specifically, let us denote an indicator of the performance of the system under study (i.e., the input reflection coefficient or the crosstalk of a waveguide) as $f(\mathbf{x})$, where $\mathbf{x}$ is a vector collecting the design parameters subject to uncertainty, which are regarded as epistemic variables. It is assumed in the following that the relation between f and the variables $\mathbf{x}$ is not know upfront; however, for each possible combination of the variables $\mathbf{x}$ it is always possible to compute the corresponding value of the function f, by adopting CAD simulations or measurements. For example, if the quantity of interest is the magnitude of the scattering parameters of a microwave filter and the variables in $\mathbf{x}$ represent the filter's design parameters, full wave simulations in suitable CAD software can be used to estimate the scattering parameters in the desired frequency range for each design configuration.

Additionally, let us assume that the PDs of the variables in $\mathbf{x}$ are known, or that they can be easily estimated, thanks to some prior knowledge of the problem at hand. This is a reasonable assumption for many practical problems: the interval of admissible values for the design parameters affected by (epistemic) uncertainty is typically known, due to the characteristics of the problem considered. For example, when manufacturing integrated circuits, the foundry can provide an estimation of the effect of the tolerances of the manufacturing process on the design parameters.

If no other information is available, a rectangular PD (see Figure 10.1(a)) can be assigned to the variables **x**. Indeed, as described in the previous section, a rectangular PD assumes that all values in the chosen interval are equally possible (situation often described as "total ignorance"). Otherwise, different PD can be more suited to describe the variability of the problem at hand. For example, the triangular PD in Figure 10.2, assumes levels of confidence which are higher around a central value, where $\pi(\mathbf{x}) = 1$, and gradually decreasing for all the other values.

In this framework, the goal is to estimate the PD of $f(\mathbf{x})$ and, consequently, the corresponding possibility Π and necessity N functions. Indeed, as indicated in Section 10.2.1, Π and N form the upper and lower bounds, respectively, for all possible CDFs of $f(x)$. The reference approach to solve this problem, based on the theory of FVs [26, 27], is described in the following. First, the interval of possibility values [0, 1] is divided into a finite set of N_α α-cuts. Note that each value α identifies a domain Ω_α, which is bounded in each dimension by the interval of the corresponding α-cut of the pertinent epistemic variable in the vector **x**. For example, let us assume that two epistemic variables are collected in the vector $\mathbf{x} = (x_1, x_2)$. For each value of $\alpha \in [0, 1]$, one interval $A_\alpha^{x_1}$ can be identified for x_1, and another $A_\alpha^{x_2}$ for x_2. $A_\alpha^{x_1}$ and $A_\alpha^{x_2}$ define the domain Ω_α. Next, for each α-cut, the minimum and maximum of $f(\mathbf{x})$ must be estimated for $\mathbf{x} \in \Omega_\alpha$. Indeed, the minimum $\inf_\alpha(f)$ and the maximum $\sup_\alpha(f)$ of $f(\mathbf{x})$ for all the N_α α-cuts define the desired $\pi(f)$ (see Section 10.2.1). Once the PD of f is computed, the corresponding possibility and necessity functions can be obtained using (10.4) and (10.5).

Hence, a series of N_α minimization and maximization problems must be solved. The standard "brute force" approach is to adopt a dense sampling of $\mathbf{x} \in \Omega_\alpha$ and to evaluate $f(x)$ for all the chosen samples in order to estimate the desired minimum and maximum. This approach, summarized in Figure 10.3, can be computationally expensive due to the large number of CAD simulations (or measurements) required. This is especially true for electromagnetic problems, where

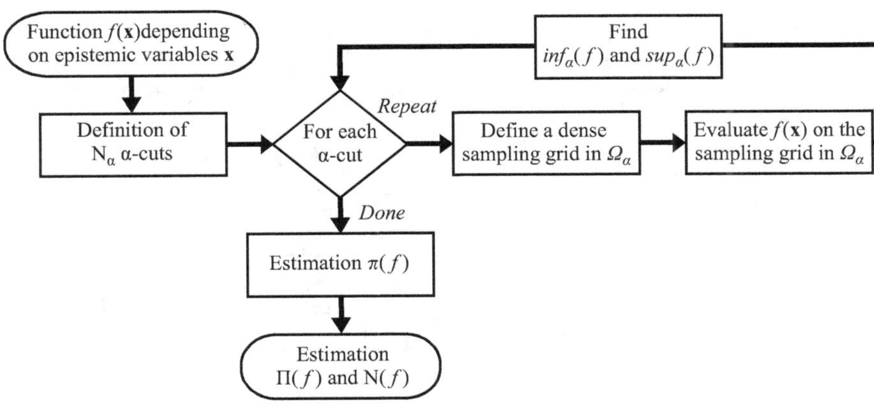

Figure 10.3 Flowchart of the "brute force" approach to estimate the possibility and necessity function of $f(x)$

numerical full wave simulations are often employed. Additionally, the approach outlined in Figure 10.3 offers a limited accuracy if the function f is non-smooth in Ω_α. Note that these issues become especially relevant when the number of epistemic variables increases.

10.2.3 Machine learning for the characterization of epistemic uncertainty sources

To alleviate the computational burden of the brute force method presented in Section 10.2.2, BO is adopted here to solve the N_α minimization and maximization problems necessary to estimate the PD of $f(\mathbf{x})$ [9]. Indeed, BO is particularly useful when the objective function is expensive to evaluate and dynamic with respect to its parameters, thus leading to several local optima [10].

In particular, BO aims to solve global optimization problems of the form:

$$f_{\min} = \min_{\mathbf{x}} (f(\mathbf{x})) \tag{10.8}$$

or

$$f_{\max} = \max_{\mathbf{x}} (f(\mathbf{x})) \tag{10.9}$$

where $f(\mathbf{x})$ is the quantity to be minimized (or maximized), which depends on a set of parameters collected in the vector $\mathbf{x}$. In the framework of BO, $f(\mathbf{x})$ is called *objective* or *cost* function, and it is considered as a "black box" function. This means that it does not have to be known *a priori*, but it can be evaluated for any combination of its parameters $\mathbf{x}$ through simulations or measurements. For epistemic UQ problems, the vector $\mathbf{x}$ collects the epistemic variables, while $f(\mathbf{x})$ represents the performance of the system under study, as defined in Section 10.2.2.

In order to solve optimization problems defined as in (10.8) and (10.9), BO adopts an iterative approach based on two crucial elements: a mathematical model and a suitable sampling strategy, as shown in Figure 10.4.

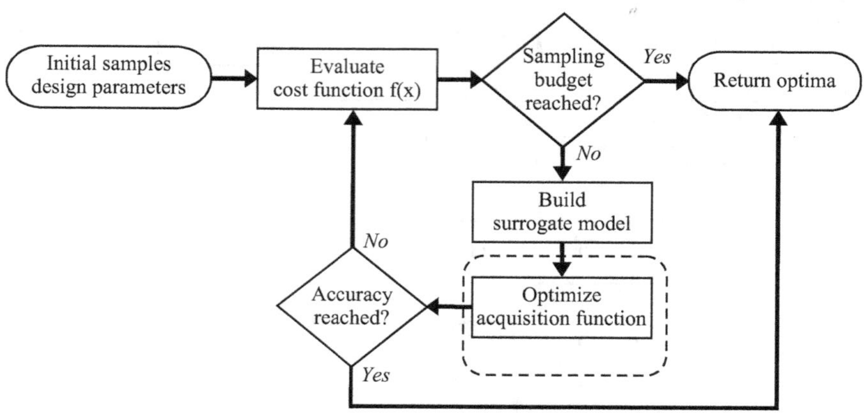

Figure 10.4 Flowchart of the BO algorithm

First, the cost function $f(\mathbf{x})$ is evaluated through simulations or measurements for an initial set of N_{in} samples $\mathbf{x}_i$, $i = 1, \ldots, N_{\text{in}}$, for example chosen according to a Latin hypercube [28]. Based on these initial samples, a suitable mathematical model of the cost function is then built. Such model, also called surrogate model, serves as a cheap substitute for $f(\mathbf{x})$, which is computationally easy to optimize. However, the model predictions are not deterministic, but *stochastic*. This is one of the main characteristics of BO: the sampling strategy can leverage on the model's uncertainty to determine the next point to evaluate. Several mathematical techniques can be adopted to build the desired surrogate model, for example Bayesian neural networks [29] and Gaussian Processes (GPs) [30,31].

In general, the sampling strategy requires to optimize a suitable function, called acquisition function. Such an acquisition function transforms the model predictions into a measure that identifies where the next evaluation should be performed in order to efficiently estimate the global optimum (see (10.8) and (10.9)). Among the most widely adopted acquisition functions there are expected improvement (EI) [32] and probability of improvement [33].

Now, if the new selected point is very close to the previously evaluated point, convergence is assumed and the optimization is finished. Alternatively, the suggested point is evaluated through a new (expensive) simulation. This procedure is repeated until the maximum computational budget is spent or the algorithm reaches convergence. For a more complete treatment of BO, the interested reader is referred to [10,32,34].

In order to apply BO to solve epistemic UQ problems, first it is necessary to choose the statistical model and the acquisition function. In this chapter, the popular GP regression model [30,31] is adopted as the stochastic surrogate model, due to its accuracy and modeling power. A GP can be considered as a generalization of the multivariate Gaussian distribution. While the latter is a distribution over a finite set of random variables, completely specified by a mean vector and covariance matrix, a GP instead considers an infinite, parameterized set of random variables. Hence, it can be interpreted as a distribution over functions that is fully characterized by its mean function and covariance function [30,31]. Typically, the mean function is chosen to be zero, as the data can be normalized, and no trend is assumed to be known beforehand. The covariance function (also called kernel) determines the types of interactions between different variables that can be captured in the final model, and must be carefully chosen among those defined in the literature. Giving a comprehensive list of the main kernel functions and their properties is outside the scope of this work. The interested reader is referred to [30,31] for a through discussion on this topic. Here the Matérn (5/2) kernel is adopted, due to its capability to model a wide class of functions (including non-differentiable ones) [30,31]. It is defined as:

$$K_{M,5/2}(\mathbf{x}, \mathbf{x}') = \sigma^2 \left(1 + \frac{\sqrt{5}d}{\rho} + \frac{5d^2}{3\rho^2}\right) \exp\left(-\frac{\sqrt{5}d}{\rho}\right) \tag{10.10}$$

where $\mathbf{x}$ and $\mathbf{x}'$ represents two points in the space of the epistemic variables, and $d = \|\mathbf{x} - \mathbf{x}'\|$.

Once the kernel is defined, the value of the corresponding hyperparameters must be estimated. For the Matérn (5/2) kernel, the hyperparameters are σ^2 and ρ. In general, the optimal value of the hyperparameters is strictly related to the specific problem: it depends on the interactions between the different variables at hand. In the application example shown in Section 10.2.4, the hyperparameters σ^2 and ρ are optimized using maximum likelihood estimation, as implemented by the GPyOpt package [35].

Regarding the sampling strategy, the acquisition function chosen here is the EI [32,36], which is defined as follows for pure minimization problems. Let us indicate the improvement at a point $\mathbf{x}$ as:

$$I(\mathbf{x}) = \begin{cases} f_{\min} - f(\mathbf{x}) \text{ } if \text{ } f(\mathbf{x}) < f_{\min} \\ 0 \text{ otherwise} \end{cases} \tag{10.11}$$

where $f_{\min}$ indicates the current evaluated point with minimum value. Since a GP is a stochastic model, its predictions are not deterministic. In particular, a GP's prediction t at location $\mathbf{x}$ is a Gaussian random variable, whose mean represents the estimated value of $f(\mathbf{x})$. Therefore, the expected value of the improvement can be written as:

$$E[I(\mathbf{x})] = E[\max(f_{\min} - t, 0)] \tag{10.12}$$

The expectation in (10.12) and the corresponding gradient can be calculated analytically, making EI optimization computationally cheap.

Once both the model and the acquisition function are defined, BO optimization can effectively replace the "brute force" sampling method, in order to solve epistemic UQ and UP problems.

The implementation of BO described so far is particularly suitable to solve a *single* minimization *or* maximization problem. However, in our problem setting, a *series* of N_α minimization *and* maximization problems must be solved in order to find $\inf_\alpha$ and $\sup_\alpha$ for each α-cut, see Figure 10.3. Hence, the formulation of BO is adapted in the following to better fit the characteristic of epistemic UQ problems [9]. First, the acquisition function adopted here is a modified version of the EI:

$$\mathrm{EI}_{\mathrm{mod}}(\mathbf{x}) = \max(E[\max(f_{\min} - t, 0)], E[max(t - f_{\max}, 0)]) \tag{10.13}$$

This corresponds to the maximum of the EI defined for both the minimization and the maximization problem. Hence, the next candidate point to be evaluated is where, according to the model, the most potential for a better optimum of either kind can be found. Thus, the proposed optimization strategy is able to find both minimum and maximum of $f(\mathbf{x})$ with a minimal amount of evaluations.

Next, the minimum and the maximum of the cost function must be computed for *each* α-cut. This is a trivial task for epistemic variables with a rectangular PD, since all α-cuts are the coincident (see Figure 10.1). However, a suitable strategy must be defined when dealing with different PDs, such as the triangular one shown in Figure 10.2. To this end, an iterative "top-down" approach is proposed in the following, which leverages on the fact that α-cuts are nested by definition. According to such an approach, BO is first carried out for the top α-cut, corresponding to $\alpha = 1$, with a small portion of the total computational budget. Then,

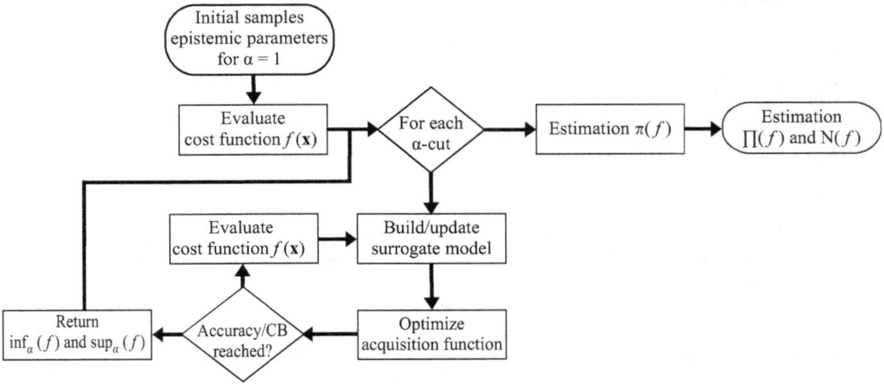

Figure 10.5 Flowchart of the BO algorithm for epistemic UQ and UP problems

one by one, the objective function is also optimized over the α-cuts corresponding to lower possibility values, making use of the samples already evaluated in the α-cuts above them, while allowing a few additional samples to be evaluated for each new α-cut. Hence, the model is refined at each new α-cut in order to accurately estimate $\inf_\alpha$ and $\sup_\alpha$. This procedure is repeated until the last α-cut is reached, corresponding to $\alpha = 0$ [9].

The approach outlined above is summarized in Figure 10.5, which illustrates the application of BO for epistemic UQ and UP problems. In particular, the surrogate model for each α-cut is updated until the desired accuracy or computational budget (indicated as CB in Figure 10.5) is reached. Note that, in post-processing, a better optimum found at lower α can replace an optimum found at higher α, as long as it falls within the bounds defined by that α-cut.

10.2.4 Application example

A dual-polarized textile patch antenna, described in [37] and shown in Figure 10.6, is considered as first case study. Such antenna operates in the $[2.4, 2.4835]$ GHz industrial, scientific, and medical (ISM) band. Let us assume that the substrate height h and permittivity ε_r can be considered as epistemic variables defined in the support $h \in [3.44, 4.44]$ mm and $\varepsilon_r \in [1.43, 1.63]$, where the substrate height is assigned a rectangular PD and the relative permittivity a triangular PD with central value $\varepsilon_{r0} = 1.53$. This choice allows us to demonstrate that the proposed approach is general and independent of the specific PDs assigned to the epistemic variables. In this framework, our goal is to estimate the uncertainty of the antenna scattering parameters for 50 frequency samples in the band $[2.4, 2.4835]$ GHz, due to the epistemic variables (h, ε_r). In particular, the scattering parameters are estimated via the Momentum electromagnetic field simulator of Advanced Design System (Momentum EEsof EDA, Agilent Technologies) as a function of the frequency and of the epistemic variables.

First, the α-cuts must be defined: $N_\alpha = 36$ α-cuts are chosen for this application, linearly spaced from the possibility level 0 to 1. Next, in order to apply the machine

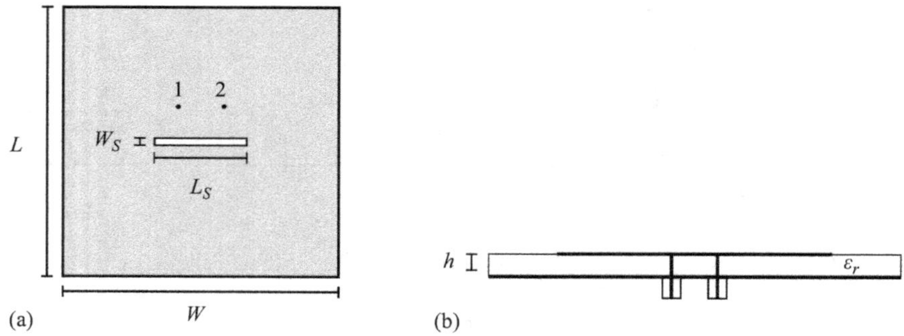

Figure 10.6 Geometry of the antenna under study. (a) Top view. The feed points are indicated as 1 and 2. (b) Cross section

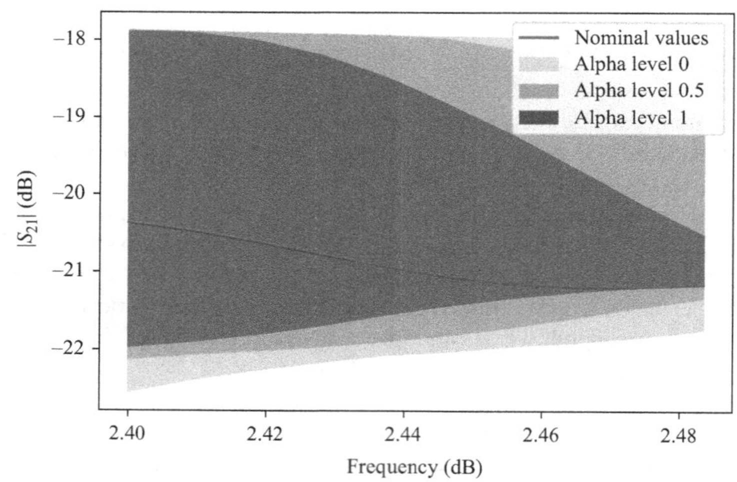

Figure 10.7 Bounds of the magnitude of S_{21} (dB) obtained with the proposed BO-based method at different α levels

learning method described in Section 10.2.3, the maximum computational budget available for the optimization must be specified, that is, the maximum number of simulations that can be used to find $\inf_\alpha$ and $\sup_\alpha$ for each α level. The maximum computational budget is set to 81 (h, ε_r) samples for this application, where 11 (h, ε_r) samples are assigned to $\alpha = 1$, and two additional samples are given for each following α-cut. Now, the technique described in Section 10.2.3 can be applied to solve the UQ problem. Figure 10.7 illustrates an example of the obtained results: the minima and the maxima of all the possible values of $|S_{21}|$ computed by the proposed BO-based method are shown in Figure 10.7, where different shades correspond to different α levels.

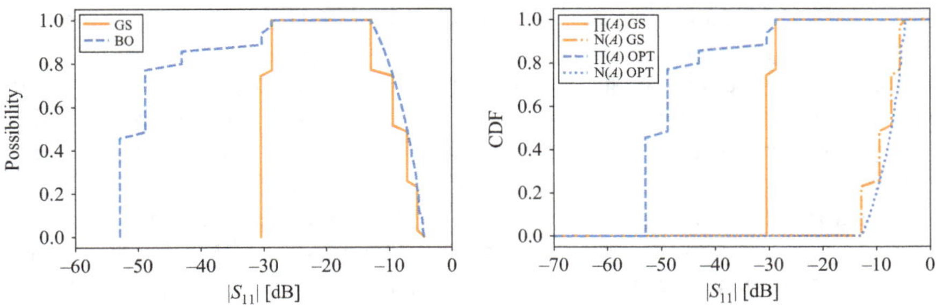

Figure 10.8 (a) PD of the magnitude of S_{11} (dB) at 2.4371 GHz estimated with the GS and the proposed BO-based method for 81 (h, ε_r) samples. (b) Corresponding possibility $\Pi(A)$ and necessity $\mathrm{N}(A)$ functions

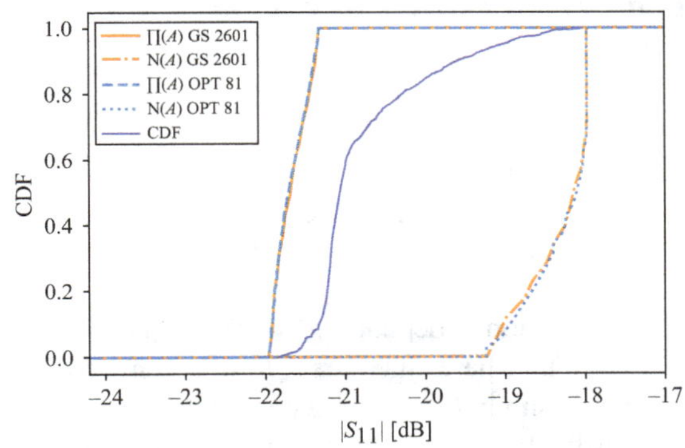

Figure 10.9 Possibility $\Pi(A)$ and necessity $\mathrm{N}(A)$ functions of the magnitude of S_{21} at 2.4557 GHz estimated via GS with 51×51 (h, ε_r) samples and the proposed BO approach with 81 (h, ε_r) samples. The CDF of the magnitude of S_{21} (dB) obtained by considering (h, ε_r) as uniform random variables is indicated as well

Next, the proposed technique is compared to the grid search (GS) method described in Section 10.2.2 for the same maximum computational budget. Hence, a uniform grid of 9×9 (h, ε_r) samples is allocated for the GS approach. Relevant results are shown in Figure 10.8(a), which compares the PD of the magnitude of S_{11} at 2.4371 GHz obtained with GS and BO. The corresponding possibility and necessity functions are shown in Figure 10.8(b). Clearly, for the same computational budget, BO offers greater accuracy than the approach based on GS.

Finally, the sampling budget of the GS method is increased until a comparable accuracy with BO can be achieved: a uniform grid of 51×51 (h, ε_r) samples is needed for the GS. Relevant simulation results are shown in Figure 10.9: the proposed method requires less computational resources compared to GS in order to reach the same accuracy. In terms of computational efficiency, the GS method requires 8 h and 11 min per single frequency point to perform 51 × 51 (h, ε_r) simulations, while the BO requires only 16 min per single frequency point. In particular, 13 min are required to perform 81 (h, ε_r) simulations and 3 min for the GP model building process and the optimization performed in the BO framework. The calculations have been performed on a computer with 8 GB and 8 cores Intel Core i7-2600 CPU @ 3.40 GHz.

It is important to remark that possibility and necessity functions individuate the boundaries that delimit the region where all the possible CDFs lie. In order to illustrate this property, let us assume that the parameters (h, ε_r) can be considered as traditional uniform random variables with support $[3.44, 4.44]$ mm and $[1.43, 1.63]$, respectively. The corresponding CDF for $|S_{21}|$, computed with 10,000 (h, ε_r) samples, is shown in Figure 10.9. As expected, the CDF is always in the domain defined by the possibility and necessity functions.

10.3 Characterization of hybrid epistemic-aleatory UQ and UP problems

10.3.1 MC-based hybrid algorithm

Several UQ and UP problems are characterized by the coexistence of aleatory and epistemic uncertainty. Hence, in this section the methodology described in Section 10.2.2 is extended to describe hybrid epistemic-aleatory uncertainty sources.

Let us consider a function f depending on a total number of M variables as $f(r_1, \ldots, r_T, e_{T+1}, \ldots, e_M)$, where the first T of them are random variables $r_1, r_2, \ldots, r_T$, and the remaining $T - M$ are epistemic variables $e_{T+1}, e_{T+2}, \ldots, e_M$. Propagation and quantification of the uncertainty of the function f can be performed by the following algorithm [13,14,38]:

1. K realizations of the random variables $\left[r_1^k, r_2^k, \ldots, r_T^k\right]_{k=1}^K$ are generated via MC sampling.
2. For a predetermined number N_α of α-cuts, f is evaluated for each α-cut on a dense grid in the space of epistemic variables.
3. The minimum and maximum of the function f at different α levels can now be determined, and denoted as $\inf_\alpha$ and $\sup_\alpha$, respectively. These minimum and maximum values identify the PD π associated with a specific random-variable realization.
4. This procedure is repeated for all random-variable realizations, leading to a collection of K PDs: $\left[\pi_f(r_1^k, r_2^k, \ldots, r_T^k)\right]_{k=1}^K$.
5. The variations of the function f due to the epistemic and random uncertainty sources can be estimated by aggregating the results obtained so far.

Note that, for each realization $\left[r_1^k, r_2^k, \ldots, r_T^k\right]$ the value of the random variables is fixed. In these conditions, the variation of the function f depends only on the $T - M$ epistemic variables. Hence, steps 2 and 3 of the methodology described above are the same as the approach presented in Section 10.2.2 to characterize epistemic uncertainty sources. The main difference here is that one PD for each realization of the random variables is estimated for hybrid probabilistic–possibilistic problems. Such PDs are eventually aggregated in step 5 to obtain a single estimation of π_f. Several approaches have been proposed in the literature to aggregate these results [13,38]: in this chapter, the methodology described in [13] is adopted, due to its accuracy and ease of implementation. Let us consider as example the following analytical function:

$$f(x_1, x_2) = \frac{2x_1^2 + x_2}{x_1 + x_2} \tag{10.14}$$

where x_1 is a Gaussian random variable with mean $\mu = 100$ and standard deviation $\sigma = 2$, while x_2 is an epistemic variable with support $[0 - 20]$ following the PD shown in Figure 10.10. Now, Figure 10.11 illustrates an example of the PDs obtained by following steps 1–4 of the methodology described above for the function (10.14). Each PD is individuated by the $\inf_\alpha$ and $\sup_\alpha$ computed for 31 α-cuts in the interval $[0 - 1]$. Note that, for each value of α, the spread between the obtained PDs is entirely a consequence of the MC random sampling. Hence, it is possible to compute an aggregated PD by building an histogram of cumulative relative frequencies of $\inf_\alpha$ and $\sup_\alpha$ for each α-cut, and extracting their final values for a specific level of probability. In this chapter, the final $\inf_\alpha$ and $\sup_\alpha$ values are taken such that there is a 5% probability of having values lower or higher, respectively. This corresponds to a 95% confidence interval. As an example, the aggregated PD for the function in (1.14) is indicated in Figure 10.11 with red circles.

Finally, once the aggregated PD is obtained, the corresponding necessity N and possibility Π functions can be estimated via (10.4) and (10.5), respectively, thus

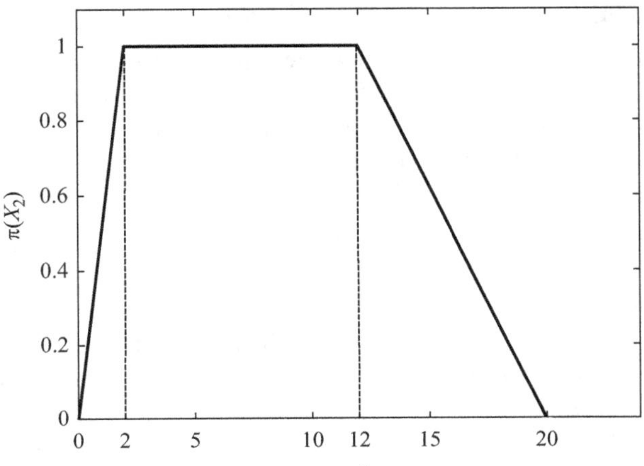

Figure 10.10 PD of the variable x_2 defined in (10.14)

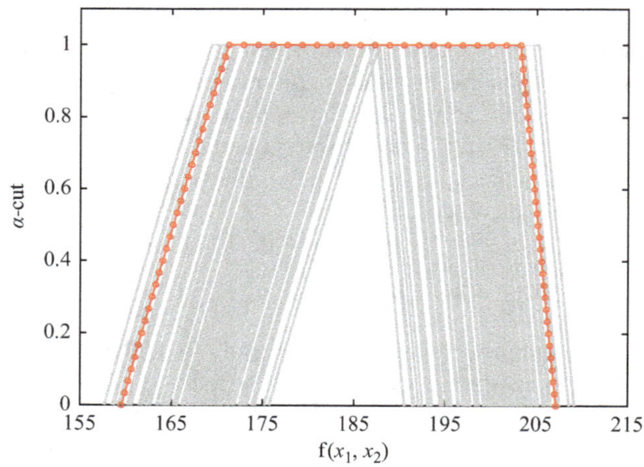

Figure 10.11 PDs of $f(x_1, x_2)$ obtained for 100 MC samples (gray lines). The corresponding aggregated PD is indicated with red circles

estimating the interval where the CDF of the function $f(r_1, \ldots, r_T, e_{T+1}, \ldots, e_M)$ lies. Clearly, this five-step approach to characterize hybrid probabilistic–possibilistic problems is computationally expensive, since it requires to perform a dense sampling of the epistemic variables *for each* realization of the random parameters. Given that thousand of realizations may be required for an accurate estimation of the uncertainty effects, the computational cost of this method can become prohibitive. This issue is especially relevant for electromagnetic problems, where the function f is often evaluated through computationally expensive full wave simulations.

10.3.2 Machine learning-based hybrid algorithm

The main drawback of the methodology described in the previous section is the large number of evaluations of the function f needed to estimate its PD π_f. A significant efficiency boost can be achieved by adopting BO to estimate the PD for each realization of the random variables, as described in Section 10.2.3. Hence, the approach described in Section 10.3.1 becomes:

1. K realizations of the random variables $\left[r_1^k, r_2^k, \ldots, r_T^k\right]_{k=1}^K$ are generated via MC sampling.
2. For a predetermined number N_α of α-cuts, ranging from $\alpha = 1$ to $\alpha = 0$, BO is adopted to estimate $\inf_\alpha$ and $\sup_\alpha$ for all the α levels. These minimum and maximum values construct the PD π associated with a specific random-variable realization.
3. This procedure is repeated for all random-variable realizations, leading to a collection of K PDs: $\left[\pi_f(r_1^k, r_2^k, \ldots, r_T^k)\right]_{k=1}^K$.
4. The variations of the function f due to the epistemic and random uncertainty sources can be estimated by aggregating the results obtained so far.

Nonetheless, the slow convergence of the MC sampling, to account for the aleatory variability of the random variables, still limits the efficiency of this novel algorithm. To overcome this limitation, the objective is to obtain a large number of PDs $\left[\pi_f(r_1^k, r_2^k, \dots, r_T^k)\right]_{k=1}^K$ depending on K different realizations of the random variables, based only on a limited set of J realizations, where $J << K$. To this end, a mathematical model $\pi_f(r_1, r_2, \dots, r_T)$ must be computed, starting from a small set of J realizations of the random variables $\left[r_1^j, r_2^j, \dots, r_T^j\right]_{j=1}^J$. Once such model is computed, it can be efficiently evaluated for an (arbitrarily) large number of K realizations of the random parameters, and the PDs of f can now be estimated by aggregating the results obtained so far [15].

More specifically, the PDs $\left[\pi_f(r_1^k, r_2^k, \dots, r_T^k)\right]_{k=1}^K$ depend on the value of $\inf_\alpha$ and $\sup_\alpha$ for all the considered α-cuts. Hence, two separate models, one for $\inf_\alpha$ and another for $\sup_\alpha$ will be computed. The PC expansion is adopted to reach this goal, given its accuracy and efficiency in characterizing stochastic variations [16–24]. Using the PC expansion, the desired models of the minimum and maximum for different α-cuts can be expressed in the form [15]:

$$F_{\min}(\alpha, \mathbf{r}) = \sum_{l=0}^{L} \beta_{\min,l}(\alpha)\phi_l(\mathbf{r}) \tag{10.15}$$

$$F_{\max}(\alpha, \mathbf{r}) = \sum_{l=0}^{L} \beta_{\max,r}(\alpha)\phi_L(\mathbf{r}), \tag{10.16}$$

where $\phi_l(\mathbf{r})$, for $l = 1, \dots, L$, are suitable orthogonal polynomials depending on the $\mathbf{r} = (r_1, r_2, \dots, r_T)$ random variables, while $\beta_{\min,l}(\alpha)$ and $\beta_{\max,l}(\alpha)$ are the corresponding PC coefficients, which depend on the α-cut.

Let us assume that models in the form (10.15) and (10.16) have been calculated for the problem at hand. These analytical models are continuous with respect to the random variables $\mathbf{r}$ and can be evaluated with great efficiency compared to a direct sampling of the function f. As such, a MC sampling of K instances of the random variables, evaluated using $F_{\min}$ and $F_{\max}$ in (10.15) and (10.16), can be used to construct the collection of PDs $\left[\pi_f(r_1^k, r_2^k, \dots, r_T^k)\right]_{k=1}^K$. Then, these PDs can be used to solve UQ and UP problems as described in Section 10.3.1.

In order to compute models (10.15) and (10.16), the approach [15] is adopted here, which is summarized in Figure 10.12 and discussed in the following. It is important to remark that the PC basis functions in (10.15) and (10.16) depend only on the distribution of the random variables r. In particular, for random variables with specific distributions, the PC basis functions are known and are formed by the polynomials of the Wiener-Askey scheme [16]. Examples include the Gaussian and Uniform distribution. In general, different methodologies have been presented in the literature to compute the basis functions for random variables with arbitrary distributions, such as [18]. Note that PC models can also be obtained if the random variables $\mathbf{r}$ are correlated [23]: for example, approaches based on suitable variable transformations (such as the copula theory [17]) can be adopted.

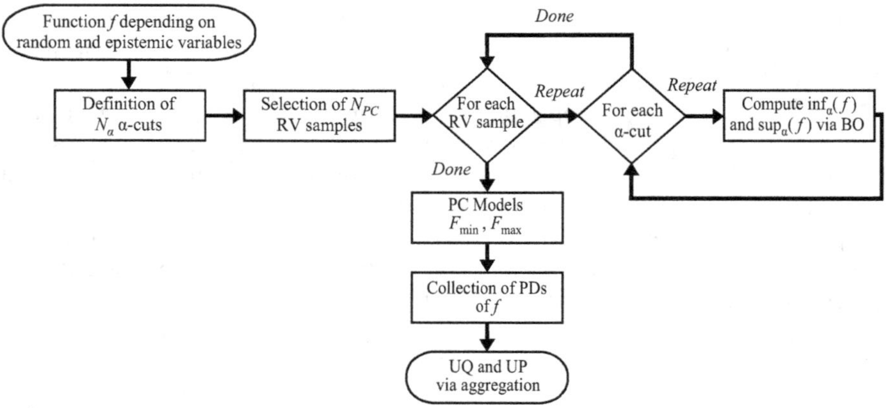

Figure 10.12 Flowchart of the proposed methodology for hybrid random-epistemic UQ and UP problems

Thus, only the coefficients $\beta_{\min,l}$ must be estimated: in this chapter, the Stochastic Testing approach presented in [19] is adopted for this purpose. Hence, a number of samples $[\mathbf{r}_i]_{i=1}^{N_{PC}}$ (also called collocation points) are chosen in the random variables' space according to the method described in [19]. At each of these points, BO is performed in order to estimate $\inf_\alpha$ and $\sup_\alpha$ of the cost function f for the different α-cuts. Next, the desired PC coefficients can be computed by solving a suitable linear system [19]. Giving a detailed description of the properties of the PC expansion is outside the scope of this contribution: the interested reader is referred to [16–24] for a comprehensive discussion.

10.3.3 Application example

The proposed framework is applied to two coupled microstrip lines with a 90 degrees bend. This layout, illustrated in Figure 10.13, was introduced in [39].

Four design parameters are considered as independent random variables in this example: the lengths of both segments, $l_1 \sim \mathcal{N}(50, 0.1)$ mm and $l_2 \sim \mathcal{N}(50, 0.1)$ mm, and the width of each line, $w_1 \sim \mathcal{N}(1.8, 0.1)$ mm and $w_2 \sim \mathcal{N}(1.8, 0.1)$ mm. The symbol $\mathcal{N}(a, b)$ indicates a Gaussian distribution with mean a and standard deviation b. The relative permittivity ε_r and the height h of the substrate are instead treated as epistemic variables with rectangular PDs defined in the supports $\varepsilon_r \in [3.36, 3.96]$ and $h \in [1.374, 1.674]$ mm, respectively. Since mode conversion is critical for bent interconnects, the goal is to estimate the effect of aleatory and epistemic parameters on the following measure of the total differential (DM) to common mode (CM) conversion [40]:

$$C = \left[\int_0^{6\text{GHz}} \left(|S_{\text{cd11}}(f)|^2 + |S_{\text{cd21}}(f)|^2 \right) df \right]^{1/2}, \qquad (10.17)$$

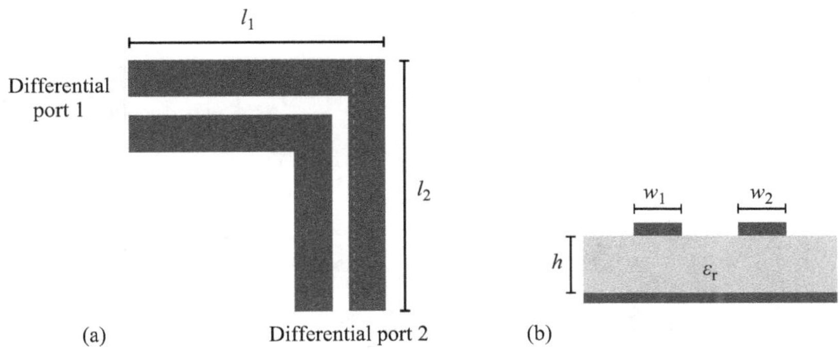

Figure 10.13 Geometry of the bended microstrips under study. (a) Top view. The differential ports are explicitly indicated. (b) Cross section

where S_{cd11} and S_{cd21} are the relevant elements of the modal S-parameters matrix. Equation (10.17) is computed into two steps: first the S-parameters are estimated in the frequency range of interest $[0, 6]$ GHz by means of full wave simulations, performed using Advanced Design System (Agilent Technologies). Then, the corresponding modal S-parameters are computed [39], and the integral in (10.17) is calculated using standard numerical techniques [40].

The methodology described in Section 10.3.2 can be applied to solve the UQ problem at hand. A maximum computational budget of 50 samples in the space of the epistemic variables was chosen. Note that, BO only needs to be carried out once for each realization of the random variables, since the epistemic variables follow rectangular PDs (i.e., all α -cut are the same, see Figure 10.1). Next, PC models with $L = 35$ (with polynomials up to third order) of the minimum and maximum of the cost function were constructed from the optima found by BO in the collocation points. Evaluating such PC models in 1,000 MC sampled points allows one to estimate the aggregated possibility and necessity functions for the total DM to CM conversion in equation (10.17), which are shown in Figure 10.14. A total of 696 full wave simulations are required to obtain these results, for a total computational cost of 17 h and 57 min.

For comparison, a GS with 10×10 points for the epistemic variables was carried out in each of these 1,000 MC points, and the corresponding aggregated results are also shown in Figure 10.14. Note that the total computational cost of the GS method is 112 day, 16 h, and 3 min, since 100,000 full wave simulations must be performed. While the possibility and necessity functions obtained with both methods are in good agreement, the proposed BO-based approach offers a significant efficiency boost compared with the "brute force" solution. Note that the improved computational efficiency is fundamental to apply a hybrid possibilistic–probabilistic approach to complex engineering problems, where "brute force" solutions become unfeasible. The calculations have been performed on a server node with a Quad-Core AMD Opteron Processor (2 GHz) and 31.4 GB DDR2 RAM.

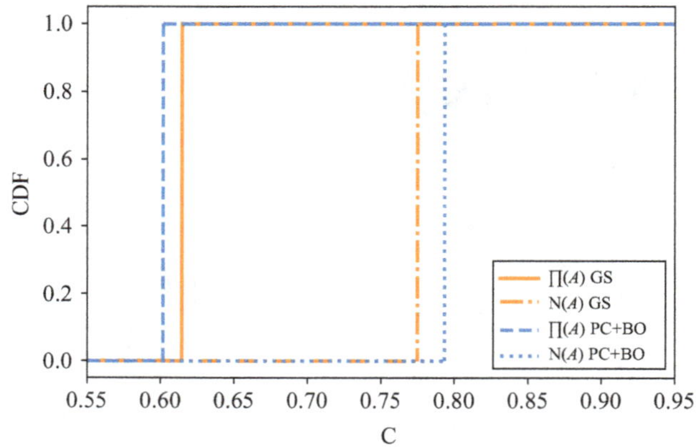

Figure 10.14 Possibility $\Pi(A)$ *and necessity* $\mathrm{N}(A)$ *functions of the total DM to CM conversion in equation (10.17) estimated via GS with 100,000 full wave simulations, and the proposed machine learning-based approach (PC + BO) with 696 full wave simulations*

10.4 Conclusions

Solving UQ and UP problems for complex engineering systems in the presence of both aleatory and epistemic uncertainty sources is of paramount importance. Possibility theory provides a common framework to describe these problems and, when coupled with data-efficient machine learning and stochastic modeling approaches, it allows one to solve UQ and UP problems with accuracy and efficiency.

In particular, a combination of BO based on GP regression and the PC expansion modeling approach is presented in this chapter. This methodology allows one to overcome the computational burden of state-of-the-art methodologies, while retaining the accuracy in estimating epistemic and aleatory variations. The proposed approach aims at minimizing the number of (expensive) simulations to characterize the system under study, and is especially useful for RF and microwave systems, where cumbersome full wave simulations are usually required.

References

[1] George F. *Monte Carlo: Concepts, Algorithms, and Applications*. Springer: New York, NY; 1996.

[2] Spina D, Ferranti F, Dhaene T, *et al.* Variability Analysis of Multiport Systems Via Polynomial-Chaos Expansion. *IEEE Transactions on Microwave Theory and Techniques*. 2012; 60(8): 2329–2338.

[3] Pham T, Gad E, Nakhla MS, *et al.* Decoupled Polynomial Chaos and Its Applications to Statistical Analysis of High-Speed Interconnects. *IEEE*

Transactions on Components, Packaging and Manufacturing Technology. 2014; 4(10): 1634–1647.

[4] Fei Z, Huang Y, Zhou J, *et al.* Uncertainty Quantification of Crosstalk Using Stochastic Reduced Order Models. *IEEE Transactions on Electromagnetic Compatibility*. 2017; 59(1): 228–239.

[5] Wang Y, Jin S, Penugonda S, *et al.* Variability Analysis of Crosstalk Among Differential Vias Using Polynomial-Chaos and Response Surface Methods. *IEEE Transactions on Electromagnetic Compatibility*. 2017; 59(4): 1368–1378.

[6] Ye Y, Spina D, Manfredi P, *et al.* A Comprehensive and Modular Stochastic Modeling Framework for the Variability-Aware Assessment of Signal Integrity in High-Speed Links. *IEEE Transactions on Electromagnetic Compatibility*. 2018; 60(2): 459–467.

[7] Boeykens F, Rossi M, Vallozzi L, *et al.* A stochastic framework to model bending of textile antennas. In: Proceedings of the IEEE Antennas and Propagation Society International Symposium; 2014.

[8] Shafer G. *A Mathematical Theory of Evidence*. vol. 42. Princeton University Press: Princeton, NJ; 1976.

[9] Kan D, Spina D, De Ridder S, *et al.* A Machine-Learning-Based Epistemic Modeling Framework for Textile Antenna Design. *IEEE Antennas and Wireless Propagation Letters*. 2019; 18(11): 2292–2296.

[10] Shahriari B, Swersky K, Wang Z, *et al.* Taking the Human Out of the Loop: A Review of Bayesian Optimization. *Proceedings of the IEEE*. 2016; 104 (1): 148–175.

[11] Knudde N, Couckuyt I, Spina D, *et al.* Data-Efficient Bayesian Optimization with Constraints for Power Amplifier Design. In: 2018 IEEE MTT-S International Conference on Numerical Electromagnetic and Multiphysics Modeling and Optimization (NEMO); 2018.

[12] Tarik MHM, Omar M, Abdullah MF, *et al.* Optimization of neural network hyperparameters for gas turbine modelling using Bayesian optimization. In: 5th IET International Conference on Clean Energy and Technology (CEAT2018); 2018.

[13] Guyonnet D, Bourgine B, Dubois D, *et al.* Hybrid Approach for Addressing Uncertainty in Risk Assessments. *Journal of Environmental Engineering*. 2003; 129(1): 68–78.

[14] Badini L, Toscani N, Spadacini G, *et al.* A possibilistic approach to radiated susceptibility of twisted-wire pairs. In: IEEE International Symposium on Electromagnetic Compatibility & Signal/Power Integrity (EMCSI); 2017.

[15] De Ridder S, Spina D, Toscani N, *et al.* Machine-Learning-Based Hybrid Random-Fuzzy Uncertainty Quantification for EMC and SI Assessment. *IEEE Transactions on Electromagnetic Compatibility*. 2020; 62(6): 2538–2546.

[16] Xiu D, Karniadakis GE. The Wiener-Askey Polynomial Chaos for Stochastic Differential Equations. *SIAM Journal on Scientific Computing*. 2002; 24(2): 619–644.

[17] Blatman G, Sudret B. An Adaptive Algorithm to Build Up Sparse Polynomial Chaos Expansions for Stochastic Finite Element Analysis. *Probabilistic Engineering Mechanics*. 2010; 25: 183–197.

[18] Oladyshkin S, Nowak W. Data-driven Uncertainty Quantification Using the Arbitrary Polynomial Chaos Expansion. *Reliability Engineering & System Safety*. 2012; 106: 179–190.

[19] Zhang Z, El-Moselhy TA, Elfadel IM, *et al.* Stochastic Testing Method for Transistor-level Uncertainty Quantification Based on Generalized Polynomial Chaos. *IEEE Transactions on Computer-Aided Design of Integrated Circuits and Systems*. 2013; 32(10): 1533–1545.

[20] Rufuie MR, Gad E, Nakhla M, *et al.* Generalized Hermite Polynomial Chaos for Variability Analysis of Macromodels Embedded in Nonlinear Circuits. *IEEE Transactions on Components, Packaging and Manufacturing Technology*. 2014; 4(4): 673–684.

[21] Spina D, Dhaene T, Knockaert L, *et al.* Polynomial Chaos-Based Macromodeling of General Linear Multiport Systems for Time-Domain Analysis. *IEEE Transactions on Microwave Theory and Techniques*. 2017; 65(5): 1422–1433.

[22] Kaintura A, Dhaene T, Spina D. Review of Polynomial Chaos-Based Methods for Uncertainty Quantification in Modern Integrated Circuits. *Electronics*. 2018; 7(3): 30. Available from: http://www.mdpi.com/2079-9292/7/3/30

[23] Cui C, Zhang Z. Stochastic Collocation with Non-Gaussian Correlated Process Variations: Theory, Algorithms and Applications. *IEEE Transactions on Components, Packaging and Manufacturing Technology*. 2019; 9(7): 1362–1375.

[24] Manfredi P, Grivet-Talocia S. Rational Polynomial Chaos Expansions for the Stochastic Macromodeling of Network Responses. *IEEE Transactions on Circuits and Systems I: Regular Papers*. 2020; 67(1): 225–234.

[25] Dubois D, Prade H. When Upper Probabilities are Possibility Measures. *Fuzzy Sets and Systems*. 1992; 49(1): 65–74.

[26] Zadeh LA. Fuzzy Sets as a Basis for a Theory of Possibility. *Fuzzy Sets and Systems*. 1978; 1: 3–28.

[27] Salicone S. *Measurement Uncertainty: An Approach Via the Mathematical Theory of Evidence*. Springer: New York, NY; 2007.

[28] Cai D, Shi D, Chen J. Probabilistic load flow computation using Copula and Latin hypercube sampling. *IET Generation, Transmission & Distribution*. 2014; 8:1539–1549.

[29] Springenberg JT, Klein A, Falkner S, *et al.* Bayesian Optimization with Robust Bayesian Neural Networks. In: *Advances in Neural Information Processing Systems (NIPS)*. vol. 29; 2016. pp. 4134–4142.

[30] Rasmussen CE, Williams CKI. *Gaussian Processes for Machine Learning*. MIT Press; 2006. Available from:http://www.gaussianprocess.org/gpml

[31] MacKay DJC. *Information Theory, Inference, and Learning Algorithms*. Cambridge University Press: Cambridge, MA; 2003.

[32] Jones DR, Schonlau M, Welch WJ. Efficient Global Optimization of Expensive Black-Box Functions. *Journal of Global optimization*. 1998; 13 (4): 455–492.

[33] Kushner HJ. A New Method of Locating the Maximum Point of an Arbitrary Multipeak Curve in the Presence of Noise. *Journal of Basic Engineering*. 1964; 86(1): 97–106.

[34] Snoek J, Hugo HL, Adams RP. Practical Bayesian Optimization of Machine Learning Algorithms. In: *Advances in Neural Information Processing Systems (NIPS)*. vol. 25; 2012. pp. 2951–2959.

[35] The GPyOpt authors. GPyOpt: A Bayesian Optimization framework in python; 2016. Available from:http://github.com/SheffieldML/GPyOpt

[36] Močkus J. On Bayesian methods for seeking the extremum. In: Marchuk GI, editor. *Optimization Techniques IFIP Technical Conference: Novosibirsk, July 1–7, 1974*. Springer: Berlin; 1975. pp. 400–404.

[37] Vallozzi L, Rogier H, Hertleer C. Dual Polarized Textile Patch Antenna for Integration Into Protective Garments. *IEEE Antennas and Wireless Propagation Letters*. 2008; 7: 440–443.

[38] Baudrit C, Dubois D, Guyonnet D. Joint Propagation and Exploitation of Probabilistic and Possibilistic Information in Risk Assessment. *IEEE Transactions on Fuzzy Systems*. 2006; 14(5): 593–608.

[39] Gazda C, Vande Ginste D, Rogier H, *et al.* A Wideband Common-Mode Suppression Filter for Bend Discontinuities in Differential Signaling Using Tightly Coupled Microstrips. *IEEE Transactions on Advanced Packaging*. 2010 Nov; 33(4): 969–978.

[40] Gazda C, Couckuyt I, Rogier H, *et al.* Constrained Multiobjective Optimization of a Common-Mode Suppression Filter. *IEEE Transactions on Electromagnetic Compatibility*. 2012 June; 54(3): 704–707.

Chapter 11
Conclusions and future directions

Sourajeet Roy[1]

In this concluding chapter, we will summarize the contents of this book and draw conclusions regarding the current state-of-the-art in uncertainty quantification. In addition, we will also discuss the most important unanswered questions and key challenges still dominating this field and try to predict the possible emerging techniques best suited to answer these challenges.

11.1 Overview of the chapters in this book

In order to summarize the contents of this book, we begin by first breaking up the term uncertainty quantification into its constituent words – "uncertainty" and "quantification." Thus, for a reader to fully comprehend what uncertainty quantification means, he/she has to start with the knowledge of what we mean by "uncertainty," especially in the context of electromagnetic devices, circuits, and systems. In particular, the reader must be familiar with what are the different types of uncertainties (e.g., epistemic, aleatory, or mixed) and how do they arise in electromagnetic devices, circuits, and systems. However, it is simply not enough to know what uncertainty means. The reader also needs to know how to quantify the effects of these types of uncertainties on the performance of the electromagnetic system in question. Both these two aspects of uncertainty quantification are introduced and discussed in Chapter 1.

Chapter 2 introduces the topic of generalized polynomial chaos (PC) theory – one of the most popular spectral approaches for uncertainty quantification [1–5]. This chapter covers the theoretical and mathematical foundations of PC theory and existing methodologies to generate the orthonormal basis functions (i.e., the basic building blocks of PC expansions).

Chapter 3 is a continuation of Chapter 2 where greater emphasis is laid on the application of the PC theory to uncertainty quantification of practical electromagnetic problems. This chapter describes the intrusive and non-intrusive approaches to evaluate the unknown coefficients of the PC expansion. From this description, the pros and cons of each approach is explained. These pros and cons are further illustrated through multiple numerical examples.

[1]Computational Modeling and Simulation Laboratory, Indian Institute of Technology, Roorkee, India

Chapters 4 and 5 cover different techniques for addressing the pervasive curse of dimensionality facing PC metamodels. In particular, Chapter 4 describes a sensitivity analysis-based approach to identify the most impactful input dimensions in any circuit and then constrict a PC metamodel only in the space enclosed by those dimensions. In effect, the unimportant dimensions of the circuit are neglected and do not contribute to the training cost. In Chapter 5, a predictor–corrector algorithm based on the notion of multi-fidelity analysis is proposed whereby the burden of training the PC metamodel is shifted from the shoulders of a high-fidelity complex circuit model to the shoulders of a more compact and approximate low-fidelity model. As a bonus benefit, the low-fidelity model can always be parlayed to perform the sensitivity analysis leading to dimension reduction in addition to utilizing a predictor–corrector algorithm. This strategy works very well for the problems where a low-fidelity model is already available for the structure or circuit under consideration, for example, multi-walled carbon nanotube interconnects.

Chapter 6 adapts the conventional PC theory for the more general and complex case where the input random dimensions exhibit a non-Gaussian correlation. It develops mathematical techniques to construct the multidimensional orthonormal PC bases directly as well as quadrature techniques to accurately solve the inner product integrals in a multidimensional correlated parameter space.

Chapter 7 presents a detailed description of the emerging machine learning based metamodeling techniques for uncertainty quantification of highly nonlinear systems. This chapter begins with a look at training conventional metamodels using regression techniques such as least-square regression, ridge regression, LASSO, etc., before moving on to the more recent techniques such as support vector regression, least-square support vector regression, and Gaussian process regression.

Chapter 8 delves into the application of artificial neural networks for the yield optimization of electromagnetic devices and circuits in the presence of uncertainties. Techniques to address the training cost for artificial neural networks using space mapping techniques have been detailed in this chapter. Besides neural networks, the utilization of PC metamodels for yield optimization has also been discussed.

Chapter 9 focuses on order reduction clustering to filter the most influential random variables involved during uncertainty quantification of electromagnetic devices and circuits. This chapter describes the foundations of the methodology and demonstrates the utility of the technique through the assessment of stochastic outputs from electromagnetic applications with high number of random and/or dependent parameters. The efficacy of the proposed methodology is illustrated through the assessment of electromagnetic composite models assuming stochastic structures within microwave frequency bandwidth.

Chapter 10 focuses on performing mixed epistemic-aleatory uncertainty quantification by combining PC and machine learning metamodels together. In particular, in this chapter, epistemic variables are characterized as fuzzy variables modeled using the Dempster-Schafer theory of evidence while aleatory variables are characterized as conventional random variables. Thereafter, the Bayesian optimization technique has been combined with PC metamodels to quantify the impact of both forms on uncertainty simultaneously on the response in a numerically efficient manner.

11.2 State-of-the-art in uncertainty quantification

With the advent of the generalized PC approach and, more recently, machine learning techniques, the idea of using surrogate models or metamodels to facilitate uncertainty quantification of electromagnetic devices, circuits, and systems has become mainstream. Indeed, it is now well accepted that metamodels are a far better alternative than rigorous numerical electromagnetic/circuit solvers for uncertainty quantification of devices and circuits because of the following reasons:

1. Metamodels, being closed-form in nature, are significantly more numerically efficient to probe at different Monte Carlo sample points than rigorous numerical electromagnetic/circuit solvers that are based on solving massive systems of coupled equations. In fact, PC metamodels and machine learning based metamodels have been demonstrated to provide three to four orders of magnitude in speedup compared to numerical electromagnetic solvers when performing Monte Carlo based statistical analysis of devices and circuits [6–8].
2. Metamodels, again on account of being closed-form in nature, can be stored at very small memory costs. This too is in contrast of numerical electromagnetic solvers that need significant memory to store the massive coupled systems of equations describing the behavior of the device or circuit under test.
3. Metamodels are much more flexible and their accuracy can be easily enhanced by tuning the order of PC expansion, the number of training samples involved, and value of the hyperparameters in machine learning techniques. Moreover, the basis functions, activation functions, and kernel functions involved in different types of metamodels can always be adapted by the user to obtain faster convergence. Such flexibility is typically not possible in numerical electromagnetic/circuit solvers. Rather, in electromagnetic solvers, the accuracy of the solution can be changed by using trial and error method to determine what is the appropriate mesh size and/or the numerical integration method required.
4. Metamodels are inherently parametric in nature. For example, in PC metamodels, the variability of the device or circuit responses with respect to the uncertain parameters is embedded in the basis functions. Thus, the regression coefficients and weights/bias terms in metamodels have to be trained only once and can then be reused for every new combinations of the input parameter values. In effect, metamodels naturally enable the fast performance assessment of the device and circuit under test across a wide range of different combinations of input geometrical, physical, and material parameter values. On the other hand, in numerical electromagnetic/circuit solvers, the coupled system of equations representing the device or circuit under test have to be reconstructed and resolved at large time and memory costs for every new combination of the input parameter values.

Given the aforementioned advantages of metamodels over numerical electromagnetic/circuit solvers, the value of this book lies in the fact that it demonstrates and explains in detail all of these advantages to the reader without requiring him/her to have any previous knowledge on the subject of uncertainty quantification. As a result, this book will enable the uninitiated reader to truly appreciate the importance of metamodels to the current state-of-the-art in automated uncertainty quantification with a specific focus on electromagnetic devices and circuits. This is of crucial importance because currently there are very few books catering to the need of electrical engineers and circuit designers on the topic of uncertainty quantification in general and the value of metamodels in particular. As a bonus benefit, this book will also ensure that the reader is aware of the present challenges in uncertainty quantification and common metamodeling techniques – in effect, what are the future topics of high importance for researchers and scientists to focus on and where do we expect the state-of-the-art in uncertainty quantification to go from here.

11.3 Key challenges facing uncertainty quantification

Currently, the key challenges facing uncertainty quantification are:

- *Curse of Dimensionality*: Although a variety of approaches such as the Karhunen–Loève (KL) transformation [9–11], dimension reduction techniques [12–21], multi-fidelity algorithms [22–24], compressed sensing techniques [25], multi-element or domain decomposition techniques [26–28], strategies based on the kernel trick [29–32], etc., have been developed to address the curse of dimensionality, most of these approaches suffer from unique problems of their own. For example, the KL transformation performs eigenvalue decomposition on only the covariance matrix of the input parameters and does not consider the impact of each input parameter on the responses of interest. Hence, this technique is not very well-informed and does not achieve the maximum possible dimension reduction for a particular problem at hand. On the other hand, other dimension reduction techniques may incur large computational overheads in order to properly gauge which dimensions can be removed and which need to be retained [12–21]. Minimizing these overheads invariably decreases the confidence in the identification of the important/unimportant dimensions of a problem [21]. Moreover, the savings provided by dimension reduction strategies is minimal if all or most of the random dimensions have large impact on the response statistics. Next, multi-fidelity algorithms work only when an appropriate low-fidelity model for the device or circuit under test is already known [22–24]. Therefore, these algorithms are not universally applicable. The multi-element PC approach attempts to decompose the random space into smaller elements where local low-order PC metamodels can be developed. By stitching together these local PC metamodels, a global low-order PC metamodel can be constructed. It has been found that training the local metamodels at low-order is far more efficient than training a single global metamodel at a much higher order [26–28]. However, in such a domain decomposition technique, the number of elements, and consequently, the number of local metamodels to be trained may increase in an exponential manner with the number of dimensions, thus diluting the very advantages of using a low-order local PC metamodel. Thus, the multi-element approach works best when the random space has to be decomposed only along very few of the dimensions. Overall, we can conclude that the curse of dimensionality remains an entrenched problem despite the various number of techniques developed to mitigate it.

- *Mixed Epistemic-Aleatory Uncertainty Quantification*: The standard approach of performing mixed epistemic-aleatory uncertainty quantification is by assuming epistemic variables to be equivalent to uniformly distributed aleatory variables [33,34]. This assumption allows us to artificially convert the mixed epistemic-aleatory uncertainty in a given problem into purely aleatory uncertainty. Thereafter, a PC metamodel can be trained for this "purely aleatory" problem. One instance of such a PC metamodel is the Chebyshev PC metamodel that was proposed in [34–38]. Unfortunately, the challenge in such techniques is that the total number of dimensions of this problem is equal to the sum of epistemic and aleatory dimensions. In effect, the total number of dimensions considered will be much larger than simply the number of aleatory dimensions. Consequently, the PC metamodel that will be constructed for such mixed problems will require far more training data and hence, far greater training time costs than if only the aleatory dimensions were considered [39]. How to mitigate this increase in training time costs remains an open question.

  Of course, one approach to mitigate the increase in training costs could be to utilize the dimension reduction techniques used for conventional PC metamodel [39]. However, the fundamental limitation of such an approach is that the metric used to distinguish the relative importance of each aleatory dimension, for example, the Sobol's sensitivity indices, will now become a function of the epistemic variables. Therefore, for different values of the epistemic variables, the sensitivity indices of each dimension will change. This will make it very difficult for the engineer to properly gauge which aleatory dimensions are globally or locally important/unimportant. The same will be true for other dimension reduction strategies, such as active subspace technique [12,13], sliced inversion [14–16], and basis adaptation methods [17,18]. Finally, even if aleatory dimensions could be reduced, how to reduce the epistemic dimensions is not known.
- *Presence of Discontinuities and High-order Nonlinearities*: The PC metamodeling approach is most effective when the uncertain device or circuit response in question exhibits a smooth variation with respect to the uncertain parameters. This is due to the fact that the variability of the response with respect to the uncertain parameters is represented by polynomial basis functions – functions that while being nonlinear, nevertheless possess a smooth trajectory. However, for many devices and circuits, the variability in the responses can be highly discontinuous and non-smooth for, for example, nonlinear semiconductor devices such as diodes and FET devices, especially at millimeter wave frequencies of operation. In such cases, the PC metamodel will display the Gibbs effect. In order to reduce the side lobes due to the Gibb's effect,

an unnaturally high order would be required leading to massive training time costs [26–28]. For these particular problems, machine learning techniques are far better because they are not restricted to polynomial nonlinear functions. Rather, they can exploit highly nonlinear activation functions and kernels such as hyperbolic tangent, arctangent, and even nonlinear functions of the users own construction to better approximate sharp discontinuities and non-smooth variations in the responses.

- *Memory Constraints due to Dynamic Response*: Electromagnetic and electronic devices and circuits are typically dynamic in the sense that their responses are time-dependent or frequency-dependent. This means that all metamodels, whether based on PC or machine learning, will have to capture the variability of the response with respect to the uncertain parameters as well as time/frequency. In effect, the training data must sweep the uncertain parameter space as well as the time/frequency window of observation. This can significantly augment the volume of training data required leading to memory issues. Of course, for each training sample point, the response may be measured at only a coarse set of time/frequency points in order to decrease the amount of memory required. However, in such cases, important features of the dynamic response will be missed. Recently, artificial neural networks have been combined with transfer functions to emulate the dependence of the poles and residues of linear circuit responses on the values of the input uncertain parameters [40,41]. In these metamodels, the poles and residues not being functions of frequency, at each training sample point there is no need to evaluate the response for different frequency points – only the poles and residues need to be known. Once the poles and residues are identified, the entire frequency-domain response can be easily described and even converted into time domain using the direct inverse Laplace transform [40]. Importantly, the same approach can also work for time-domain data. Unfortunately, this approach will not work for nonlinear circuits. Thus, more generalized techniques for nonlinear circuits are required.
- *Arbitrary PC*: The concept of generalized PC theory is predicated on the assumption that the marginal probability density function (PDF) of the uncertain parameters are well-approximated by known distributions such as Gaussian, uniform, gamma, or beta distributions. However, this is not true for many real-world electronic devices and circuits. Thus, there has been a growing demand of the PC theory to be adapted to more general input distributions. Recently, the concept of arbitrary PC (aPC) was developed in [42]. The aPC was based on the observation that continuous/discrete PDFs could be well-approximated using the dominant statistical moments of the underlying random process even if complete knowledge of the shape of the PDFs was unavailable. Thereafter, based on the knowledge of the dominant statistical moments, orthogonal basis functions of the random dimensions could be numerically extracted and these functions be used to construct an aPC metamodel [43]. Indeed, to construct the orthogonal bases of a random dimension up to an order m, the first $2m + 1$ statistical moments of the dimension must be known [43]. While this approach is highly robust and can construct PC

metamodels even for PDFs of imperfectly known or multi-modal shapes, it is important to realize that in practice, it is not always possible to know *a priori* the leading $2m + 1$ statistical moments of each and every dimension of a high-dimensional problem. Unfortunately, if this condition is not satisfied for even one dimension, then the construction of the multidimensional orthonormal bases becomes a major challenge.

All of the above points are key questions that remain partially or completely unanswered in the current state-of-the-art of uncertainty quantification. Thus, these topics are anticipated to be the major focus areas for the coming years.

References

[1] D. Xiu, *Numerical Methods for Stochastic Computations: A Spectral Method Approach*. Princeton, NJ: Princeton University Press, 2010.

[2] O. P. Le Maitre and O. M. Konio, *Spectral Methods for Uncertainty Quantification*. New York, NY: Springer-Verlag, 2010.

[3] D. Xiu, "Fast numerical methods for stochastic computations: a review," *Commun. Comput. Phys.*, vol. 5, no. 2–4, pp. 242–272, Feb. 2009.

[4] D. Xiu and G. E. Karniadakis, "The Wiener-Askey polynomial chaos for stochastic differential equations," *SIAM J. Sci. Comput.*, vol. 24, no. 2, pp. 619–644, July 2002.

[5] R. G. Ghanem and P. Spanos, *Stochastic Finite Elements: A Spectral Approach*. New York, NY: Springer-Verlag, 1991.

[6] I. S. Steivano, P. Manfredi, and F. Canavero, "Parameters variability effect on multiconductor transmission lines via Hermite polynomial chaos," *IEEE Trans. Comp. Packag. Manuf. Technol.*, vol. 1, no. 8, pp. 1234–1239, Aug. 2011.

[7] M. Larbi, R. Trinchero, F. Canavero, P. Besnier, and M. Swaminathan, "Analysis of parameter variability in an integrated wireless power transfer system via partial least-squares regression," *IEEE Trans. Comp. Packag. Manuf. Technol.*, vol. 10, no. 11, pp. 1795–1802, Nov. 2020.

[8] R. Trinchero and F. Canavero, "Combining LM-SVM and GP regression for the uncertainty quantification of the EMI of power converters affected by several input parameters," *IEEE Trans. Electromagn. Compat.*, vol. 62, no. 5, pp. 1755–1762, Oct. 2020.

[9] D. Xiu and G. E. Karniadakis, "Modeling uncertainty in steady sate diffusion problems via generalized polynomial chaos," *Comput. Methods Appl. Mech. Eng.*, vol. 191, no. 43, pp. 4297–4248, Sept. 2002.

[10] Y. M. Marzouk and H. N. Najm, "Dimensionality reduction and polynomial chaos acceleration of Bayesian inference in inverse problems," *J. Comput. Phys.*, vol. 228, no. 6, pp. 1862–1902, Apr. 2009.

[11] A. Doostan, R. G. Ghanem, and J. Red-Gorse, "Stochastic model reduction for chaos representations," *Comput. Methods Appl. Mech. Eng.*, vol. 196, no. 37–40, pp. 3951–3966, Aug. 2007.

[12] P. G. Constantine, E. Dow, and Q. Wang, "Active subspace methods in theory and practice: applications to kriging surfaces," *SIAM J. Sci. Comput.*, vol. 36, no. 4, pp. A1500–A1524, 2014.

[13] P. G. Constantine, *Active Subspaces: Emerging Ideas for Dimension Reduction in Parameter Studies*. Philadelphia, PA: SIAM, 2015.

[14] W. Li, G. Lin, and B. Li, "Inverse regression-based uncertainty quantification algorithms for high-dimensional models: theory and practice," *J. Comput. Phys.*, vol. 321, no. 3, pp. 259–278, Sept. 2016.

[15] K.-C. Li, "Sliced inverse regression," *J. Am. Stat. Assoc.*, vol. 86, no. 414, pp. 316–327, June 1991.

[16] Q. Pan and D. Dias, "Sliced inverse regression-based sparse polynomial chaos for reliability analysis in high dimensions," *Reliab. Eng. Sys. Saf.*, vol. 167, pp. 484–493, Nov. 2017.

[17] R. Tipireddy and R. G. Ghanem, "Basis adaptation in homogeneous chaos spaces," *J. Comput. Phys.*, vol. 259, no. 2, pp. 304–317, Feb. 2014.

[18] P. Tsilifis and R. Ghanem. "Reduced Wiener chaos representation of random fields via basis adaptation and projection," *J. Comput. Phys.*, vol. 341, pp. 102–120, July 2017.

[19] J. Jiang, Y. Chen, and A. Narayan, "A goal-oriented reduced basis methods-accelerated generalized polynomial chaos algorithm," *SIAM J. Uncertain. Quantif.*, vol. 4, no. 1, pp. 1398–1420, Dec. 2016.

[20] Z. Zhang, X. Yang, I. V. Oseledets, G. E. Karniadakis, and L. Daniel, "Enabling high-dimensional hierarchical uncertainty quantification by ANOVA and tensor-train decomposition," *IEEE Trans. Comput.-Aided Des. Integr. Circuits Syst.*, vol. 34, no. 1, pp. 63–76, Jan. 2015.

[21] A. K. Prasad and S. Roy, "Accurate reduced dimensional polynomial chaos for efficient uncertainty quantification of microwave/RF networks," *IEEE Trans. Microw. Theory Tech.*, vol. 65, no. 10, pp. 3697–3708, Oct. 2017.

[22] L. W. T. Ng and M. S. Eldred, *"Multifidelity uncertainty quantification using non-intrusive polynomial chaos and stochastic collocation,"* in *Proc. 53rd AIAA structures, structural dynamics, and materials conference*, Apr. 2012, pp. 1–17.

[23] P. S. Palar, T. Tsuchiya, and G. T. Parks, "Multi-fidelity non-intrusive polynomial chaos based on regression," *Comput. Methods Appl. Mech. Eng.*, vol. 315, pp. 579–606, June 2016.

[24] A. Narayan, C. Gittelson, and D. Xiu, "A stochastic collocation algorithm with multifidleity models," *SIAM J. Sci. Comput.*, vol. 36, no. 2, pp. A495–A521, Mar. 2014.

[25] J. Hampton and A. Doostan, "Compressive sampling of polynomial chaos expansions: convergence analysis and sampling strategies," *J. Comput. Phys.*, vol. 280, pp. 363–386, Jan. 2015.

[26] X. Wan and G. E. Karniadakis, "An adaptive multi-element generalized polynomial chaos method for stochastic differential equations," *J. Comput. Phys.*, vol. 209, no. 2, pp. 617–642, Nov. 2005.

[27] X. Wan and G. E. Karniadakis, "Multi-element generalized polynomial chaos method for arbitrary probability measures," *SIAM J. Sci. Comput.*, vol. 28, no. 3, pp. 901–928, June 2006.

[28] J. Foo, X. Wan, and G. E. Karniadakis, "The multi-element probabilistic collocation method: error analysis and applications," *J. Comput. Phys.*, vol. 227, no. 22, pp. 9572–9595, Nov. 2008.

[29] V. N. Vapnik, *Statistical Learning Theory*. Hoboken, NJ: John Wiley and Sons Inc., 1998.

[30] J. A. K. Suykens, T. V. Gestel, J. D. Brabanter, B. D. Moor, and J. Vandewalle, *Least Squares Support Vector Machines*. Singapore: World Scientific Publishing, 2002.

[31] R. Trinchero, M. Larbi, H. M. Torun, F. Canavero, and M. Swaminathan, "Machine learning and uncertainty quantification for surrogate models of integrated devices with a large number of parameters," *IEEE Access*, vol. 7, pp. 4056–4066, Jan. 2019.

[32] R. Trinchero, P. Manfredi, I. S. Steivano, and F. Canavero, "Machine learning for the performance assessment of high-speed links," *IEEE Trans. Electromagn. Compat.*, vol. 60, no. 6, pp. 1627–1634, Dec. 2018.

[33] A. Monti, F. Ponci, and F. Valtorta, "Extending polynomial chaos to include interval analysis," *IEEE Trans. Instrum. Meas.*, vol. 59, no. 1, pp. 48–55, Jan. 2010.

[34] J. Wu, Z. Luo, H. Li, and N. Zhang, "A new hybrid uncertainty optimization method for structures using orthogonal series expansion," *Appl. Math. Model.*, vol. 45, pp. 474–490, May 2017.

[35] J. L. Wu, Y. Q. Zhang, and Z. Luo, "A new method for hybrid of probability and interval analysis," in *Proceedings on International Conference on Computational Methods*, July 2014, pp. 1–8.

[36] J. Wu, Z. Luo, N. Zhang, and Y. Zhang, "A new interval uncertain optimization method for structures using Chebyshev surrogate models," *Comput. Struct.*, vol. 146, pp. 185–196, Jan. 2015.

[37] S. Yin, D. Yu, H. Yin, and B. Zai, "Interval and random analysis for structure–acoustic systems with large uncertain-but-bounded parameters," *Comput. Methods Appl. Mech. Eng.*, vol. 305, June 2016, pp. 910–935.

[38] C. Jiang, J. Zheng, and X. Han, "Probability-interval hybrid uncertainty analysis for structures with both aleatory and epistemic uncertainties: a review," *Struct. Multidiscip. Optimiz.*, vol. 864, no. 4, pp. 1–18, Nov. 2017.

[39] A. K. Prasad and S. Roy, "Reduced dimensional Chebyshev-polynomial chaos approach for fast mixed epistemic-aleatory uncertainty quantification of transmission line networks," *IEEE Trans. Comp. Packag. Manuf. Technol.*, vol. 9, no. 6, pp. 1119–1132, June 2019.

[40] Y. Cao, G. Wang, and Q. J. Zhang, "A new training approach for parametric modeling of microwave passive components using combined neural networks and transfer functions," *IEEE Trans. Microw. Theory Tech.*, vol. 57, no. 11, pp. 2727–2742, Nov. 2009.

[41] F. Feng, J. Zhang, W. Zhang, Z. Zhao, J. Jin, and Q. J. Zhang, "Recent advances in parametric modeling of microwave components using combined neural network and transfer function," *Int. J. Numer. Model. Electron. Netw. Devices Fields*, vol. 33, no. 6, Nov./Dec. 2019.

[42] S. Oladyshkin and W. Novak, "Data-driven uncertainty quantification using the arbitrary polynomial chaos expansion," *Reliab. Eng. Syst. Saf.*, vol. 106, pp. 179–190, Oct. 2012.

[43] R. Ahlfeld, B. Belkouchi, and F. Montomoli, "SAMBA: Sparse approximation of moment based arbitrary polynomial chaos," *J. Comput. Phys.*, vol. 320, pp. 1–16, Sept. 2016.

Index

www.ingramcontent.com/pod-product-compliance
Lightning Source LLC
LaVergne TN
LVHW011849200926
848484LV00002B/427